ENERGY-EFFICIENT MOTOR SYSTEMS:

A Handbook on Technology, Program, and Policy Opportunities

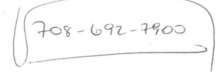

American Council for an Energy-Efficient Economy
Series on Energy Conservation and Energy Policy

Series Editor, Carl Blumstein

ENERGY-EFFICIENT MOTOR SYSTEMS:

A Handbook on Technology, Program, and Policy Opportunities

STEVEN NADEL

MICHAEL SHEPARD

STEVE GREENBERG

GAIL KATZ

ANIBAL T. DE ALMEIDA

American Council for an Energy-Efficient Economy
Washington, D.C., and Berkeley, California

in cooperation with:
Universitywide Energy Research Group
University of California

revised edition
1 9 9 2

Energy-Efficient Motor Systems:
A Handbook on Technology, Program, and Policy Opportunities

Published by the American Council for an Energy-Efficient Economy
1001 Connecticut Avenue, N.W., Suite 801, Washington, D.C. 20036.
2140 Shattuck Avenue, Suite 202, Berkeley, California 94704.

Cover art copyright © 1991 M.C. Escher Heirs/Cordon Art - Baarn - Holland

Cover design by Wilsted & Taylor

Book design by Paula Morrison

Library of Congress Cataloging-in-Publication Data
Energy-efficient motor systems: a handbook on technology, program, and policy
   opportunities / Steve Nadel . . . [et al.].
      408 p.   23 cm.
   Includes bibliographical references and index.
   ISBN 0–918249–10–4: $27.00
   1. Electric motors—Energy conservation—Handbooks, manuals, etc.
I. Nadel, Steve, 1957–   .   II. American Council for an Energy-Efficient
Economy.   III. University of California (System), Universitywide Energy
Research Group.
TK2781.E54   1992
333.79'32—dc20   92–26265
                                                                              CIP

NOTICE

The submitted manuscript has been authored by a contractor (grantee) of the U.S. Government under contract (grant) No. DE-AC02-76CH00016. Accordingly, the U.S. Government retains a nonexclusive, royalty-free license to publish or reproduce the published form of this contribution, or allow others to do so, for U.S. Government purposes.

Printed on recycled paper.

Acknowledgments

A book of this scope requires the help of many individuals and organizations. Thanks to the dozens of people, too numerous to list here, who provided us with data and information, much of it not previously published. It is due to their assistance that this book is able to break new ground.

Thanks also to our project sponsors—organizations who recognized the usefulness of this project and agreed to provide funds before a single word was written. Project sponsors are: American Public Power Association, Bonneville Power Administration, California Energy Commission, North Carolina Alternative Energy Corporation, Northeast Utilities, U.S. Department of Energy (both the Least-Cost Utility Planning Program and the Office of Industrial Programs), and the Western Area Power Administration.

A number of people reviewed drafts, and provided insightful comments, including Sam Baldwin, Office of Technology Assessment; Joseph Flores, Bonneville Power Administration; Howard Geller, ACEEE; Bill Gilmore, Walco Electric; Chuck Glaser and Dietrich Roesler, DOE; Walter Johnston, North Carolina Industrial Extension Service; Jon Leber and Michael Messenger, California Energy Commission; Dan Lewis, APPA; Peter Morante, Northeast Utilities; Gary Morgan and Susan De Belle, Western Area Power Administration; Barbara Pierce, Brookhaven National Laboratory; Marc Ross, University of Michigan; Tim Stout, New England Power Service Company; Todd Litman, Washington State Energy office; Paul Willis, Willis Energy Services.

Other important contributors include Carl Blumstein, the series editor, who helped conceive the project, the production team at Home Energy magazine, including Karina Lutz, managing editor, and Michelle Stevens, production manager, and the freelance copy editor, Stephen Frantz.

Finally, we would like to thank our families, who put up with many evenings and weekends of meeting deadlines and reviewing manuscripts.

Dedication

This book is dedicated to our valued colleague and friend, Gail Katz, whose practical perspective enriched this book and helped keep us on track. Her tragic death near the end of this project was a blow to all who have worked with her.

Gail was nationally recognized for her pioneering work in energy conservation engineering in industrial and commercial facilities. A memorial scholarship fund has been established in Gail's name to benefit women engineering students working in the fields of energy conservation, renewable energy, and appropriate technology. Contributions and inquiries should be directed to the Gail Katz Memorial Fund, Society of Women Engineers, 345 East 47th St., New York, N.Y. 10017.

Preface

Motor-driven systems use more than half of all electricity in the United States and many other countries, and over 70% of the electric power in many industries. The cost of powering motors is immense—roughly $90 billion a year in the United States.

Increasing the efficiency of drivepower systems can save vast amounts of energy and money. The technologies to realize this potential include material and design advances, power electronic and microelectronic instruments that can control motor systems more precisely than ever before, and more durable and reliable equipment. Several of these technologies, especially control systems, improve not only energy efficiency but also product and process quality. Furthermore, many utilities are now offering incentives for drivepower efficiency improvements.

This handbook is intended to help utility customer-service and demand-side management personnel, energy policymakers, analysts, and planners to design and implement programs that will foster more efficient and reliable motor systems. While the technical press has published much on technologies to improve motor system efficiency, this information is scattered and usually too technical for program and policy planners. Information on drivepower program and policy experience is also dispersed and often unpublished. This handbook pulls together the relevant literature and fills the gaps.

The handbook is divided into six sections:

1. Overview and summary (chapter 1). An introduction to the book and a summary for the busy executive.
2. Technologies (chapters 2 to 5). An explanation of motor system technologies and of energy-saving opportunities and economics.

This section should be particularly useful for semi-technical readers, such as staff who need to have a basic understanding of the technologies in order to implement programs in the field.

3. Data (chapters 6 and 7). An examination of the motor population by type, size, and operating characteristics, and an estimate of the national savings possible from drivepower system improvements. This section should be useful for utility and government planners who need data to incorporate motor efficiency improvements into integrated resource plans.

4. Program and policy constraints and opportunities (chapters 8 and 9). A look at the motor marketplace, and at programs and policies designed to overcome market barriers and promote motor system efficiency improvements. This material should help planners who are designing programs and policies and want to know where to start, or who want specific information on a particular program or policy option.

5. Summary and recommendations (chapter 10). A brief summary of our findings and a compilation of recommendations to be addressed by manufacturers, end-users, utilities, trade associations, and governments.

6. Appendices. Additional information and worksheets on the economics of efficiency improvements, listings of manufacturers and trade associations, a glossary, and a list of references, including an annotated listing of the most useful references.

We recognize that not all readers will want to read all sections of the book. We encourage everyone to read chapter 1, but thereafter the reader should feel free to jump around. The text will let you know when you need to refer to previous or subsequent sections. The book covers a lot of ground but is necessarily incomplete. More detailed technical information is available elsewhere (see the annotated references), and experience with programs and policies is continually evolving. While we believe this book contains the most complete information currently available, we look to our readers to advance the state of knowledge in the important and rapidly emerging field of drivepower efficiency.

Contents

Illustrations

FIGURE

FIGURE

FIGURE

FIGURE

Tables

TABLE

TABLE

Chapter 1

Overview and Summary

This book describes how to save 9–23% of U.S. electricity (calculated in chapter 7) by optimizing the performance of electric motors and their associated wiring, power conditioning equipment, controls, and transmission components—networks of devices we refer to as motor systems.

Electric motors are remarkable machines: rugged, reliable, and far more efficient than the animals and steam powered equipment they replaced over the past century. A well-designed and well-maintained electric motor can convert over 90% of its input energy into useful shaft power, 24 hours a day, for decades. The popularity of motors attests to their effectiveness: they provide more than four-fifths of the nonvehicular shaft power in the United States and use as input upwards of 60% of the nation's electricity. It is their popularity that makes electric motor systems such an important potential source of energy savings: because more than half of all electricity flows through them, even modest improvements in their design and operation can yield tremendous dividends.

Touring a Motor System

The key to making motor systems more efficient and economical is to take advantage of high-performance technologies and the synergisms among the various system components. To illustrate, let's take a brief tour of a system, starting from where electricity enters the facility; moving downstream through the wiring, power-conditioning equipment, and controls to the motor; and then continuing through transmission hardware to the driven devices. Along the way we will identify some of the major opportunities for savings.

In theory, electricity arrives at a customer's facility as perfectly balanced and synchronized single- or three-phase power of constant voltage, free of harmonics and other kinds of distortion. In the real world, this ideal condition is almost never reached. Phases are often slightly out of balance, voltages may dip and rise, and various kinds of distortion are common. This less-than-perfect power is subject to further distortions from equipment—such as welders, arc furnaces, and variable-frequency motor controls—inside customers' facilities. Sometimes problems can arise from a poor arrangement of equipment, such as the uneven distribution of single-phase and three-phase devices on a circuit.

Such deviations from the pure, ideal electric waveform can reduce the efficiency, performance, and life of motors and other electric equipment. Avoiding and correcting such problems requires careful monitoring of power quality, repair of faulty devices, and, in some cases, installation of specialized power-conditioning equipment. Some analysts believe that such tune-ups may be among the largest reservoirs of untapped drivepower savings, although scanty data allow only rough estimates of the overall potential. Field studies suggest that the effort and expense of electrical tune-ups can be worthwhile in terms of reduced energy costs, better equipment performance, improved process control, and reduced downtime from damaged equipment. Details of some major opportunities in this area are discussed in chapter 3.

Just as it pays to clean up the power flowing through the wires, so too it is important to optimize the efficiency of the wires themselves. In most facilities, distribution wiring is sized according to the National Electrical Code, which principally addresses safety, not energy efficiency. Larger wires have lower resistance to the flow of electricity—hence lower losses—than the sizes required by the code. Therefore, in new installations or major renovations, it often pays to exceed code standards. Unfortunately, the benefits of doing so are not widely appreciated by architects, designers, electricians, and facility managers, so considerable amounts of energy and money are being wasted through in-plant distribution losses, before the electricity even does any work. Details on wire sizing are covered in chapter 3.

Many motor-driven processes require some form of control over the motor's speed, start-up, or torque (rotational force). For example, fan-, compressor-, and pump-driven systems moving gaseous or liquid loads require frequent changes in the rate of flow, as in the cases of fans and chillers for ventilation and cooling of commercial buidings, pumps for hydronic heating and/or cooling systems, fans and feed pumps for industrial and power plant boilers, and municipal water and wastewater pumps. Rather than varying the speed of the motor, which until recently was difficult to do with the predominant alternating current (AC) motor, most systems use mechanical devices such as inlet vanes, outlet dampers, or throttling valves to control fluid flow while the motor continues to run at full speed. These techniques are analagous to driving a car with the accelerator pushed to the floor while controlling the vehicle's speed with the brake. Such methods yield imprecise control and waste a lot of energy.

In other kinds of loads requiring varying speed or torque— winders, mills, conveyors, elevators, cranes, and servodrivers—motor users have employed various kinds of mechanical, electro- mechanical, or hydraulic speed controls in conjunction with AC motors, or have used DC motors, whose speed can be easily con- trolled. Most of these speed control options, however, have pitfalls, including high cost, low efficiency, or poor reliability.

The speed-control problem has been largely solved by the development of the electronic adjustable-speed drive (ASD). This device precisely controls the speed of AC motors, eliminating the need for wasteful throttling devices in fluid flow applications and rendering many traditional controls and uses for DC motors obsolete. ASDs yield sizable energy savings (15–40% in many cases) and extend equipment life by allowing for gentle start-up and shut-down.

The electronic ASD is not the only new control technology, although it may be the most important one. Other technologies include sequencing controls for pumps and fans; lead-lag control systems for compressors; feedback control systems that regulate rather than bypass flow; and power-factor controllers that can trim the energy use of small motors driving grinders, drills, and other devices that idle at nearly zero loading most of the time.

Other developments enlarge the range of control applications. For instance, advanced sensors are allowing ASDs to be used in applications (lumber drying kilns, for example) where they previ- ously would not work because of limitations in sensing or in match- ing the response time required by a control loop. Electronic advances also are allowing lumber mills to control cuts better and to mill more product from raw stock without increasing energy use. These developments and others in the controls area represent the largest slice of the drivepower savings pie and are discussed in chapter 4.

High-efficiency motors, also known as energy-efficient motors or EEMs, are available for about 80% of applications. EEMs are typically two to ten percentage points more efficient than standard-efficiency motors, with smaller motors at the high end of this range and larger motors at the low end. Due principally to their better materials, high-efficiency units cost 10–30% more and tend to last longer than standard models. While a few percentage points of efficiency do not sound like much, such an improvement can add up to sizable savings over the life of a motor. A heavily used motor can easily have an electricity bill ten times its purchase price each year. If cars were comparable, a $10,000 car would use $100,000 worth of gasoline annually. With so much of the life-cycle cost in operating expense, each increment of efficiency is extremely valuable. Hence the payback on the added cost of high-efficiency motors is often very attractive.

EEMs are still a small part of the market—about 11% of current national sales of motors 1 hp and larger, roughly 3% of the existing motor stock. Great savings are thus possible from applying EEMs wherever these motors are cost-effective. Motor types, their applications, and economics are discussed in chapter 2.

As we replace standard-efficiency motors with more efficient models we can capture savings bonuses by correcting for two problems endemic to the existing motor stock: oversizing and rewind damage. Many motors are oversized for their applications, and because motor efficiency drops off sharply below about 40% of rated load, oversized motors often run far below their nameplate efficiency. In addition, many motors are rewound at least once and often several times before they are discarded. Standard rewind practices are thought to damage the magnetic properties of the motor core in many instances, thus reducing efficiency. Proper sizing of new motors and either the use of nondamaging rewind practices or the adoption of replace-instead-of-rewind policies can thus add significant savings. These matters are covered in chapters 2 and 3.

Energy enters a motor as electricity and emerges as mechanical power in the form of a rotating shaft. To put that energy to use often requires a transmission, provided typically by belts, gears, or chains. Such devices are often overlooked in efficiency analyses and typically receive fairly unsophisticated installation and maintenance. This neglect is unfortunate, because, as discussed in chapter 3, the proper selection, installation, and maintenance of transmission hardware can profoundly affect the performance and efficiency of a motor system. For example, a belt that is too loose will slip, wasting energy. A belt that is too tight can place extreme loads on a bearing, causing it to fail prematurely and lead to costly downtime. Such problems can be avoided in some applications by using synchronous belts, which run on toothed sprockets and are generally more efficient than V-belts, which run on smooth pulleys.

Optimized drivetrains are also important because they are far downstream in the drivepower system, where even modest improvements can ripple back through the system to yield significant savings. For instance, a unit of energy saved in the drivetrain means the motor doesn't have to work as hard, so it draws less energy, which reduces losses in the distribution wiring, and so on back to the power plant. An additional, potentially large bonus comes in the form of indirect savings from reduced building cooling load due to lower current flow and less heat dissipation from the more efficient equipment.

The need for careful, ongoing monitoring and maintenance applies to the entire motor system. A high-efficiency system will only stay that way if given proper care, from simple cleaning and lubrication to sophisticated troubleshooting of power quality problems. While the energy savings from top-notch maintenance are substantial, the greatest dividend comes in the form of more reliable, trouble-free operation and extended equipment life. Where equipment downtime can mean thousands of dollars per hour in lost production, quality maintenance pays.

We have completed our tour of the motor system and touched on some of the major technical areas that later chapters will deal with in greater detail. If nothing else, this brief survey is designed to emphasize the notion of a motor *system*, and to underscore the critical importance of the interactions and synergisms among the various system components.

A Note on Lost Opportunities

Most of the efficiency options discussed here are more economical in new installations than in retrofits. These options are termed "lost opportunity" resources, because if not implemented during new construction or renovation, they are much more costly to install later. In some cases, however, it makes economic sense to replace and upgrade operating equipment rather than to wait for it to fail. Where load factors are very high, for instance, it often pays to scrap standard-efficiency motors and replace them with high-efficiency models. As chapter 2 describes, Stanford University recently did this with 73 motors, with average paybacks of under three years. Energy conservation program planners and facility managers should remember this distinction between new and retrofit efficiency opportunities as they implement programs.

Barriers to Drivepower Savings

If the potential savings are so large, why are so few motor users aggressively pursuing them? The answer lies in a maze of barriers to investment in energy efficiency in general and to drivepower improvements in particular. Some of the most important of these barriers are highlighted below and discussed in detail in chapters 8, 9, and 10.

Payback Gap

The "payback gap" is a key impediment to investment in more efficient equipment. The concept is simple. Consumers, including businesses, typically will invest in efficiency improvements only if the resulting energy savings will yield a simple payback of under two or three years (even though longer-payback investments can yield significant life-cycle savings) while utilities regularly invest in power plants that take at least 10 years, and sometimes more than 20 years, to pay back. This gap in payback requirements causes society to invest too little in savings and too much in power supply.

Lack of Information

One of the principle causes of the payback gap is the lack of readily available, accurate information on the financial and operating benefits of efficiency investments. This book will help motor users become aware of the opportunities and make wise decisions with regard to drivepower improvements. Other information programs are reviewed in chapter 9.

Limited Access to Capital

Even an informed consumer who wishes to invest in more efficient equipment may have limited access to capital, for which energy savings must compete with many other potential investments. Various financial incentives, like rebates, are being employed in a number of jurisdictions to help overcome this barrier. Details are covered in chapter 9.

Lack of Institutional Incentives

Often the person in the best position to implement an energy-saving program has little to gain from doing so. Facility managers whose energy bills are treated as a flow-through operating expense are less likely to invest in efficiency improvements than managers who get to keep in their budget some share of the money they save, savings they can plow back into further conservation measures or use, say, for staff bonuses.

Market Structure

Roughly 75% of electric motors are purchased by original equipment manufacturers (OEMs), who incorporate them into products sold through various intermediate distributors and retailers to the ultimate users, who will have to pay the electricity bills. Among the many OEMs are refrigerator, air conditioner, pump, and fan manufacturers. OEMs buy more than 75% of motors 20 hp and smaller but only 15% of the motors larger than 125 hp. Because OEMs compete largely on the basis of price, they tend to avoid the use of more costly, high-efficiency components. And because they are several steps removed from end-users, they rarely receive direct requests for efficiency improvements.

Aversion to Downtime and to Innovation

In many businesses, particularly in industry, shutting down equipment for upgrading or replacement can mean losing thousands of dollars per hour in forgone production. Such penalties may induce an understandable aversion to downtime and cause many facility managers to shy away from new, unfamiliar technology they fear might be less reliable than the equipment they are used to. Furthermore, if a high-efficiency substitute for a failed motor is not stocked by the distributor, in order to save time, the user is likely to buy a standard replacement or simply repair the old motor.

Repair Shops Compete on Speed and Price

Efforts to reduce downtime also lead to problematical attitudes toward motor repair. When a motor fails, the user generally wants it repaired as fast as possible. This fix-it-by-yesterday pressure leads motor repair shops to compete not only through price but also through speed. One of the most effective ways to hasten a motor rewind is to increase the temperature of the oven used to burn out the motor windings. Yet the higher the temperature, the more likely that the motor core will be damaged. This invisible damage makes the motor less efficient.

Standardized Inventories

For the sake of simplicity, many motor-using facilities will keep in inventory a few standard motor sizes to serve a wide range of loads. This practice can lead to the installation of motors oversized for their loads. While this approach might save on inventory costs and thus keep certain accountants happy, it is likely to harm the firm by increasing energy costs. A related problem is that distributors often do not keep a full range of EEMs in stock.

Purchase Decision Criteria

As with inventory managers, personnel in charge of purchasing drivepower equipment are typically uninvolved and untrained in the energy-use consequences of their choices. For example, maintenance staff with limited technical training often make motor replacement decisions. At OEMs, engineers often provide a list of acceptable products to purchasing departments, which then select products primarily on the basis of price and delivery terms.

This listing of the barriers to motor system improvement is by no means exhaustive. It does, however, cover enough of the major impediments to clarify the nature of the challenge. Fortunately, there are many ways to remove or lower these hurdles to sound investment. Some of the more important options are outlined briefly below and are covered in greater detail in chapter 9.

Overcoming the Hurdles

To improve the quality and availability of drivepower information, utilities, energy agencies, manufacturers, universities, and private organizations should produce more and better publications, videos, seminars, and design and calculational aids; provide on-site technical assistance; and improve engineering curricula. Information programs should address several audiences, including policymakers, utility staff, motor users, electrical distributors, original equipment manufacturers, design engineers, and engineering faculty and students. Chapter 8 discusses the perspectives and needs of these various players in the motor market.

As chapter 9 discusses, efficiency data on all motors, and particularly in all motor catalogs, is critical for users to make informed purchases. Manufacturers stamp nominal and minimum motor efficiency on the nameplates of 1–125 hp NEMA Design A and B motors, but not on other sizes or types. Labeling should be expanded to additional categories and it should be made more accurate. Published efficiency ratings are often rounded down to one of 43 preset values. This practice can result in a discrepancy of up to two percentage points between the published and "actual" nominal efficiency of a motor. More accurate efficiency information is needed to give consumers a more precise indication of the performance to expect from a given motor.

Because most motors are purchased from catalogs, not off a showroom floor, complete efficiency information in catalogs is critical. In addition, efficiency data on the motor nameplate are important because they offer a permanent record of the motor's design efficiency, no matter how old the motor is and how many times it has changed hands. Over time, of course, degradation from poor rewinds and other damage will decrease many motors' efficiencies to levels lower than those inscribed on their nameplates.

In addition to better labeling, national minimum efficiency standards for motors are needed. With appliances, home heating and cooling systems, and cars, such standards have proven to be a relatively low-cost way of removing the worst products from the market, increasing efficiency, and saving the nation billions of dollars. As this handbook goes to press, legislation establishing minimum efficiency standards is making its way through the U.S. Congress.

Testing programs also are needed, to help motor users know exactly how their equipment is performing and how well it is sized, rather than relying on nominal nameplate ratings. Before-and-after testing of motors bound for the rewind shop can also play a big role in raising consumer awareness and improving motor repair practices. Testing of motors in the field and at the repair shop is covered in chapter 2.

Financial incentives may be useful in certain instances to overcome the perverse effects of the payback gap and motor users' limited access to capital. As of mid-1990, some two dozen utilities in the United States and Canada offered rebates for the purchase of efficient motors. The impacts of these programs have been modest, but a number of important lessons have been learned that should lead to greater success in future programs: rebates should be targeted to dealers as well as customers; high rebate levels tend to increase customer interest; extensive one-on-one contact with dealers and customers is necessary; and certification and educational measures, such as providing dealers with marketing support, lists of complying EEMs, and worksheets for calculating the savings, are useful. Chapter 9 covers the experience to date with drivepower rebate programs as well as other financial options, including loans, leasing, performance contracting, bidding, and tax credits.

Financial incentives should not be limited to programs between utilities and their customers. Internal policies can do a lot too by making incentives work for, rather than against, efficiency. Allowing individuals and departments to keep a share of the energy savings they generate for the firm will often stimulate impressive results.

Demonstrations of new technologies are very important for hastening their adoption. The Electric Power Research Institute, for example, has funded a series of carefully documented applications of ASDs in large industrial and power plant settings. But more of these demonstrations and further dissemination of their findings are needed to overcome equipment users' reluctance to innovate.

The complexities of energy analysis can often be simplified by rules of thumb. For example, the billion-dollar Southwire Company halved its energy use per unit of output in less than ten years. Southwire buys only high-efficiency motors and thus can negotiate with its motor distributor to get EEMs for only 5% more than the company would pay for standard motors. A failed motor under 125 hp is not rewound but is automatically replaced with a new EEM. When stringing wire to new loads under 100 amps, Southwire uses wire one size larger than required by code, and often does so for larger loads. These policies probably result in some uneconomic installations, but these instances are undoubtedly outnumbered by applications that save energy and money; this approach also eliminates the time required to analyze each energy-related installation.

We need research to develop improved hardware, better end-use data, and more insightful program and policy options. There are many opportunities to improve motors, controls, diagnostic equipment, and other hardware. Although motor systems use more than half of all electricity, the end-use data are sketchy and inconsistent. And because drivepower efficiency programs are uncommon and most are young, we have much to learn about what programs and policies work and why.

Finally, the programs for increasing drivepower efficiency need to be broader in scope. Most drivepower efficiency programs have focused solely on high-efficiency motors, although a few have included electronic ASDs. Virtually all programs have ignored other efficiency-related topics, such as motor sizing, rewinding, and controls other than ASDs. And no programs that we know of have addressed the savings available from electrical tune-ups, better selection and maintenance of drivetrains and bearings, better system monitoring, and the upsizing of distribution wires in new installations.

Chapter 2

Motor Technologies

otors produce useful work by causing a shaft to rotate.
The twisting force (torque) applied to the shaft is produced
by the interaction of two magnetic fields, one produced
by the fixed part of the motor (stator) and the other produced by
the rotating part of the motor (rotor). The forces developed in a
motor resemble the force between two magnets held close together—
similar poles repel each other, and dissimilar poles attract each
other. If one of the magnets is mounted on a shaft, the attracting
and repelling forces create torque (Figure 2-1).

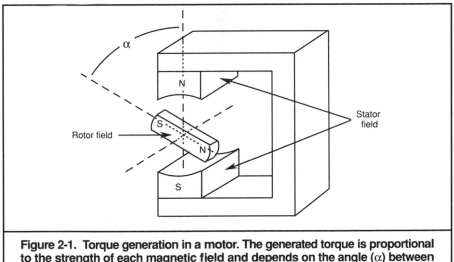

Figure 2-1. Torque generation in a motor. The generated torque is proportional
to the strength of each magnetic field and depends on the angle (α) between
the two fields. Mathematically, torque equals $B_{rotor} \times B_{stator} \times sine\ \alpha$ where B
refers to a magnetic field.

A magnetic field can be generated either by a permanent magnet, in which case the field is constant, or by a winding in which electric current flows. In the latter case, the magnetic field is generally proportional to the number of turns of wire in the winding and to the amount of current. The iron in the motor provides an easy path for the magnetic field in the same way that copper provides a low-resistance path for electric current. A wire with a low resistance to current flow has high conductivity; a material, like iron, with low resistance to a magnetic field has high permeability. Using high-permeability material in the magnetic circuits of the rotor and stator reduces the amount of current required to produce a given magnetic field.

There are three basic types of electric motors: alternating-current (AC) induction/asynchronous motors, AC synchronous motors, and direct-current (DC) motors. A detailed breakdown of motor types by horsepower and end use appears in chapter 6. Figure 2-2 shows the relative shares of electrical input used by different motor types. Because more than 90% of energy input goes to AC induction motors, this type is discussed in more detail than the others.

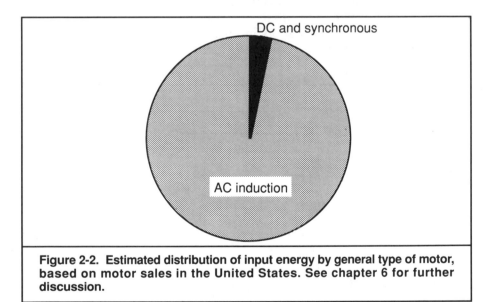

Figure 2-2. Estimated distribution of input energy by general type of motor, based on motor sales in the United States. See chapter 6 for further discussion.

Principles of Induction Motors

Induction motors can be categorized by whether they run on single- or three-phase power. Houses are usually supplied with single-phase electricity. As a result, household appliances such as refrigerators, washers, dryers, heat pumps, and furnaces use single-phase motors. Utilities provide commercial and industrial facilities with three-phase service, which is used to run most motors larger than 1 hp. The overwhelming majority of motors are single phase. Because of their relatively small size, however, these motors account for less than 20% of the total drivepower input in the United States.

Many single-phase motors are integrated with the equipment they drive, so when the motor fails, the equipment must be replaced. Three-phase motors are typically separate from equipment and can be replaced without changing the equipment. Three-phase motors are emphasized in this section because they use more energy and can be more readily replaced with high-efficiency models.

Rotating Field and Synchronous Speed

Three-phase induction motors, also called polyphase asynchronous motors, have three stator windings symmetrically arranged 120° apart in a cylinder surrounding the rotor. When supplied with three-phase power, also offset by 120°(Figure 2-3), the windings act as electromagnets and create a rotating magnetic field, which starts and then drives the motor.

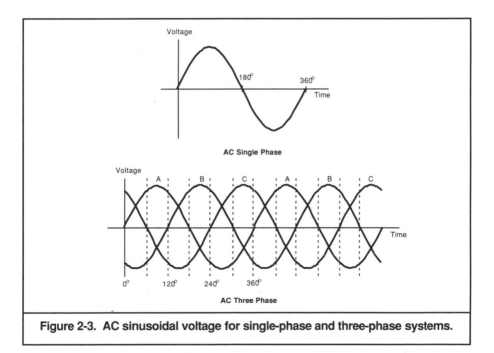

Figure 2-3. AC sinusoidal voltage for single-phase and three-phase systems.

Because a single-phase motor does not have a three-phase field, it needs a special starting system employing an auxiliary winding, offset 90° from the main winding and normally connected in series with a capacitor. In capacitor-start designs, the auxiliary winding and capacitor are disconnected after the motor starts. Motors that do not disconnect the capacitor are known as permanent split-capacitor (PSC) designs. The basic circuitry of a single-phase motor is shown in Figure 2-4.

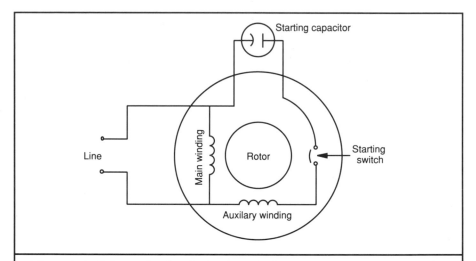

Figure 2-4. Schematic of a single-phase induction motor. In addition to the main winding there is an auxiliary winding offset by 90°, normally connected in series with a capacitor. The sum of the magnetic fields generated by the main and the auxiliary windings is a rotating field of north and south magnetic poles that revolves or "moves around" the stator. The changing magnetic field from the stator induces current in the rotor conductors, in turn creating the rotor magnetic field. Magnetic forces in the rotor tend to follow the stator magnetic fields, producing rotary motor action. In some designs (as shown) the auxiliary winding and capacitor are disconnected after start, by a centrifugal or thermal switch; such machines are commonly known as capacitor-start motors. Motors that do not disconnect the capacitor are known as permanent split-capacitor (PSC). There are also motors that combine the two designs, using one capacitor for starting, another for normal operation; these are known as capacitor-start, capacitor-run motors (Andreas 1982).

The speed of the rotating magnetic field in an induction motor, known as the synchronous speed, depends on the frequency of the supply voltage and the number of pole pairs in the motor, according to the following equation:

$$\text{synchronous speed (rpm)} = \frac{\text{frequency of the applied voltage (Hz)} \times 60}{\text{number of pole pairs}}$$

Thus, when a motor with two poles (one pole pair) is supplied by a 60 cycle-per-second (Hz) supply, the synchronous speed is 3,600 rpm. A four-pole motor supplied with 60 Hz power has a synchronous speed of 1,800 rpm.

Induction Motor Slip

Induction motors are referred to as asynchronous motors because they operate slightly below synchronous speed. For example, a motor with four poles and a synchronous speed of 1,800 rpm will actually spin between 1,725 rpm and 1,790 rpm.

The difference between the synchronous and actual speeds of an induction motor is called the motor slip. Slip is expressed either as a percentage of synchronous speed or as rpm. For example, a four-pole induction motor (synchronous speed of 1,800 rpm) operating at 1,750 rpm has a slip of 2.8% or 50 rpm. The full load motor slip ranges from 4% in small motors to 1% in large motors (see Figure 2-5).

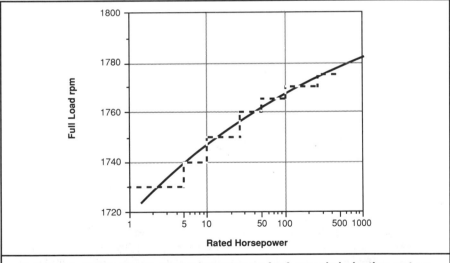

Figure 2-5. Full load rpm versus horsepower for four-pole induction motors. The stepped curve is data from typical motors; the smooth curve is fitted to the data. For four-pole induction motors, 1,800 rpm is the synchronous speed (approximately the speed under no load). The full load speed is less than the synchronous speed; this difference (or "slip") is smaller for larger motors (Nailen 1987).

Power Factor

The current in an induction motor has two components: active and reactive. The active component is responsible for the torque and work performed by the motor; the active part of the current is small at no load and rises as the load grows. The reactive component creates the rotating magnetic field; this component is almost constant from no load to full load, as is the magnetic field.

Although the reactive component does not perform useful work, it is required to excite the motor and has to be supplied by the power network. The ratio of active to total current is called the power factor (see Figure 2-6).

When the motor is operating at no load, the energy absorbed by the motor is limited to the power losses (motor inefficiencies), so the active component is small and the power factor can be as low as 10%. At full load the active component is at its maximum, so the power factor is typically 70–95% for a three-phase motor. A high power factor is desirable because it implies a low reactive power component. (Power factor is commonly expressed as a percentage or a decimal fraction.)

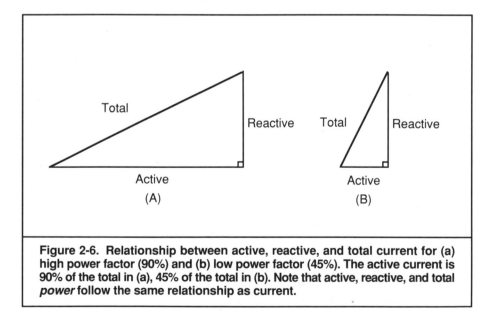

Figure 2-6. Relationship between active, reactive, and total current for (a) high power factor (90%) and (b) low power factor (45%). The active current is 90% of the total in (a), 45% of the total in (b). Note that active, reactive, and total *power* follow the same relationship as current.

A poor power factor has the following effects:

- Higher losses in the cables and transformers, and thus higher energy bills for a given amount of useful work output.

- Reduced available capacity of transformers, circuit breakers, and cables, whose capacity depends on the total current. The capacity falls linearly as the power factor decreases: a 1,000-kVA transformer supplying loads with a 70% power factor is only able to supply 700 kW.
- Higher voltage drops, yielding problems associated with undervoltage, as discussed in chapter 3.

These effects have caused most utilities to penalize consumers whose power factor is below a threshold level, typically in the range of 85–95%. Thus, when consumers improve the power factor, they reduce both the energy bill and the reactive power bill. As discussed in chapter 3, the savings from avoiding utility penalties are typically larger than the energy savings from power factor correction. Measures to improve power factor are discussed in chapters 3 and 4.

Types of Induction Motors

According to the rotor configuration, induction motors can be classified as either squirrel cage or wound rotor. Squirrel cage induction motors are the most common, and can be either three phase or single phase.

Squirrel Cage Induction Motors

Most induction motors contain a rotor in which the conductors, made of either aluminum or copper, are arranged in a cylindrical format resembling a "squirrel cage" (Figure 2-7, next page). Squirrel cage induction motors are used in the vast majority of commercial and industrial applications because they are relatively simple, inexpensive, reliable, and efficient.

Squirrel cage induction motors have no external electrical connections to the rotor, which is made of solid, uninsulated aluminum or copper bars short-circuited at both ends of the rotor with solid rings of the same metal. The rotor and the stator are connected by the magnetic field that crosses the air gap. This simple construction leads to relatively low maintenance requirements.

The relationship between torque and speed in squirrel cage motors depends largely on rotor resistance. As the rotor resistance decreases, the performance near synchronous speed improves, and the starting torque decreases. The smaller the slip for a given load, the higher the efficiency, because the induced rotor currents and their associated rotor losses also are smaller.

Three-phase squirrel cage induction motors predominate for

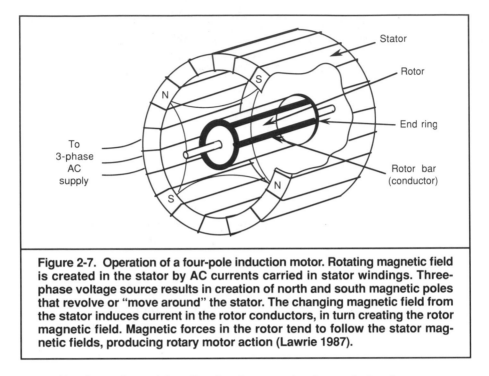

Figure 2-7. Operation of a four-pole induction motor. Rotating magnetic field is created in the stator by AC currents carried in stator windings. Three-phase voltage source results in creation of north and south magnetic poles that revolve or "move around" the stator. The changing magnetic field from the stator induces current in the rotor conductors, in turn creating the rotor magnetic field. Magnetic forces in the rotor tend to follow the stator magnetic fields, producing rotary motor action (Lawrie 1987).

applications above 1 hp. Single-phase squirrel cage induction motors are more common in sizes below 1 hp and in large home appliances. Single-phase motors are larger and more expensive and have lower efficiency than three-phase motors with the same power and speed ratings. For example, the efficiency of a 1/2-hp, 1,800-rpm, three-phase, standard-efficiency motor is 72.0% with a power factor of 62.0%, while the efficiency of a 1/2-hp single-phase motor from the same manufacturer is 66.2% with a power factor of 62.1%. Additionally, three-phase motors are more reliable, since they do not need special starting equipment. They thus are typically used whenever a three-phase supply is available. In commercial and industrial installations involving a large number of small motors, single-phase models have the further disadvantage of causing voltage unbalance if they are not evenly distributed on the three phases (see chapter 3 discussion of voltage unbalance).

Shaded-Pole Motors

Another type of induction design, the shaded-pole motor, is used most commonly in packaged equipment applications below 1/6 hp, such as computers, small fans found in portable heaters, and small condensing units for air conditioning and refrigeration. Although shaded-pole motors are cheaper than single-phase squirrel cage motors, their efficiency is poor (below 20%), and their use should be restricted to low-power applications with a limited number of operating hours. For low-power applications with higher operating hours, higher-efficiency single-phase motors should be used, such as permanent split-capacitor (a type of squirrel cage motor—see caption, Figure 2-4) or permanent magnet units (discussed later in this chapter).

Wound-Rotor Induction Motors

Wound-rotor induction motors are sometimes used in industrial applications—typically 20 hp or larger—where the starting current, torque, and speed need to be precisely controlled. As the name suggests, these motors feature insulated copper windings in the rotor similar to those in the stator. The rotor windings are fed with power using slip rings and brushes. This rotor construction is substantially more expensive and requires more maintenance than the squirrel cage type.

Factors to Consider in Selecting Induction Motors

Some of the factors described in this section apply to all kinds of motors, including noninduction designs. Others, like the NEMA design classes, apply exclusively to squirrel cage induction motors, particularly the three-phase versions.

NEMA Designs

The type of load determines the type of motor chosen to drive it. The National Electrical Manufacturers Association (NEMA) has defined standards for different types of squirrel cage designs to meet the needs of different operating conditions. The NEMA standard designs fall into four categories: A, B, C, and D.

- NEMA Design B motors are the dominant type of motor on the market and are used for most applications including fans, pumps, some compressors, and many other types of machinery. "Normal" torque is that produced by a Design B motor. The torque peaks at approximately 80% of the synchronous speed. Design B units have a normal starting current of approximately five times the full load current. In manufacturers' literature, general purpose motors are design B motors.

- NEMA Design A motors are similar to Design B motors except that the maximum torque is 15% to 25% higher. The starting current for these motors is six to seven times the full load current because of the design trade-offs necessary to increase the peak torque.
- NEMA Design C motors are high-torque motors. The high starting torque makes these motors useful for machines that can start with a full load (such as conveyors or some compressors).
- NEMA Design D motors have high starting torque and high slip. High slip allows the motor speed to vary somewhat from the rated speed. As a result, these motors are normally used where there is the potential for a shock load on the motor, as in punch presses and shears, because the motor's ability to adjust its speed will act as a shock absorber and protect the driven equipment. Because of the high slip, these motors have the lowest efficiency for a given size and speed.

In general, Design B motors are available in both standard and energy-efficient versions. Other NEMA designs are manufactured in small quantities and generally only in standard efficiency, except by special order.

Figure 2-8 shows the available torque as a function of speed for NEMA Design A, B, C, and D motors.

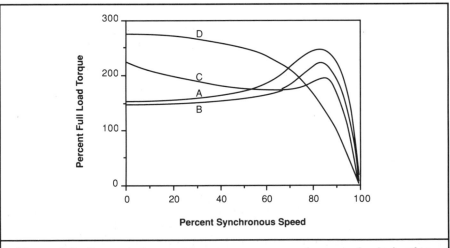

Figure 2-8. General shape of torque-versus-speed curves for induction motors with NEMA Designs A, B, C, and D. The torque at zero speed is the starting torque. The full load torque occurs at speeds somewhat below the synchronous speed (i.e. at the full load speed). See glossary for further discussion of torque (Smeaton 1987).

Available Speeds

As mentioned above, the number of pole pairs determines the synchronous speed of the motor. Induction motors are available with synchronous speeds of 3,600, 1,800, 1,200, 900, 720, 600, 450, and 300 rpm. Actual speeds are slightly lower because of the motor slip. For a given horsepower, as the speed decreases, the cost increases, and the efficiency and power factor go down (Figure 2-9).

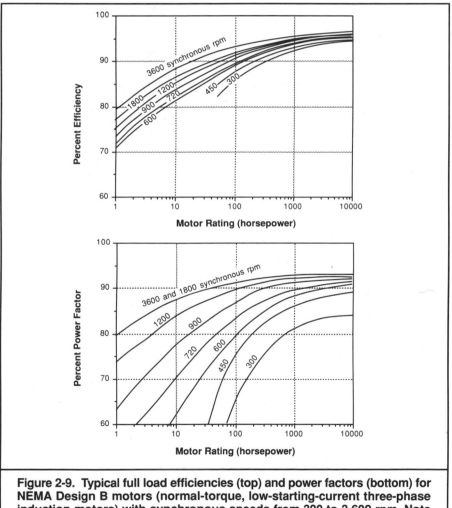

Figure 2-9. Typical full load efficiencies (top) and power factors (bottom) for NEMA Design B motors (normal-torque, low-starting-current three-phase induction motors) with synchronous speeds from 300 to 3,600 rpm. Note that these are general trends; values for particular motors may vary (Smeaton 1987).

The 1,800-rpm motor probably accounts for more than 50% of the motor population. Both 1,200 and 3,600-rpm motors are popular enough to be stocked by distributors and manufactured in large quantities. Motors slower than 1,200 rpm are often treated as special orders. For fixed speeds lower than 300 rpm, it is more economical to use a motor with a speed of 300 rpm or more, combined with a mechanical transmission system like gears or belts, to achieve the speed reduction.

Enclosures

Motor enclosures are designed to match the motor to the surroundings. An open enclosure allows heat to dissipate readily, leading to better motor cooling. It offers less protection, however, against the entry of potentially damaging foreign objects such as dirt, metal pieces, and water. NEMA Standard MG 1-1978 defines 20 types of enclosures clustered into 2 basic groups: open and totally enclosed. The most common enclosures used in commercial and industrial facilities are open drip-proof (ODP); totally enclosed fan-cooled (TEFC); and explosion proof (EXP), a type of TEFC motor (see Figure 2-10, page 24). Each of these three basic types of enclosures has subsets, listed in catalogs, for special environments (agricultural, corrosive, or wet conditions, for example).

Temperature Ratings and Classes of Insulation

Motor losses are transformed into heat, which increases the motor temperature. Table 2-1 shows the insulation class required to withstand different temperature rises, according to NEMA Standard

Table 2-1. Allowable temperature increases (°C) for various insulation classes (NEMA standards).

| | Insulation Class | | | |
	A[a]	B	F	H
Open or TEFC motors with 1.0 service factor, rise at rated load (°C)	60	80	105	125
All motors having 1.15 service factor, rise at 115% rated load (°C)	70	90	115	135[b]

[a]Of historical interest only. Class A insulation is no longer used in integral horsepower motors. Classes C, D, E, and G were never used.
[b]Not NEMA standard but common industry practice.
Source: Nailen 1987.

Motor Enclosures

The open enclosure types normally used are as follows:

- Drip-proof (ODP), in which the ventilation openings are positioned to keep out liquid or solid particles falling at any angle from 0° to 15° from the vertical. This enclosure is not adequate for harsh environments, but is very common in driving the fans and pumps in indoor HVAC applications.

- Splash-proof, which is the same as drip-proof except that the downward angle of the vents is increased to 100° so that liquid or solid particles arriving at a slightly upward angle will not enter the motor.

- Guarded, in which all openings giving direct access to rotating or electrically live metal parts are limited in size by screens, baffles, grilles, or other barriers to the entry of objects larger than 0.75" in diameter. Thus, insects and dirt are not prevented from entering. The purpose of guarding fittings is more to protect personnel than to protect the motor.

- Weather protected, for outdoor use, include guarded enclosures and ventilation passages designed to minimize the entrance of rain, snow, and airborne particles.

The totally enclosed machines are designed to prevent the free exchange of air between the inside and the outside. The most common types are as follows:

- Totally enclosed fan-cooled (TEFC), in which the motor is equipped with a fan for external cooling. Normally the external fan is mounted on the shaft opposite the load and equipped with a guard to improve safety and aerodynamics.

- Explosion-proof (EXP), where the enclosure is designed to withstand the explosion of a specified gas or vapor within and prevent ignition of a specified external gas or vapor by sparks, flashes, or explosions that may occur within the motor casing. These enclosures may be fan-cooled (EXPFC) or nonventilated (EXPNV).

- Dust-ignition-proof, designed to exclude dust, ignitable or not, that might affect performance or rating, and prevent arcs, sparks, or heat generated within from igniting external dust.

(a)

(b)

Figure 2-10. Motor enclosure types. The most common motor enclosures are: (a) Open Drip-Proof (ODP) and (b) Totally Enclosed Fan-Cooled (TEFC). In ODP motors such as the one shown, internal fans bring cooling air into the motor through openings in one or both ends, then discharge it through openings in the side. Another ODP design brings in air through one end and discharges it through the other end. In either case, the cooling air flows directly through the motor. In TEFC designs, there is no air exchange between the inside and outside of the motor. A fan, driven by an extension of the motor shaft, and located here in the smooth housing on the left end of the motor, pulls air through slots in its housing, then blows it over the exterior of the motor, which is usually made with fins (as shown) for cooling. Figure 2-20 is a cutaway view of a TEFC motor; the internal arrangement of both motor types is basically the same (courtesy of Toshiba).

MG1-1978. Class A is no longer made. Class B is the most common. Classes C, D, E, and G were never used. Classes F and H are used in applications with high ambient temperatures or to allow a larger reserve margin for overload conditions or to enable the design of smaller, less expensive motors for intermittent duty operation. Most energy-efficient motors use Class F insulation as part of the general package to upgrade the motor performance.

Service Factor

The service factor specifies the capacity of the motor to withstand prolonged overload conditions. When the service factor is 1.0, prolonged operation above full load can damage the insulation and cause the motor to fail. If the service factor is 1.15, the motor can work at 1.15 times its rated horsepower without failing, although the insulation life may be reduced (typically by 50% when compared with the same motor working at full load). Standard service factors for 3,600- and 1,800-rpm motors range as high as 1.35 for 0.5 hp and smaller, 1.25 for 0.5 and 0.75 hp, and 1.15 for 1 hp and above. Motors running at 1,200 and 900 rpm generally have lower service factors. However, service factors of 1.5 or more are available on special order.

In general, the class of insulation on the motor windings determines the service factor. Motors above 1 hp with class B insulation have a service factor of 1.0 while motors with class F insulation have a service factor of 1.15.

Frame Size

The frame size defines the shape and size of the motor and depends upon horsepower, speed, voltage, and duty requirements. Motors built prior to 1952 did not use industry-wide standard frame sizes. In 1952 the industry standardized the U-frame, so all motors with the same code, such as 254U, had the same frame size. With the advent of new, high-temperature insulation, NEMA in 1964 authorized smaller, lighter T-frames, which remain the prevalent frame type today for new three-phase motors. Most small (under 1.0 hp) and very large (over 300 hp) motors use frames other than T- or U-designs. Standard-efficiency U-frame designs are still made for replacing worn-out or damaged U-frames, because replacing a U-frame motor with a standard- or high-efficiency T-frame unit typically requires modification of the mounting hardware and is not practical in all applications. Thus, not all existing U-frame motors can be replaced with high-efficiency T-frame models.

Frame size is an important determinant of motor efficiency and performance. To reduce production cost, manufacturers often try to fit a motor into the smallest possible frame, but, in so doing, they must limit the service factor to insure that the motor will not overheat under the reduced cooling of the smaller frame. Thus, some manufacturers build 5-hp motors in frames typically used for 3-hp motors; to meet cooling requirements, such motors will have a service factor no higher than 1.0 (Gilmore 1990).

Supply Voltage

Most three-phase motors are designed to operate at 460 volts, which allows for some voltage drop from the nominal 480-V supply

---

How to Read a Motor Catalog

The specific information presented in a motor catalog and the format for that information vary among manufacturers. Figures 2-11 and 2-12 show sample pages from one catalog.

Most catalogs cluster information on specific motor types. For example, all ODP standard-efficiency motors are listed in a single table that contains generic information on the type of mounting system, the housing, the materials of construction, the insulation class, the service factor, and the design rating. In most catalogs, the general listing also will specify whether the motor line is standard or premium efficiency.

In addition to generic information, other tables will list size, speed, frame number, full load amps, and list price for individual motors. Most motors actually sell for 30% to 70% of the list price.

Most catalogs list in a separate table motor efficiencies and power factors at full, 3/4, and 1/2 load. This table also typically contains data on motor torque.

Finally, most motor catalogs include a table of dimensional data organized by the motor frame number. In the NEMA system, all motors with the same frame number are the same size. Dimensional data are generally used to determine the changes required in the mounting system or drive shaft when downsizing a motor or converting from a U-frame motor to a T-frame motor.

---

commonly used in newer large commercial and industrial facilities. Smaller commercial facilities often use either 230-V three-phase or 208-V three-phase power. Older facilities (both commercial and industrial) sometimes use 575-V three-phase power. To decrease the distribution losses in cables and transformers, large motors (200 hp and above) can be specified with a supply voltage over 600V. Fractional horsepower single-phase motors most commonly run on 120–V power. Specialty motors are available to run on two voltages (230/460 V, for example) or on a range of voltages and some other, typically doubled, value (such as 208–230/460 V).

Other Types of Motors

Although induction models use more than 90% of all motor input energy, they are not appropriate for all applications. The following varieties of motors are also important.

Figure 2-11. Typical motor application data from a major motor manufacturer's catalog (General Electric 1991).

Totally Enclosed Fan Cooled - Severe Duty

HP	RPM	Volts	NEMA Frame	SF	Brgs.	Encl.	Prot.	Base	Full Load Amps @ NP Volts	Catalog No.	List Price	Multiplier Symbol	Shpg. Wt.	"C" Dim.	Notes
1/3	1725	230/460	56	1.15	Ball	TENV	None	WB	1.5/.8	K264	$294	GO-1	25	10.0	
1/2	1725	230/460	56	1.15	Ball	TENV	None	WB	1.9/1.0	K265	320	GO-1	27	10.9	
3/4	1725	230/460	56	1.00	Ball	TENV	None	WB	2.6/1.3	K266	346	GO-1	38	11.9	CI 29
3/4	1200	230/460	143T	1.15	Ball	TEFC	None	Ftd	3.0/1.5	K347	407	GO-1	56	12.4	CI 29
• 3/4	1140	208-230/460	143T	1.15	Ball	TEFC	None	WB	2.8-2.8/1.4	E34S6	509	GO-1	35	13.0	E$
1	1800	230/460	143T	1.15	Ball	TEFC	None	Ftd	3.4/1.7	K348	346	GO-1	56	12.4	CI 29
1	1800	575	143T	1.15	Ball	TEFC	None	Ftd	1.4	K1439	346	GO-1	56	12.4	CI
• 1	1725	208-230/460	143T	1.15	Ball	TEFC	None	Ftd	3.6-3.4/1.7	E10S1	433	GO-1	39	13.0	E$
1	1200	230/460	145T	1.15	Ball	TEFC	None	Ftd	4.2/2.1	K349	436	GO-1	63	13.4	CI 29
1	1200	575	145T	1.15	Ball	TEFC	None	Ftd	1.7	K1440	436	GO-1	63	13.4	CI
• 1	1140	208-230/460	145T	1.15	Ball	TEFC	None	WB	3.8-3.6/1.8	E10S6	545	GO-7XSD	48	14.3	E$
• 1	900	200-230/460	182T	1.15	Ball	TEFC	None	Ftd	4.2/2.1	E9871	646	GO-7XSD	109	15.9	E$ CI
• 1	900	460	182T	1.15	Ball	TEFC	None	Ftd	2.1	E9837	646	GO-7XSD	109	15.9	CI
• 1	900	575	182T	1.15	Ball	TEFC	None	Ftd	1.7	E9872	646	GO-7XSD	109	15.9	CI
1 1/2	3600	230/460	143T	1.15	Ball	TEFC	None	Ftd	4.0/2.0	K350	355	GO-1	56	12.4	CI 29
1 1/2	3600	575	143T	1.15	Ball	TEFC	None	Ftd	1.0	K1441	355	GO-1	56	12.4	CI
• 1 1/2	3450	208-230/460	143T	1.15	Ball	TEFC	None	WB	4.6-4.2/2.1	E15S2	444	GO-1	35	13.0	E$
1 1/2	1800	230/460	145T	1.15	Ball	TEFC	None	Ftd	5.2/2.6	K351	377	GO-1	63	13.4	CI 29
1 1/2	1800	575	145T	1.15	Ball	TEFC	None	Ftd	2.1	K1442	377	GO-1	63	13.4	CI
• 1 1/2	1725	208-230/460	145T	1.15	Ball	TEFC	None	WB	4.6-4.2/2.1	E15S1	471	GO-1	44	13.8	E$
• 1 1/2	1200	200-230/460	182T	1.15	Ball	TEFC	None	Ftd	4.6/2.3	E9702	507	GO-7XSD	109	15.9	E$
1 1/2	1200	200-230/460	182T	1.15	Ball	TEFC	None	Ftd	4.8/2.4	N253	474	GO-7KW	109	15.9	CI
• 1 1/2	1200	460	182T	1.15	Ball	TEFC	None	Ftd	2.3	E9943	507	GO-7XSD	109	15.9	E$ CI
• 1 1/2	1200	575	182T	1.15	Ball	TEFC	None	Ftd	1.8	E9703	507	GO-7XSD	109	15.9	E$ CI
1 1/2	1200	575	182T	1.15	Ball	TEFC	None	Ftd	1.9	N245	474	GO-7KW	109	15.9	CI
• 1 1/2	900	200-230/460	184T	1.15	Ball	TEFC	None	Ftd	6.0/3.0	E9873	764	GO-7XSD	120	15.9	E$ CI
• 1 1/2	900	460	184T	1.15	Ball	TEFC	None	Ftd	3.0	E9888	764	GO-7XSD	120	15.9	E$ CI
• 1 1/2	900	575	184T	1.15	Ball	TEFC	None	Ftd	2.4	E9874	764	GO-7XSD	120	15.9	CI
2	3600	230/460	145T	1.15	Ball	TEFC	None	Ftd	5.4/2.7	K353	438	GO-1	63	13.4	CI 29
2	3600	575	145T	1.15	Ball	TEFC	None	Ftd	2.2	K1443	438	GO-1	63	13.4	CI
• 2	3450	208-230/460	145T	1.15	Ball	TEFC	None	WB	5.8-5.4/2.7	E20S2	548	GO-1	40	13.8	E$
2	1800	230/460	145T	1.15	Ball	TEFC	None	Ftd	5.8/2.9	K354	410	GO-1	63	13.4	CI 29
2	1800	575	145T	1.15	Ball	TEFC	None	Ftd	2.3	K1444	410	GO-1	63	13.4	CI
• 2	1725	208-230/460	145T	1.15	Ball	TEFC	None	WB	6.2-5.8/2.9	E20S1	513	GO-1	53	14.3	E$
• 2	1200	200-230/460	184T	1.15	Ball	TEFC	None	Ftd	5.8/2.9	E97C4	566	GO-7XSD	120	15.9	E$ CI
2	1200	200-230/460	184T	1.15	Ball	TEFC	None	Ftd	7.0/3.5	N254	530	GO-7KW	120	15.9	CI
• 2	1200	460	184T	1.15	Ball	TEFC	None	Ftd	2.9	E9944	566	GO-7XSD	120	15.9	E$ CI
• 2	1200	575	184T	1.15	Ball	TEFC	None	Ftd	2.4	E9706	566	GO-7XSD	120	15.9	E$ CI
2	1200	575	184T	1.15	Ball	TEFC	None	Ftd	2.6	N246	530	GO-7KW	120	15.9	CI
• 2	900	200-230/460	213T	1.15	Ball	TEFC	None	Ftd	7.2/3.6	E9875	985	GO-7XSD	192	20.1	E$ CI
• 2	900	460	213T	1.15	Ball	TEFC	None	Ftd	3.6	E9705	985	GO-7XSD	200	20.2	E$ CI
• 2	900	575	213T	1.15	Ball	TEFC	None	Ftd	2.9	E9876	985	GO-7XSD	192	20.1	CI

• E$ Energy $aver, premium efficiency motor.
CI Cast iron frame and endshields.
29 Usable at 200 volts at 1.00 service factor.

Figure 2-12. Typical motor performance data from a major motor manufacturer's catalog (General Electric 1991).

Totally Enclosed Fan Cooled, Frames 182-449

Horse power	Full-load RPM	Amperes: Full-load @ 460v	Amperes: Full-load @ 200v	Amperes: NEMA Locked Rotor (max.)	NEMA Code Letter	Torque: Full-load Lb.-ft.	Torque: NEMA ST %FL (avg.)	Torque: NEMA BD %FL (avg.)	Efficiency ⑩: Nom-inal ①	Efficiency ⑩: Full Load Guar. Min. TEFC	Efficiency ⑩: Full Load Guar. Min. Cast-iron X$D	Efficiency ⑩: 3/4 load ①	Efficiency ⑩: 1/2-load ①	Power Factor ①: Full-load	Power Factor ①: 3/4-load	Power Factor ①: 1/2-load	Max. KVAR ③	dBA Sound Press. ① ⑨
1	885	2.1	4.0	15.0	N	6.1	175	240	75.5	-	74.0	75.9	71.4	60.5	53.7	41.9	1.2	-
1 1/2	1170	2.3	4.8	20.0	M	6.7	248	347	87.5	85.5	86.5	88.4	86.4	72.0	66.2	53.5	1.0	51
	865	3.0	5.9	20.0	M	9.1	193	253	80.0	-	78.5	80.8	77.4	60.0	53.1	41.1	1.7	-
2	1165	2.9	6.3	25.0	L	9.0	242	336	87.5	85.5	86.5	88.6	87.2	74.0	68.4	55.8	1.3	51
	880	4.1	-	25.0	L	11.9	130	210	85.5	-	81.5	84.8	80.9	54.0	47.4	36.1	2.2	-
3	3520	3.7	8.2	32.0	K	4.5	218	335	88.5	86.5	87.5	90.4	89.2	87.0	86.6	78.8	0.9	53
	1765	4.0	8.7	32.0	K	8.9	267	382	89.5	87.5	88.5	90.3	88.8	80.0	75.8	64.6	1.5	59
	1175	4.3	9.1	32.0	K	13.4	243	333	89.5	87.5	88.5	91.0	89.3	74.5	69.2	57.1	1.8	57
	875	5.2	-	32.0	K	18.0	130	205	86.5	-	82.5	86.8	84.5	62.5	55.9	43.8	2.5	-
5	3515	6.0	13.9	46.0	J	7.5	210	330	89.5	87.5	88.5	90.2	88.9	88.0	87.9	81.2	1.4	73
	1755	6.3	14.3	46.0	J	15.0	247	335	90.2	88.5	89.5	91.3	90.7	83.0	70.1	70.7	2.0	59
	1170	7.0	15.1	46.0	J	22.4	261	352	89.5	87.5	88.5	91.6	90.5	75.0	70.8	58.3	2.9	57
	875	7.4	-	46.0	J	29.9	130	205	89.5	-	84.0	89.9	88.7	70.5	64.9	52.5	3.1	-
7 1/2	3530	8.7	20.2	63.5	H	11.2	152	273	91.7	90.2	91.0	93.1	92.4	88.0	89.1	83.9	1.7	64
	1765	9.3	21.9	63.5	H	22.3	212	252	91.7	90.2	91.0	92.4	91.6	82.5	81.0	73.4	2.5	65
	1180	10.7	23.5	63.5	H	33.4	202	244	91.7	90.2	91.0	92.3	91.0	72.0	67.6	56.2	4.4	58
	875	10.9	-	63.5	H	44.9	125	200	89.5	-	85.5	90.5	89.6	72.0	66.4	53.8	4.5	-
10	3525	11.6	26.8	81.0	H	14.9	161	284	91.7	90.2	91.0	93.3	92.9	88.5	89.1	83.7	2.3	64
	1760	12.7	29.4	81.0	H	29.8	216	252	91.7	90.2	91.0	92.6	92.1	81.0	79.6	71.7	3.6	65
	1175	14.3	31.4	81.0	H	44.6	207	245	91.7	90.2	91.0	92.4	91.4	71.5	67.3	56.0	5.9	58
	885	13.2	-	81.0	H	59.5	125	200	91.0	-	90.2	91.5	90.8	78.0	73.9	63.2	4.4	-
15	3545	17.3	39.6	116.0	G	22.2	169	298	91.7	90.2	91.0	92.8	91.6	88.5	88.8	83.2	3.6	71
	1770	18.7	43.6	116.0	G	44.5	183	230	92.4	91.0	91.7	93.4	92.7	81.5	80.2	72.3	5.3	69
	1180	20.1	45.8	116.0	G	66.9	169	195	91.7	90.2	91.0	92.1	91.3	76.5	74.8	65.7	6.6	60
	880	19.8	-	116.0	G	89.3	125	200	91.0	-	90.2	91.9	91.6	78.0	74.5	63.9	6.4	63
20	3540	22.5	52.1	145.0	G	29.7	181	292	92.4	91.0	91.7	93.8	92.9	90.0	91.2	87.4	3.7	71
	1765	24.4	58.5	145.0	G	59.5	178	211	93.0	91.7	92.4	93.9	93.6	82.5	82.1	75.7	6.0	69
	1175	26.7	63.1	145.0	G	89.5	156	186	92.4	91.0	91.7	93.4	93.3	76.0	75.2	67.4	8.2	60
	880	26.3	-	145.0	G	118.7	125	200	91.7	90.6	91.0	92.5	92.0	77.5	74.6	64.5	8.3	64
25	3560	27.9	64.8	217.5	G	36.9	197	279	92.4	91.0	91.7	93.2	92.1	91.0	91.8	88.2	4.6	75
	1770	30.0	70.7	182.5	G	74.2	162	205	93.6	92.4	93.0	94.4	94.1	83.5	83.6	78.0	6.9	64
	1175	30.4	-	182.5	G	111.2	135	200	92.4	91.4	91.7	93.3	93.1	83.5	83.2	76.9	6.7	64
	890	32.8	-	182.5	G	148.6	125	200	91.7	90.6	91.0	92.7	92.4	78.0	75.6	66.0	9.8	64
30	3555	33.4	77.8	182.5	G	44.3	194	274	92.4	91.0	91.7	93.4	92.6	91.0	91.8	88.0	5.6	75
	1765	36.0	85.8	217.5	G	89.1	165	206	93.6	92.4	93.0	94.4	94.3	83.5	83.6	78.1	8.2	64
	1175	36.2	-	217.5	G	133.5	135	200	93.0	92.0	92.4	93.6	93.5	83.5	83.0	76.6	8.1	66
	890	39.2	-	217.5	G	176.9	125	200	93.6	92.7	93.0	93.9	93.3	76.5	73.0	62.5	12.6	66

Figure 2-12 (continued).

Totally Enclosed Fan Cooled, Frames 182-449 (cont'd)

Horse power	Full-load RPM	Amperes: Full-load @ 460v	Amperes: Full-load @ 200v	NEMA Locked Rotor (max.)	NEMA Code Letter	Torque: Full-load Lb.-ft.	NEMA ST %FL (avg.)	NEMA BD %FL (avg.)	Efficiency: Nominal	Full Load Guar. Min. TEFC	Full Load Guar. Min. Cast-iron X$D	3/4 load	1/2 load	Power Factor: Full-load	Power Factor: 3/4-load	Power Factor: 1/2-load	Max. KVAR	dBA Sound Press.
40	3565	43.7	-	290.0	G	58.9	130	200	93.6	92.7	93.0	93.5	92.5	92.0	93.4	91.1	5.5	80
	1780	45.5	-	290.0	G	118.0	140	200	94.1	93.3	93.0	94.6	94.3	87.5	86.7	81.2	9.1	74
	1185	46.8	-	290.0	G	177.2	135	200	93.6	92.7	93.0	94.2	93.9	85.5	84.4	77.9	10.1	66
	890	53.3	-	290.0	G	236.1	125	200	93.0	92.0	92.4	93.7	93.1	75.5	71.6	60.8	17.8	66
50	3560	54.7	-	362.5	G	73.7	120	200	93.0	92.0	92.4	93.7	93.2	92.0	93.6	92.1	6.1	80
	1775	57.5	-	362.5	G	147.5	140	200	94.1	93.3	93.6	94.6	94.6	86.5	85.9	80.1	12.0	74
	1185	58.5	-	362.5	G	221.5	135	200	93.6	92.7	93.0	94.4	94.2	85.5	84.6	78.2	12.5	66
	890	60.3	-	362.5	G	295.6	125	200	93.6	93.0	93.0	94.5	94.5	84.0	82.6	75.5	13.6	68
60	3570	65.3	-	435.0	G	88.5	120	200	94.1	93.3	93.6	94.4	93.8	92.0	93.7	91.6	7.6	80
	1785	69.9	-	435.0	G	176.0	140	200	95.0	94.3	94.5	95.4	94.7	84.5	83.2	75.8	16.7	80
	1190	69.9	-	435.0	G	264.7	130	200	94.1	93.3	93.6	94.7	94.5	85.5	84.5	78.2	14.9	68
	885	71.9	-	435.0	G	354.9	125	200	93.6	92.7	93.0	94.6	94.7	83.5	82.5	75.4	16.4	68
75	3570	81.2	-	542.5	G	110.3	105	200	94.5	93.7	94.1	94.9	94.4	91.5	92.8	90.1	10.7	80
	1785	86.1	-	542.5	G	220.6	140	200	95.4	94.7	95.0	95.7	95.3	85.5	84.4	78.0	18.8	80
	1190	86.0	-	542.5	G	330.9	135	200	95.0	94.3	94.5	95.2	94.8	86.0	85.0	78.8	18.1	68
	885	87.8	-	542.5	G	444.0	125	200	94.1	93.3	93.6	94.8	94.8	85.0	83.9	77.5	18.5	74
100	3570	109.0	-	725.0	G	147.1	105	200	94.1	93.3	93.6	94.0	93.0	91.5	92.7	90.0	14.7	80
	1790	115.0	-	725.0	G	293.3	125	200	95.4	94.7	95.0	95.9	95.3	85.0	83.0	75.0	28.8	80
	1190	111.0	-	725.0	G	441.0	125	200	95.0	94.3	94.5	95.4	95.3	88.5	88.6	84.3	18.8	74
	885	119.8	-	725.0	G	592.3	125	200	94.1	93.3	93.6	94.8	94.8	84.0	82.6	75.4	27.1	74
125	3575	132.0	-	907.5	G	183.8	100	200	94.5	93.7	94.1	94.5	93.7	93.5	95.1	93.8	12.9	80
	1785	137.0	-	907.5	G	367.6	110	200	95.4	94.7	95.0	95.7	95.3	89.0	88.1	84.4	27.1	80
	1190	139.0	-	907.5	G	551.0	125	200	95.0	94.3	94.5	95.6	95.4	88.5	88.6	84.4	23.3	74
	885	150.0	-	907.5	G	740.5	120	200	94.5	93.7	94.1	95.1	95.0	82.5	80.4	71.8	39.0	80
150	3575	158.0	-	1085.0	G	220.3	100	200	94.5	93.7	94.1	94.8	94.2	93.5	95.0	93.7	15.6	80
	1790	166.0	-	1085.0	G	439.9	110	200	95.8	95.2	95.4	96.1	95.6	88.0	86.9	81.0	34.0	80
	1190	171.0	-	1085.0	G	661.0	120	200	95.8	95.2	95.4	96.0	95.7	85.5	84.0	76.8	39.1	74
	885	180.0	-	1085.0	G	889.0	120	200	94.5	93.7	94.1	95.1	95.3	82.5	79.9	70.8	49.0	80
200	3575	210.0	-	1450.0	G	293.7	100	200	95.0	94.3	94.5	94.7	93.7	94.0	95.0	93.5	22.0	80
	1785	216.0	-	1450.0	G	588.2	110	200	95.8	95.2	95.4	96.1	96.0	90.0	89.7	85.4	36.0	80
	1190	227.0	-	1450.0	G	882.0	120	200	95.4	94.7	95.0	96.1	96.0	86.5	85.8	79.7	46.3	80
250	3575	274.0	-	1812.0	G	367.1	70	175	95.4	94.7	95.0	95.5	95.6	89.5	90.6	86.6	45.1	91 ⑦
	1790	293.0	-	1812.0	G	733.2	80	175	96.2	95.7	95.8	96.5	96.1	83.0	80.4	71.7	81.0	80
	1190	285.0	-	1812.0	G	1102.0	100	175	95.4	94.7	95.0	96.1	96.1	86.0	84.6	77.7	63.8	80
300	1785	348.0	-	2200.0	G	882.0	80	175	95.8	95.2	95.4	96.3	96.3	84.0	82.7	75.1	84.1	80

① Average expected values – do not use as guaranteed values. Efficiency, speed, torque, power factor and sound values are the same for 200, 230, or 575 volts. Current values vary inversely with voltage.

② Recommended maximum capacitor rating when capacitor and motor are switched as a unit.

③ Motor with sound pressure of 80 dBA can be provided at 1.0 service factor.

④ Sound Power dBA - re 10⁻¹² watts; Sound Pressure (dBA) measured in a free field with a reference pressure of 0.00002 pascals, average reading at three feet.

⑤ Tested in accordance with IEEE Standard 112, Test Method B, using accuracy improvement by segregated loss determination including stray load loss improvement as specified in NEMA standard MG1-12.53a.

Synchronous Motors

Synchronous motors have a stator similar to that of induction motors, with three windings that produce a rotating field. The rotor contains a winding to produce the rotor field and a starting winding similar to the rotor of a squirrel cage induction motor; the connection from the power supply to the rotor field winding is made through slip rings and brushes. Because of their complex rotors, synchronous motors are more expensive to build and maintain than induction motors.

The starting winding makes the motor act like an induction motor up to about 95% of the synchronous speed. At that point, the rotor field winding is switched on and the rotor quickly catches up to the rotating field, reaching the synchronous speed. For further information on the operation of synchronous motors refer to Nailen 1987, Fitzgerald 1983, or Smeaton 1987.

Synchronous motors can run at lower speeds and are slightly more efficient than induction motors, especially at low speeds. They also have the virtue of being able to generate or absorb reactive power, whereas induction motors only absorb reactive power. A large synchronous motor can thus correct the overall power factor of an entire plant by generating the reactive power absorbed by the induction motors in the plant.

Figure 2-13 shows the typical speed and horsepower ranges of induction and synchronous motors. Synchronous motors are used in applications where fixed constant speed is required, such as in the textile fiber industry or in large-power, low-speed applications, where the motor's additional cost is offset by its higher efficiency and capability for power factor compensation. Synchronous motors tend to be large and in operation most of the time. There are not many of them in use, and their number has been decreasing in recent years. Their percentage contribution to energy and power demands is therefore small.

Direct-Current Motors

Direct-current (DC) motors normally have windings in both the stator and the rotor. As the name implies, DC motors are fed by a DC voltage, which may change in magnitude but not in polarity. The magnetic field produced by the stator thus has a constant orientation, but its size may change as a function of the voltage applied to the terminals. DC motors are often used for applications where speed control is required, since varying the voltage varies the motor speed.

Electricity reaches the windings in the rotor via a ring of electrically isolated copper bars, a device known as a commutator

(Figure 2-14). Corresponding contacts known as brushes are connected to the power supply and ride against this commutator. DC motor rotors are complex, expensive to manufacture, and unreliable because of wear on the brushes and commutator caused by sparking and friction as the rotor turns. The wear creates the need for

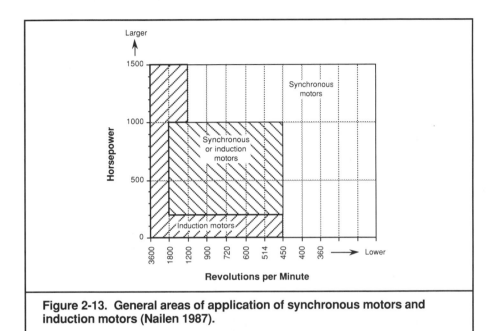

Figure 2-13. General areas of application of synchronous motors and induction motors (Nailen 1987).

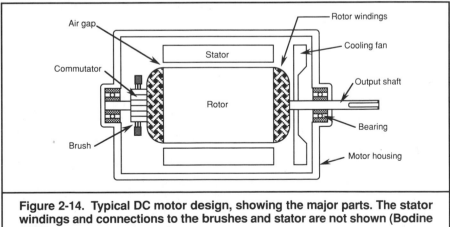

Figure 2-14. Typical DC motor design, showing the major parts. The stator windings and connections to the brushes and stator are not shown (Bodine 1978).

frequent inspection and replacement of the brushes. In addition, the commutator must be machined or replaced at longer intervals.

DC motors have other drawbacks as well. Especially in larger sizes, they are bulky, cannot sustain high speeds, and are less efficient than AC motors of similar size.

A type of DC motor, in which the rotor and stator windings are connected in series, can also be used with AC voltages (because the torque in series motors maintains the same direction even if the polarity of the voltage is reversed). These motors are known as universal motors, which are generally found in small portable appliances and power tools. For fractional horsepower sizes the universal motor has a superior power to weight ratio. Universal motors generally operate for very limited periods, so their energy use is not significant.

Besides being used in low energy-use applications such as small appliances and power tools, DC motors are still being sold for industrial applications requiring very high starting torque or inexpensive speed regulation. However, their market share is dwindling to below 5%, with the advent of high performance AC drives.

Permanent-Magnet Motors

In some small DC models, a permanent magnet replaces the stator winding, but the rotor is still fed by a conventional brush-and-commutator system. A more important type of permanent-magnet (PM) motor has a stator with three windings, producing a rotating field as in induction and synchronous motors. The rotor consists of one or more permanent magnets that interact with the rotating field so as to align the poles in the rotor with the poles of the rotating field. Thus the speed of the motor is the speed of the rotating field. Because there is no rotor current, and the rotor magnetic field is constant, there are no losses in the rotor, helping to make PM motors more efficient (by five to ten percentage points in small sizes) than induction motors.

The most common form of PM motor is the brushless DC motor, also known as an electronically commutated motor (see Figure 2-15). This motor is often used with electronic speed controls. It runs either directly off a DC supply or from an AC/DC converter. The function of the commutator and brushes in conventional DC motors is replaced in permanent magnet designs by power transistors that sequentially switch the voltages applied to the three windings, in order to create a rotating field. Figure 2-16 shows the comparative performance of AC induction motors with ASD and brushless DC motors, rated at 10 hp.

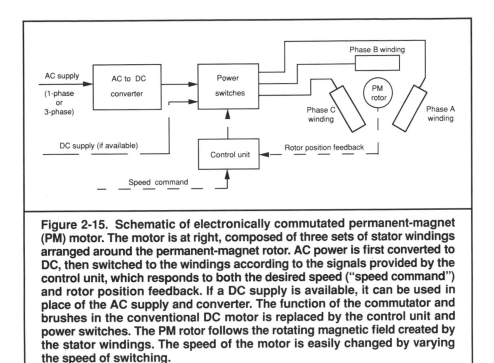

Figure 2-15. Schematic of electronically commutated permanent-magnet (PM) motor. The motor is at right, composed of three sets of stator windings arranged around the permanent-magnet rotor. AC power is first converted to DC, then switched to the windings according to the signals provided by the control unit, which responds to both the desired speed ("speed command") and rotor position feedback. If a DC supply is available, it can be used in place of the AC supply and converter. The function of the commutator and brushes in the conventional DC motor is replaced by the control unit and power switches. The PM rotor follows the rotating magnetic field created by the stator windings. The speed of the motor is easily changed by varying the speed of switching.

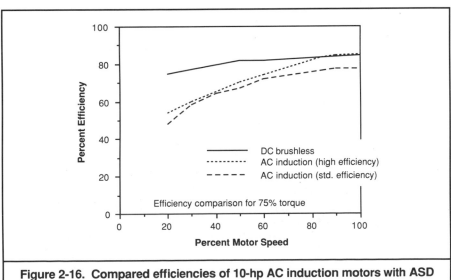

Figure 2-16. Compared efficiencies of 10-hp AC induction motors with ASD and brushless DC motors as a function of speed (Lovins et al. 1989).

Other kinds of permanent-magnet motors include (1) small DC motors that have brushes, commutator, and wound rotor plus a permanent-magnet stator, and (2) a type of AC synchronous motor available on special order from Siemens U.S. and Reliance Electric (Lovins 1989).

Permanent-magnet motors have generally been limited to fractional-horsepower sizes because they are bulkier and much more expensive than induction motors. The magnets in most fractional-horsepower PM motors, such as the motors used in residential appliances, are made from ferrite, primarily because of its low cost. In recent years, however, the performance of PM materials has improved dramatically (see Figure 2-17). In particular, neodymium-iron-boron alloys feature high energy density at moderate cost. Such PM materials allow the design of compact and high-efficiency motors in larger sizes.

One U.S. firm, Powertec, sells off-the-shelf PM motors of up to 300 hp. A custom-built 1-MW permanent-magnet motor was reportedly developed by Siemens for ship propulsion (Leonard 1988). Improved materials can also yield very high efficiencies. A 50-hp

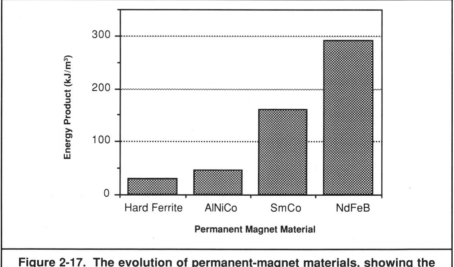

Figure 2-17. The evolution of permanent-magnet materials, showing the increasing magnetic energy density ("energy product"). Ferrites were developed in the 1940s; AlNiCos (aluminum, nickel, and cobalt) in the 1930s. The "rare-earth" magnets were developed beginning in the 1960s (samarium-cobalt) and in the 1980s (neodymium-iron-boron). The higher the energy density, the more compact the motor design can be for a given power rating (Krupp-Widia 1987, Baldwin 1989).

PM motor with an efficiency of 97% was recently developed (EPRI 1989).

Since PM motors have neither rotor windings nor slip rings, they can be as robust and reliable as induction motors. They must be totally enclosed, however, to avoid attracting iron particles. And because the magnets can be demagnetized by high temperature, the motor cannot be overheated. For example, the neodymium-iron-boron magnets can be demagnetized if the temperature exceeds 302°F (150°C). However, the low losses of PM motors mean that their operating temperature is well below 300°F at the rated power.

PM motors coupled with electronic speed controls are already being used in cordless power tools, as well as residential air conditioners, furnaces, and heat pumps. Refrigerators and freezers are likely candidates for PM motor application. Due to their high efficiency and reliability, and to the recent availability of high-performance magnetic materials at reasonable cost, PM motors are likely to become increasingly prevalent.

Reluctance Motors

Reluctance designs are another promising family of motors, which are synchronous but do not require electrical excitation of the rotor.

PM Motors in Residential Appliances

An ever-increasing number of residential appliances use PM motors. A particularly innovative example is the Carrier Weathermaker SXi gas-fired, forced-air furnace (also sold under the Bryant, Day and Night, and Payne brand names). Like many top-of-the-line furnaces, it uses a condensing heat exchanger for high thermal efficiency, and an induced-draft exhaust blower to vent the flue gases. Unlike most others, this furnace is also electrically efficient, due to the use of two electronically commutated motors: one for the main ventilation fan, the other for the induced-draft blower.

The burner operates in two stages, allowing the furnace to operate at low output for most of the required heating hours. The corresponding low speed of the main fan results in a power consumption of less than 100 W, compared to 600–1,000 W in a typical furnace. The low-speed capabilities, combined with sophisticated controls, also bring increased comfort and reduced noise. The installed cost of the furnace is about $2,500, roughly $300 to $700 more than other condensing furnaces (Nisson 1988).

Because no current is induced in the rotor, losses are lower and efficiencies are generally higher than in induction motors. The shaft power of reluctance motors is smaller than similar-sized PM motors. They are currently used in very low power applications such as clocks, timers, and turntables, where an inexpensive, low-power, constant-speed motor is required. A variation known as the switched-reluctance motor shows promise as a future competitor to the induction motor, especially in adjustable-speed applications, but further development and commercial experience are needed to fully evaluate its potential.

We have now discussed the principal types of commercially available motors. Their major characteristics are summarized in Table 2-2. Other types of motors exist in various stages of commercialization. For an overview, see Lovins et al. 1989. We now focus on the elements of and trends in motor efficiency.

Motor Efficiency

From the end of World War II to the early 1970s (a period of inexpensive energy), manufacturers built relatively inefficient motors that minimized the use of copper, aluminum, and steel. These motors had lower initial costs than their predecessors but used more energy due to their inefficiency.

These less efficient and more compact motors were made possible by the development of insulation materials that could withstand higher temperatures. Paper, cotton, enamels, and varnishes used in the 1940s deteriorate rapidly above 210°F (99°C), while the synthetic insulating materials developed since then can tolerate operating temperatures up to 390°F (199°C). Thus, it was possible to design motors with higher losses, as the temperature rise due to the losses could be accommodated without damaging the insulation and reducing motor lifetime.

Using less material in the magnetic and electrical circuits led to more compact designs. The development of improved steel with higher permeability allowed further reduction of the magnetic circuits. The superior electrical characteristics of the new insulation materials allowed windings to be packed tighter in the slots, reducing volume requirements even further. A combination of these factors led to smaller motor frames, as shown in Figure 2-18 (page 38). This size and efficiency reduction occurred mainly in motors smaller than 100 hp. In large motors, the amount of heat that could be dissipated determined the motor size and limited the minimum efficiency that could be tolerated.

Table 2-2. Classification of common motor types, with their general applications and special characteristics.

A.C.	Induction	Squirrel Cage	3-phase (general purpose; >0.5 hp; low cost; high reliability)
			1-phase (low [typ. <5] hp range; high reliability)
		Wound Rotor	(special purpose for torque and starting current regulation; typ. >20 hp; greater maintenance requirement than squirrel cage)
	Synchronous	Wound Rotor	(high efficiency and reliability; very large sizes; greater maintenance requirement than squirrel cage)
		Reluctance	(small motors; high reliability; synchronous speed)
		Brushless Permanent Magnet	(high efficiency; high performance applications; high reliability)
D.C.	Wound Rotor	(limited reliability; relatively high maintenance requirements)	Series (traction and high torque applications)
			Shunt (good speed control)
			Compound (high torque with good speed control)
			Separated (high performance drives [e.g. servos])

Note that brushless permanent magnet motors overlap between AC and DC types.

By the mid-1970s, electricity prices started escalating rapidly. Most of the large motor manufacturers started offering a line of energy-efficient motors (EEMs) in addition to their standard-efficiency models. Some now offer three lines—standard, high, and premium efficiency. Until recently, however, there has been no industry-wide definition of what constitutes an energy-efficient motor. Indeed, the "standard-efficiency" models from some manufacturers have in some cases had higher efficiencies than "high-efficiency" offerings from other firms. The industry's progress in defining "high efficiency" is discussed later in this chapter in the "Motor Efficiency Index" section and again in chapter 9. EEMs feature optimized design, more generous electrical and magnetic circuits, and high-quality materials. These kinds of improvements are best understood in the context of the motor losses they aim to reduce.

Motor Losses

There are four basic kinds of loss mechanisms in a motor: electrical, magnetic (core), mechanical (windage and friction), and stray. Their relative contributions vary with motor load and are depicted in Figure 2-19.

Whenever current flows through a conductor, some power is dissipated. These electrical losses are a function of the square of the current times the resistance and thus are termed I^2R losses, where I is the symbol for current and R is the symbol for resistance. In a motor, I^2R losses occur in the stator and rotor. Because they rise with the square of the current, such losses increase rapidly with the motor's load. By using more and sometimes better, lower-resistance materials (switching from aluminum to copper, for instance), manufacturers have reduced the I^2R losses in energy-efficient motors.

Magnetic losses occur in the steel laminations of the stator and rotor. These losses are due to eddy currents and hysteresis (see

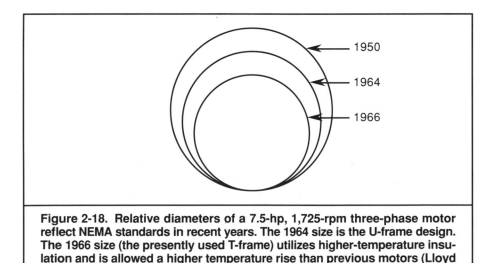

Figure 2-18. Relative diameters of a 7.5-hp, 1,725-rpm three-phase motor reflect NEMA standards in recent years. The 1964 size is the U-frame design. The 1966 size (the presently used T-frame) utilizes higher-temperature insulation and is allowed a higher temperature rise than previous motors (Lloyd 1969).

glossary). The magnetic losses can be decreased by using larger cross-sections of iron in the stator and rotor, thinner laminations, and improved magnetic materials. As Figure 2-19 illustrates, magnetic losses in a given motor decrease slightly as the load increases.

Mechanical losses occur in the form of bearing friction and "windage" created by the fans that cool the motor. Windage losses can be decreased through improved fan design. Mechanical losses are relatively small in open, low-speed motors, but may be substantial in large high-speed motors or in totally enclosed fan-cooled motors.

Stray losses are a collection of miscellaneous terms, resulting from leakage flux, nonuniform current distribution, mechanical imperfections in the air gap, and irregularities in the air gap flux density. They typically represent 10–15% of total losses and increase with the load. Stray losses also can be decreased by optimal design and careful manufacturing.

Figure 2-20 (next page) shows a cutaway of an EEM with the areas of efficiency improvement and the kind of loss minimized by each measure. In the integral horsepower sizes, EEMs are two to ten percentage points more efficient than standard motors, with the low end of this range applying to large motors. In fractional-horsepower, single-phase models, the spread between standard- and high-efficiency units can be greater than ten points. Table 2-3 (page 41) shows current data on efficiencies of standard and energy-efficient three-phase motors by manufacturer, size, and enclosure.

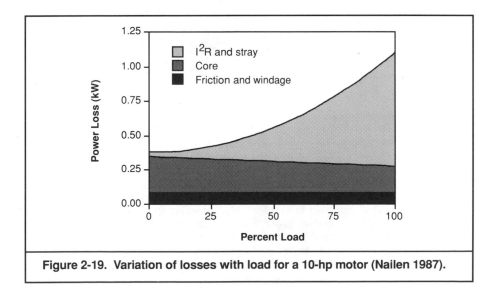

Figure 2-19. Variation of losses with load for a 10-hp motor (Nailen 1987).

Availability of EEMs

Energy-efficient three-phase induction motors are available from most manufacturers in T-frame ODP and TEFC enclosures; in speeds of 1,200, 1,800, and 3,600 rpm; and in sizes from 1–200 hp. A few manufacturers make EEMs as small as 1/2 hp and as large as 350 hp; some make EEMs that run at 900 rpm and have explosion-proof enclosures or C-face frames or both. Single-phase motors also are available in standard and efficient lines.

Standard single-phase motors have extremely poor efficiencies. For example, 1/4-hp motors from different manufacturers range in efficiency from 52% to 60% with power factors of 53% to 62%. Energy-efficient single-phase motors in this size can achieve efficiencies of 75%.

Efficient single-phase motors currently are available through most distributors that stock small motors. They also are appearing as

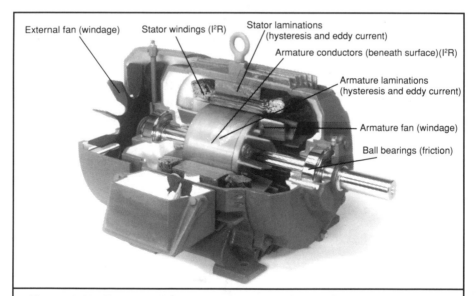

Figure 2-20. Energy-efficient induction motor cutaway view shows important features and construction. Labeling indicates major components that contribute to motor losses and (in parentheses) the type of loss that takes place. Note that "armature" is another name for rotor. Windage losses can be reduced through improved fan design. Hysteresis and eddy current (magnetic losses) can be reduced through the use of larger cross sections, thinner laminations, and special steel alloys. I²R losses can be reduced through the use of conductors with larger cross sections (e.g. bigger wire) (reprinted with permission from Reliance Electric).

Table 2-3. Nominal full load efficiencies for three-phase 1,800-rpm NEMA Design B induction motors.

Standard-Efficiency Open Drip-Proof (ODP)

Manufacturer	1	2	3	5	7.5	10	25	50	75	100	200
Baldor	77.0	78.5	81.5	80.0	84.0	84.0	89.5	93.0	93.0	91.7	none
GE	72.0	77.0	80.0	85.5	88.5	88.5	90.2	90.2	91.7	91.7	93.6
Lincoln	77.0	80.0	84.0	84.0	85.5	86.5	87.5	91.7	93.0	none	93.0
Marathon	77.0	81.6	81.5	85.5	84.0	86.5	88.5	90.2	91.7	93.0	94.5
MagneTek	78.5	80.0	81.5	84.0	84.0	86.5	88.5	89.5	92.4	93.0	93.6
Reliance	77.0	77.0	80.0	82.5	85.5	86.5	87.5	90.2	90.2	90.2	92.4
Toshiba	77.0	80.0	82.5	82.5	86.5	87.5	90.2	91.0	92.4	92.4	93.6
US	78.5	78.5	80.0	81.5	84.0	84.0	89.5	91.0	91.0	91.0	94.1
Average	76.3	78.5	80.6	83.2	85.3	86.3	88.9	90.9	91.9	91.9	93.5

High-Efficiency Open Drip-Proof (ODP)

Manufacturer	1	2	3	5	7.5	10	25	50	75	100	200
Baldor	82.5	84.0	86.5	87.5	88.5	89.5	92.4	94.1	94.1	94.1	none
GE	84.0	84.0	89.5	89.5	91.7	91.7	94.1	94.5	95.4	96.2	96.2
Lincoln	none	none	none	none	none	none	none	none	none	93.6	none
Marathon	82.5	84.0	86.5	86.5	88.5	89.5	92.4	93.0	94.1	94.1	95.0
MagneTek	82.5	84.0	89.5	89.5	91.7	91.0	93.6	94.5	95.4	95.4	96.2
Reliance	82.5	84.0	87.5	88.5	89.5	90.2	93.0	94.1	95.0	95.4	95.8
Toshiba	85.5	86.5	88.5	87.5	90.2	91.0	92.4	93.6	94.5	94.5	95.0
US	85.5	85.5	86.5	88.5	89.5	90.2	none	93.6	95.0	95.4	95.4
Average	83.6	84.6	87.8	88.2	89.9	90.4	93.0	93.9	94.8	94.8	95.6

Standard-Efficiency Totally Enclosed Fan-Cooled (TEFC)

Manufacturer	1	2	3	5	7.5	10	25	50	75	100	200
Baldor	77.0	82.5	82.5	84.0	85.5	86.5	91.0	93.0	93.0	93.0	93.6
GE	72.0	80.0	81.5	85.5	87.5	88.5	91.0	91.7	91.7	91.7	94.1
Lincoln	77.0	78.5	82.5	84.0	81.5	84.0	none	none	none	none	94.1
Marathon	77.0	82.5	82.5	82.5	85.5	86.5	90.2	91.0	93.0	92.4	95.0
MagneTek	80.0	81.5	78.5	86.5	81.5	86.5	88.5	91.7	91.7	90.2	92.4
Reliance	74.0	81.5	78.5	82.5	84.0	85.5	87.5	90.2	90.2	90.2	94.5
Toshiba	77.0	80.0	81.5	84.0	86.5	86.5	89.5	90.2	92.4	93.0	93.6
US	80.0	82.5	84.0	82.5	85.5	87.5	90.2	92.4	93.0	93.0	94.5
Average	76.8	81.1	81.4	83.9	84.7	86.4	89.7	91.5	92.1	91.9	94.0

High-Efficiency Totally Enclosed Fan-Cooled (TEFC)

Manufacturer	1	2	3	5	7.5	10	25	50	75	100	200
Baldor	none	none	89.5	90.2	91.0	91.7	94.1	95.0	95.0	95.0	none
GE	84.0	86.5	89.5	90.2	91.7	91.7	93.6	94.1	95.4	95.4	95.8
Lincoln	none	none	none	none	none	none	91.0	92.4	93.0	93.6	none
Marathon	82.5	84.0	88.5	88.5	91.7	91.7	93.6	94.5	95.4	95.8	95.0
MagneTek	82.5	84.0	88.5	89.5	91.0	91.0	93.0	94.1	94.5	95.4	95.8
Reliance	84.0	85.5	88.5	88.5	89.5	90.2	93.0	93.6	95.0	95.4	96.2
Toshiba	85.5	86.5	88.5	87.5	90.2	90.2	91.7	93.0	93.6	94.1	95.0
US	85.5	85.5	88.5	88.5	90.2	90.2	93.0	93.6	95.0	95.4	95.8
Average	84.0	85.3	88.8	89.0	90.8	91.0	92.9	93.8	94.6	95.0	95.6

NEMA nominal values for standard- and high-efficiency motors are listed for the ODP and TEFC enclosures of eight major motor manufacturers. The high-efficiency motors listed are the highest efficiency motor of each manufacturer; they are not listed if they do not meet the NEMA definition of high efficiency (see Table 9-1). The lowest-efficiency motor of each manufacturer is listed as the standard efficiency, even though some of these motors meet the NEMA definition of high efficiency. If a manufacturer makes only one line of motors, it is listed as high efficiency if it meets the NEMA definition, otherwise as standard. Note that these numbers are subject to change as product lines change.

an option in some packaged equipment, including commercial refrigeration cases.

Benefits of EEMs

The most obvious benefit of an EEM is energy savings. Even in the largest motor sizes, where the efficiency improvement between standard-efficiency and high-efficiency models is small in percentage terms, a small relative improvement can yield substantial energy savings. For example, a 1% improvement for a 500-hp motor operating at 75% load is 2.8 kW. If the motor operates almost continuously, as many large motors do, that 1% improvement could yield annual energy savings of nearly $1,500 at $.06/kWh. If the utility has a monthly demand charge of $6/kW, the demand savings will be $202/yr. In contrast, improving the efficiency of ten 1-hp motors by 10% (from 75% to 83%) saves 0.72 kW, which yields annual energy savings of $380 and demand savings of $52.

Energy-efficient motors not only save energy in themselves and contribute to reduced demand but also save energy in the cables and transformers that feed the motor. Most EEMs have a higher power factor than standard-efficiency motors. EEMs also are likely to last longer. Because EEMs have 20–40% lower losses, they run cooler. The decrease in operating temperature is not as dramatic as one would expect, however, since EEMs have downsized ventilation to decrease the ventilation losses. Manufacturers no doubt vary in how they make the trade-off between ventilation (hence cooler, longer operation) and efficiency. One manufacturer, General Electric, states that its EEMs' bearings run 10°C cooler than those in its standard motors, and that this advantage should double lubricant life. G.E. also asserts that its EEM windings run 20°C cooler, thus quadrupling insulation life (General Electric 1989).

Along with temperature effects, lubrication practice is another critical factor affecting the lifetime of a motor. Many motor failures are caused by bearing failures, many of which are in turn caused by under- or over-lubrication. Lubrication is a function of the plant maintenance practices, not motor design, but EEMs often use heavier duty bearings that are presumably more resilient in the face of poor lubrication. On the basis of cooler operation and better bearings, then, EEMs should tend to last longer than standard-efficiency motors. How much longer is a matter of speculation, however, since limited data are available on this issue.

Because of their lower losses, EEMs suffer less thermal stress when starting and operating at small overload intervals than standard motors. This makes EEMs attractive in some duty cycling applications because they can withstand a higher on-off cycle rate.

Because EEMs have lower losses, their slip is smaller and their speed is slightly higher than their standard-efficiency counterparts. When an EEM is driving loads where the power increases with the cube of the speed (such as many pumps and fans), the increased speed causes the power drawn by the motor to increase, and some of the savings associated with the EEM can be lost. This cube law phenomenon and ways to mitigate losses from faster rotation are discussed further in chapter 5.

EEMs do have a few potential drawbacks. Their small slip causes them to have lower starting torque than standard-efficiency units. EEMs thus should not be used in some applications where starting torque is critical. Some EEMs also have lower power factors than standard-efficiency motors.

Later in the chapter, we discuss the economics of EEMs. But first we address another important issue, motor rewinding.

Motor Rewinding

Most motor failures are caused by either bearing failure or winding failure. Windings fail when their insulation degrades, usually due to some combination of overheating, aging, and overvoltage transients. Minor insulation failure can lead to poor motor performance, shock, and fire hazard. Major insulation failure will trip the overcurrent protection devices. When the damage is restricted to the windings and bearings, only those parts need replacement. The rebuilt motor can use the same rotor and stator iron and case, leading to considerable savings in raw materials. The process of stripping out old windings and replacing them with new wire is called rewinding.

Because it is economic in terms of initial cost, rewinding is very common. Commercial or industrial facilities that purchase motors at the trade price rewind most motors over 10 hp. Larger users that receive a volume discount on motor purchases may restrict rewinding to motors larger than 25 hp. It is generally cheaper to replace standard motors under 10 hp with EEMs than to rewind them.

Users need to consider two key economic criteria when considering a motor rewind. The first is the cost difference between rewinding and buying a new EEM (see Tables A-1 and A-2 in appendix A). The second is that the motor might not be as efficient as the user expects when it returns from the repair shop, either because some preexisting damage is not detected and corrected during the repair or because the repair itself damages the motor. The possibility of such performance degradation is often overlooked in the effort to minimize initial cost. An efficiency loss of only 1% in a large motor can cost $1,500 per year in energy bills, yet many motors run several percentage points below nameplate efficiency (Montgomery 1989).

Severely damaged motors are typically identified as such either by the repair shop or by the user when the motor fails prematurely after being repaired. Slightly damaged units, which look fine but are running, say, 1–5% below nameplate efficiency, can run up thousands of dollars in excess losses over the years. For instance, in the course of severe bearing failure, the rotor may hit the stator and damage the magnetic properties of the iron core. If the bearing is replaced but the magnetic damage is ignored, the repaired motor will appear to be as good as new but it will in fact be sustaining excess operating losses. The only way to quantify these losses is to test the motor. Such testing is discussed later. First we address the question of how poor rewind practices can damage motors.

Impact of Rewinding on Motor Losses

In theory, rewinding can produce a motor with the same efficiency rating as it had when it was new. In practice, motor efficiency is often degraded through normal rewind practices, making the initial low cost a potentially poor investment (Montgomery 1989).

The insulation materials used in the last few decades in the stator copper windings are solvent-resistant and very hard to remove. The conventional approach to softening the windings for removal is to bake the stator in an oven. If the stator gets too hot, however, its magnetic properties can be damaged, leading to increased core losses. These increased losses are primarily due to damaged insulation between the laminations in the core. Very high temperatures can distort the iron and the air gap, also resulting in increased losses.

There is some uncertainty regarding the exact temperature at which the iron core properties start to deteriorate. The Electric Apparatus Service Association (EASA), a trade group representing motor repair shops, published a study recommending that oven setpoints not exceed 650°F (343°C) (EASA 1985). This recommendation takes into account the fact that stator cores sometimes reach temperatures 150°F (66°C) higher than the oven setpoint, due largely to heat released by the combustion of the insulation materials. Other analysts, citing EASA's own test data and reports from motor manufacturers, suggest that oven setpoints should not exceed 500°F (260°C) (Lovins et al. 1989). This uncertainty and the high potential rewards of resolving it suggest the need for further research on optimal rewind practices.

Even if the EASA recommendations are correct, many rewind shops do not follow them. A study performed by the Bonneville Power Administration (BPA) on rewind practices in the Pacific Northwest found that about half of the motors are baked at oven setpoint temperatures above 650°F and are thus likely subject to efficiency-degrading core damage (Seton, Johnson & Odell 1987b). Such damage not only can increase operating costs but also can reduce the motor's service life by increasing the heat released by the motor during operation, degrading insulation and lubricants.

There are not enough data available to precisely estimate the decline in efficiency caused by poor rewinds. Few motors are tested for either core loss or overall efficiency, in part because many rewinds are done hastily. The EASA test results shown in Figure 2-21 indicate a general correlation between oven temperature and efficiency degradation, although the sample size is small, the results vary widely among individual motors, and most motors in the sample inexplicably gained efficiency on the second rewind.

The only sizeable published survey on measured rewind losses was conducted by General Electric in the early 1980s, based on tests made over a one-year period in the company's repair facilities (Montgomery 1984). The study found an increase in core losses,

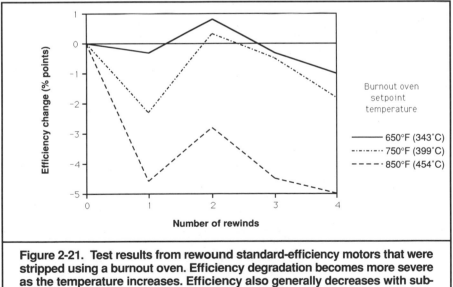

Figure 2-21. Test results from rewound standard-efficiency motors that were stripped using a burnout oven. Efficiency degradation becomes more severe as the temperature increases. Efficiency also generally decreases with subsequent rewinds (EASA 1985, Lovins et al. 1989).

compared to manufacturer ratings for these motors when they were new, ranging from 0–400%, with an average of 32%. This survey further concluded that 8–10% of the motors had at least doubled core losses, that 50–60% of the rewound motors had core loss increases ranging from 10–100%, and that the losses were most pronounced in large motors, which are most likely to be rewound repeatedly. Table 2-4 shows the distribution of core losses for different horsepower ranges. Because the 32% increase in core loss was reportedly not an input-weighted average but simply based on the number of units (Lobodovsky 1989), the tendency toward higher losses in bigger motors suggests that the size-weighted average effect in the G.E. sample was significantly greater than 32%.

As Figure 2-19 (page 39) shows, core losses range from about 60% of total losses at low load to about 20% at full load. At the typical loading of about 75%, core losses represent about one-quarter of total losses. This means that for a motor with 90% efficiency at 75% load, about 2.5 percentage points of the total 10-point efficiency loss is due to core loss. Increasing that core loss by 40%, roughly the input-weighted average effect of rewinds found in the General Electric survey, would reduce that motor's efficiency by one percentage point. Based on the presumably representative G.E. sample, we estimate that poor rewinds cause a 1% efficiency penalty across the entire motor stock.

Larger core losses cause larger energy bills and reduced motor lifetimes. For example, consider a 50-hp continuous duty standard motor that has the core losses increased by 50%, 100%, 150%, and 200% in the rewinding process. Table 2-5 shows the increase in the losses as well as the extra cost of the losses for an electricity price of

---

Table 2-4. Percent increase in the core losses of rewound motors.

% Increase in core loss	Motor size		
	5–20 hp	25–60 hp	75–200 hp
0–9%	36%	39%	21%
10–50	47	48	58
51–100	9	4	11
101–400	8	9	10
Total	100%	100%	100%

The distribution of loss-increases is shown for different hp ranges. The data are from tests of motors made over a one-year period in General Electric repair facilities.

Source: Montgomery 1989.

$.06/kWh. If there is an increase of 100% in the motor core losses, the extra annual operating cost will be similar to the rewinding cost.

The increased losses also increase motor temperature, which decreases insulation lifetime. Table 2-5 also shows the corresponding temperature increases and associated reduction in the insulation lifetime when the core losses go up. A large increase in the core losses dramatically affects the lifetime of the rewound motor. Higher motor temperatures also affect the lifetime of the lubricant in the bearings, although to avoid premature failure the user can perform more frequent regreasing, incurring additional costs. Thus the poorer the quality of the rewind job, the shorter the motor life.

Given the apparent prevalence of damaging rewinds, this topic deserves further attention. Utilities could provide a valuable service to motor users by providing low- or no-cost testing of motor performance before and after rewinding. Further research and development is needed as well, to evaluate the effects of and find ways to improve current practice and to explore alternative methods.

One alternative stripping method widely used in Europe (Dreisilker 1987, Lovins et al. 1989) uses moderate temperatures up to 300°F (149°C) to soften the insulation, so that it can be safely removed with mechanical pulling. The low temperatures used in this process are much less likely to damage the core. Unfortunately, this technique is not well known or widely used in the United States. Its main champion in the United States is a firm in Glen Ellyn, Illinois, called Dreisilker Electric Motors, Inc. See appendix B for sources of motor testing and repair equipment.

Some rewind shops use chemicals to loosen the windings. However, chemical solvents pose health and environmental problems due to their toxicity and are not able to dissolve modern epoxy-based varnishes. High-pressure water jets also can be used for stripping, but the equipment is very costly.

Table 2-5. The effect of increased core loss on motor operating cost and insulation life for a 50-hp, 3,600-rpm, open drip-proof motor.

Core loss increase		Increase in annual operating cost		Temp. rise increase	Approx. % of normal insulation life
		Dollars	As % of rewind cost		
50%	515 Watts	$ 271	28%	7°C	62%
100	1,030	542	55	14	38
150	1,545	813	83	21	24
200	2,060	1,084	110	29	14

Source: Montgomery 1989.

In addition to the methods used to remove old windings, the materials and techniques used in reassembling a motor affect its subsequent performance. Most motor rewind shops install materials in the motor equal to or better than those in the original motor. For example, most rewind shops use wire with at least a class F insulation, even on motors that were originally equipped with class B insulation. In some cases, however, this attempt to improve the motor can backfire and result in lower operating efficiency.

For example, based on the data from the BPA study on rewind practices, approximately 20% of the rewound motors are U-frame motors. These motors were originally equipped with older style wiring with bulky insulation. When wire with modern, thinner insulation is installed in the motor, most rewind shops will use more wire than in the original design. Using the same number of turns of thicker wire will improve motor efficiency, since extra copper will reduce resistance losses. However, if a larger number of turns is used (as is common), resistance losses will increase and magnetic losses will substantially increase, resulting in decreased operating efficiency.

Specifying Rewind Practices

With all the damage that rewinding potentially can cause, motor users need to know how to specify the rewind practices to be performed on their motors and the evaluation criteria needed to weed out damaged motors.

Before a motor is rewound, its iron should be tested for mechanical damage as well as for core losses. Fairly primitive loop ring tests wrap a cable around the stator core and energize it. This heats the core, and the operator then passes a hand near the stator surface to search for hot spots. A far more accurate approach uses an integrating electronic core-iron tester. Apparently, the latter testing device is made in the United States by only one firm, Lexseco, Inc., in Louisville, Kentucky. The firm's 10-, 25- and 125-kVA models cost in 1988 $8,400, $10,200, and $21,400 respectively (Lovins et al. 1989). The test is fairly easy to administer and takes about 30 minutes. A British line of electronic core-loss spot meters starting at around $8,000 is distributed in North America by Multi-Amp Corporation of Dallas, Texas.

If the core is in good condition, the rewinding operations can start. The test should be repeated after stripping the winding, to ensure that the core was not damaged in the process. If testing reveals previous damage, either the core should be scrapped or, in the case of large motors, it may be worth rebuilding.

Given the potential for costly damage from poor repairs and the fact that motor downtime can be very costly, there are strong incentives for carefully choosing a rewind shop. The motor owner should find out about the rewind shop's practices. Does it use an oven or chemical solvents to soften the windings? If an oven is used, what is the temperature setpoint? Whatever the technique, the user should specify before-and-after core iron loss tests. The specification should also state that the winding have the same number of turns of wire of a diameter at least equal to the original winding.

In the absence of before-and-after core testing, poor rewinds can be detected by efficiency testing in the field (described later in this chapter), or by measuring the no-load losses. Core losses represent about 60% of the no-load losses. Excess losses of at least a few percent can be identified by comparing measured core losses with published manufacturer's data or with the losses of a similar motor. A badly rewound motor should generally be replaced by an EEM, although a standard-efficiency motor may be appropriate for applications with very low operating hours.

Wanlass and Unity Plus Rewinds

Proponents of two novel rewind techniques have claimed that their methods can lead to substantial improvements in motor efficiency (Wanlass 1978 and 1980). In both cases, the modified windings are coupled with capacitors, which may improve power factor but do not improve motor efficiency (Umans 1983 and 1988). Umans tested the Unity Plus method on two sizes of motors from three manufacturers and concluded that the rewind reduced motor efficiency in all cases. Moreover, other motor characteristics including peak power were adversely affected (Umans 1988). Motor users should view all claims about such rewind "inventions" with caution.

Economics of Energy-Efficient Motors

The economics of EEMs must be evaluated separately for three distinct situations: (1) installing an EEM instead of a standard-efficiency motor in a new application, (2) installing a new EEM instead of rewinding a failed motor, and (3) installing a new EEM as a retrofit for an existing operational motor.

Economics of EEMs versus New Standard Motors

EEMs typically carry a price premium of 10–40%, or $5–$15/hp in the 5–250-hp range and $15–$40/hp in sizes below 5 hp (see

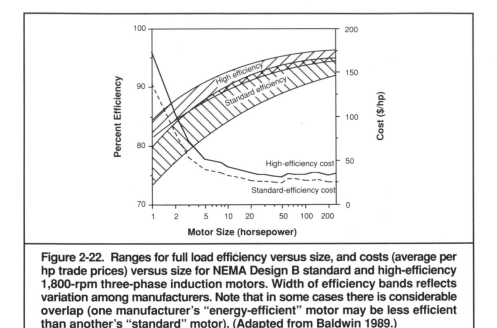

Figure 2-22. Ranges for full load efficiency versus size, and costs (average per hp trade prices) versus size for NEMA Design B standard and high-efficiency 1,800-rpm three-phase induction motors. Width of efficiency bands reflects variation among manufacturers. Note that in some cases there is considerable overlap (one manufacturer's "energy-efficient" motor may be less efficient than another's "standard" motor). (Adapted from Baldwin 1989.)

Tables A-1 and A-2 for specific values). This higher price is due to increases in the quantity and quality of materials, the cost of designing the motor, and the cost of tooling. Figure 2-22 illustrates the efficiency and trade price ($/hp) ranges of both standard- and energy-efficient motors. The market ("trade") prices of electric motors can be substantially lower than the list prices. A medium-sized user can get a 25–50% discount off list price, and a large user can get a discount as high as 50–70% (Seton, Johnson & Odell 1987a). Some large users have been able to negotiate bulk motor purchase contracts in which they pay only a 5% premium for EEMs (Lovins et al. 1989).

In most new installations, the extra cost of an EEM is justified by the energy and demand savings.

Consider a new application of a 50-hp motor with the following specifications:

4,000 hours of annual use at 75% load

Cost of electricity	=	$.06/kWh
Demand charge	=	$70/kW-yr
Efficiency of standard motor	=	90% at 75% load
Efficiency of EEM	=	94% at 75% load
Extra cost of EEM	=	$500

The yearly savings afforded by the EEM are as follows:

Demand Savings =
    50hp × (1/0.90 − 1/0.94) × 0.75 × 0.746 kW/hp = 1.32 kW

Energy Savings = 1.32 kW × 4,000 hr/yr = 5,280 kWh

Cost Savings =
    $.06/kWh × 5,280 kwh + $70/kW-yr × 1.32 kW = $409/year

Simple payback period = $500/$409 = 1.2 years

Economic calculation methods other than simple payback are discussed in appendix A.

The payback decreases linearly with the number of operating hours. Therefore, when buying a motor for a new application, an EEM can be a very attractive investment if the motor has high operating hours.

Figures 2-23 and 2-24 (next page) show payback periods for ODP and TEFC motors in new applications. The efficiencies used in these figures are drawn from Tables A-1 and A-2. The figures assume a trade price actually paid by most commercial and industrial customers. Very large customers can often purchase motors at a larger discount. For other values of annual operating hours or electricity costs, the paybacks can be adjusted linearly. For example, a motor operating 3,000 hrs/yr would take twice as long to pay back as one operating 6,000 hrs/yr.

Life-cycle cost analysis can determine the threshold number of annual operating hours at which a high-efficiency motor becomes cost-effective. The results of such an analysis are illustrated in Figure 2-25 (page 53). For most motor sizes, under the assumptions used in this analysis, high-efficiency motors are cost-effective at 250 or more operating hours per year. For the largest motors, the breakeven point is approximately 1,500 hours per year.

Because no one manufacturer has the most efficient motor in every style and size, users should comparison shop across brands. However, manufacturers change their designs from time to time, so it is important to obtain current efficiency information on the specific motors under consideration. The Washington State Energy Office (WSEO), with funding from Bonneville Power Administration, has compiled a database of three-phase motors 5 hp and larger sold in the United States. It lists efficiency and power factor of each motor at 50%, 75%, and 100% load. WSEO plans to keep the data base current and is making it available through an electronic bulletin board or a report listing all motors ranked by full-load efficiency.* BC Hydro

---

*Contact Electric Ideas Clearinghouse, (206) 586-8588, or (800) 872-3568, within Washington, Oregon, Montana, and Idaho. Further access options may be developed in the future.

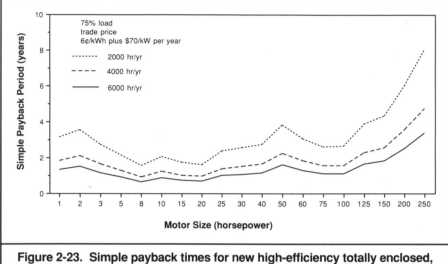

Figure 2-23. Simple payback times for new high-efficiency totally enclosed, fan-cooled (TEFC) motors versus new standard TEFC motors as a function of motor size and annual operating hours. The efficiencies and motor costs listed in Table A-1 are used (average of eight manufacturers' efficiencies for 1,800-rpm motors).

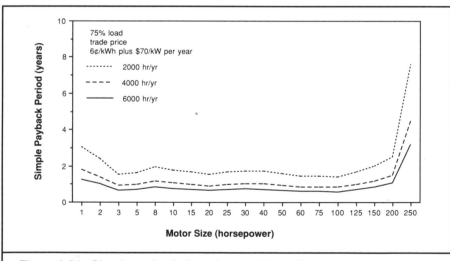

Figure 2-24. Simple payback times for new high-efficiency open drip-proof (ODP) motors versus new standard ODP motors as a function of motor size and annual operating hours. The efficiencies and motor costs listed in Table A-2 are used (average of eight manufacturers' efficiencies for 1,800-rpm motors).

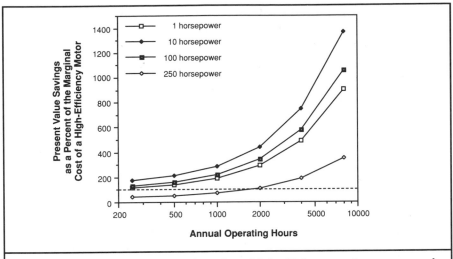

Figure 2-25. Present-value savings from high-efficiency motors compared to the marginal cost of high-efficiency motors relative to standard-efficiency motors as a function of motor size and annual operating hours. Assumptions: 1,800-rpm TEFC motors; values for efficiency and cost from Table A-1; 75% load; 15-year life and 7% real discount rate; energy and power savings are valued at $0.06/kWh and $70 per kW year; demand coincidence factor of 0.75. High-efficiency motors are cost-effective from a life-cycle cost perspective whenever the present-value savings exceed the marginal cost of a high-efficiency motor (i.e., wherever present-value savings exceed the 100% line in the graph). Note that the structure of the demand charge makes for a favorable present value even for motors with low operating hours.

has a database of three-phase motors available in British Columbia (BC Hydro 1989b).

Economics of EEMs versus Rewinding

As discussed above, rewinding can reduce the efficiency of a motor. In such cases, the energy savings from installing a new EEM instead of rewinding an existing motor can be more attractive than nameplate comparisons would suggest.

Consider an application where the economics of rewinding a motor are compared with the economics of purchasing a new EEM. This application uses a 50-hp motor with the following specifications:

4,000 hours of annual use at 75% load

Cost of electricity = $.06/kWh

Demand charge = $70/kW-yr

Nominal efficiency of standard motor when new = 90% at 75% load

Increased losses due to rewind = 1 efficiency percentage point
    Efficiency of EEM = 94% at 75% load
    Extra cost of EEM = $1,874 (EEM) − $1,175 (rewind) = $699

The yearly savings afforded by the EEM are as follows:

Demand Savings =
    50 hp × (1/0.89 − 1/0.94) × 0.75 × 0.746 kW/hp = 1.67 kW

Energy Savings = 1.67 kW × 4,000 hr/yr = 6,680 kWh

Cost Savings =
    $.06/kWh×6,680 kwh+$70/kW-yr×1.67 kW =$518/yr

Simple payback period = $699/$518 = 1.3 years

Note that because of the 1% decrease in motor efficiency due to rewinding, the savings are 21% higher than if a new standard motor is compared with a new EEM under the same operating conditions. However, the economics of replacing motors instead of rewinding them does not hinge on the 1% damage assumed from rewinds. Ignoring the 1% damage in the above example, the replacement motor still has a very attractive 1.7-year payback.

Figures 2-26 and 2-27 show the generally very favorable economics of replacing rewound motors with EEMs. The figures assume a 1% degradation in efficiency from rewinding and that new standard motors are 1% more efficient than the older versions of the same models were when they were new.

Economics of Replacing Operating Motors with EEMs

EEMs are clearly economic for most new applications and in comparison to rewinding in many instances. But what about the economics of replacing operating motors with EEMs? The incremental cost of the EEM in such instances is generally its full purchase and installation cost, not the marginal difference between it and a standard motor or a rewind. Thus the paybacks of such replacements are considerably longer than with new applications or rewinds. The right combination of conditions, however, can make replacements cost-effective. Some of these are listed below.

• The existing motor has an efficiency below its nameplate rating, a likely condition if the motor has been damaged in rewinding or if it is grossly oversized.
• The number of operating hours near full load is high.
• The replacement motors can be purchased in bulk at a large discount.
• The price of electricity is high.

    A motor replacement program conducted at Stanford University

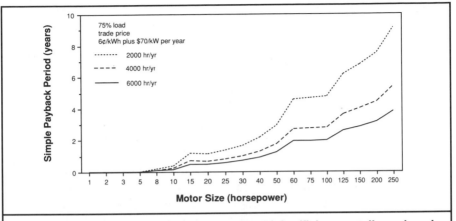

Figure 2-26. Simple payback times for new high-efficiency totally enclosed fan-cooled (TEFC) motors versus rewound standard TEFC motors as a function of motor size and annual operating hours. The efficiencies and motor costs listed in Table A-1 are used (average of eight manufacturers' efficiencies for 1,800-rpm motors). The rewound motor is assumed to be two percentage points lower in efficiency than the same-size new standard motor (approximately one point for degradation during rewind; one for the fact that older motors when new were about one point less efficient than current new motors).

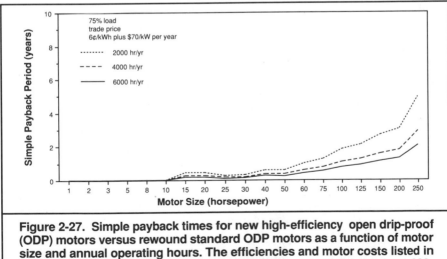

Figure 2-27. Simple payback times for new high-efficiency open drip-proof (ODP) motors versus rewound standard ODP motors as a function of motor size and annual operating hours. The efficiencies and motor costs listed in Table A-2 are used (average of eight manufacturers' efficiencies for 1,800-rpm motors). As for Figure 2-26, the rewound motor is assumed to be two percentage points lower efficiency than the same-size new standard motor.

several years ago offers an interesting example (Wilke and Ikuenobe 1987). Seventy-three standard-efficiency motors in HVAC applications were replaced in a group retrofit with EEMs. Motor sizes ranged from 7.5 hp to 60 hp; annual operating hours from 2,000 to continuous (8,760). Because of the size of the purchase, the university received a substantial cost break on the motors. Based on field tests, many of the motors were operating below their nominal efficiency ratings, possibly due to previous rewind damage. In addition, many of the motors were oversized for their loads: 32 of the replacements involved downsizing (see section on motor oversizing in chapter 3). The overall payback period on the project, counting a utility rebate, was under three years. The program's payback without the rebates would have been under five years.

This case study is instructive on several counts. First, there is no reason to believe that this fleet of motors was specified or maintained with anything but typical skill. This suggests that many such motors in institutional and commercial buildings are operating well below their published efficiencies, due to oversizing, rewind damage, or both. The attractive paybacks achieved in the Stanford program further suggest that group replacement of standard-efficiency motors might be cost-effective in many settings.

Even without credit for downsizing, replacement is cost-effective in many cases, particularly in motors between 5 hp and 100 hp. Figures 2-28 and 2-29 show paybacks for retrofitting operating motors assuming that the user pays the trade price and that the existing motor efficiency is two percentage points lower than that of its replacement (one point from rewind damage and one point because new motor offerings have become more efficient over time). In practice, large users should be able to negotiate larger price breaks on group purchases.

For further discussion of the economics of motor efficiency options, see appendix A.

Motor Testing Standards

Different standards for testing and labeling motors have been developed by various organizations in several countries. Because of these differences, the test results for a given motor may vary depending on the procedure that is used. Users need to be aware of these differences so that they will compare motors as much as possible on the basis of uniform test methods. The principal testing methods are briefly described below, with comments on the relative efficiency rating produced by each procedure.

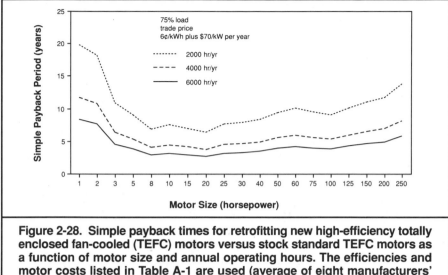

Figure 2-28. Simple payback times for retrofitting new high-efficiency totally enclosed fan-cooled (TEFC) motors versus stock standard TEFC motors as a function of motor size and annual operating hours. The efficiencies and motor costs listed in Table A-1 are used (average of eight manufacturers' efficiencies for 1,800-rpm motors). The stock motor is assumed to be two percentage points lower in efficiency than the same size new standard motor (for the same reasons as in Figure 2-26).

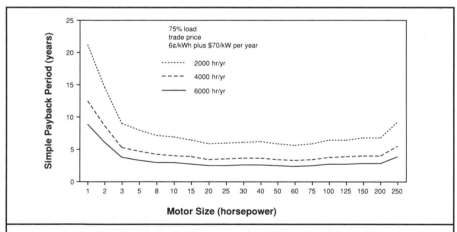

Figure 2-29. Simple payback times for retrofitting new high-efficiency open drip-proof (ODP) motors versus stock standard ODP motors as a function of motor size and annual operating hours. The efficiencies and motor costs listed in Table A-2 are used (average of eight manufacturers' efficiencies for 1,800-rpm motors). As for Figure 2-28, the stock motor is assumed to be two percentage points lower efficiency than the same size new standard motor.

As mentioned at the outset of this chapter, motor efficiency is the ratio of the mechanical output and the electrical input. Although the definition is simple, there are difficulties associated with its accurate measurement. In the United States, the basic motor test standard is IEEE Standard 112-1984, entitled "Standard Test Procedure for Polyphase Induction Motors and Generators," which comprises five testing methods. IEEE Standard 112-Method B is the most accurate but also the most time-consuming and expensive. Method B uses the basic definition of motor efficiency and directly measures the mechanical output and the electrical input to determine the efficiency.

The Canadian Standards Association (CSA) has developed the CSA Standard C(390)-M1985. This is similar to NEMA Standard MG1-12.53, which is a tighter specification of IEEE Standard 112-Method B. The additional specifications minimize the room for error introduced by operator interpretation of the data.

Both NEMA and CSA standards account for stray losses, measuring them indirectly. Not surprisingly, CSA and NEMA standards give very similar efficiency results.

Other standards used in the international market provide a less accurate estimate of motor efficiency. The International Electrotechnical Commission (IEC) Standard 34.2 (used in Europe) allows a tolerance in the efficiency and does not calculate the stray losses but assumes they are fixed at 0.5% of full load power. British Standard (BS) 269 uses similar methodology and gives similar results to the IEC-34.2 Standard. The Japanese Electrotechnical Commission (JEC) Standard 37 ignores stray losses, giving even less credible results. The assumptions of IEC-34.2 and JEC-37 are especially optimistic in small- and medium-horsepower motors. Considering that stray losses represent typically 10–15% of the motor losses at full load, in a motor whose efficiency is 85%, stray losses represent 1.5–2.25% of the full load power, not the 0.5% as assumed by the IEC or 0% as assumed by the JEC.

Table 2-6 shows the efficiency of two motors tested according to the different standards.

Motor Efficiency Index

With the exception of special-purpose or large units (above approximately 300 hp), electric motors are made to each manufacturer's standard specifications. Even so, there are variations in efficiency in units of the same model. These variations can be attributed to changes in the quality of raw materials and to multiple random factors in the manufacturing process. A 10% variation in the iron core

Table 2-6. Comparison of the efficiencies of typical motors, tested according to different standards.

Standard	Full load efficiency (%)	
	7.5 hp	20 hp
CSA C390	80.3	86.9
NEMA MG-1	80.3	86.9
IEC-34.2	82.3	89.4
JEC-37	85.0	90.4

CSA is the Canadian Standards Association; NEMA the National Electrical Manufacturers' Association; IEC the International Electrotechnical Commission (in Europe); JEC the Japanese Electrotechnical Commission.

Source: BC Hydro 1988.

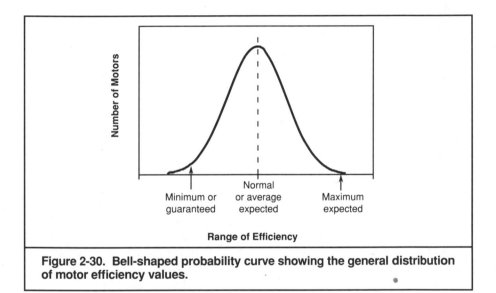

Figure 2-30. Bell-shaped probability curve showing the general distribution of motor efficiency values.

losses, which is within the tolerance of magnetic steel manufacturers, can by itself produce a change of 0.3% in the efficiency of a 10-hp motor. Mechanical variations can also affect efficiency by changing the size of the air gap (a 10% difference in air gap size is not uncommon) and, consequently, in the stray losses. Precision machining of motor parts is costly and motor manufacturers settle for a trade-off between precision and cost when purchasing the equipment used in the production line.

Under these conditions a population of motors of the same model can be statistically characterized by the average efficiency and the minimum efficiency of that population. Figure 2-30 shows a bell-shaped curve (normal distribution), which represents the efficiency distribution of a motor population. The average is considered the nominal efficiency of the motor and is used to predict the power requirements for a given installation of motors. The minimum efficiency represents a near-worst-case combination of raw materials and manufacturing tolerances. However, 5% of the motors in a population may, depending on the manufacturer, have efficiencies lower than the minimum efficiency (NEMA 1989a). In a study of different motors of the same model, motor losses varied often by 10% and sometimes by as much as 19%, corresponding to efficiency reductions of one to two percentage points (NEMA 1989a).

Despite the variations in efficiency among supposedly identical motors, NEMA has adopted a standard for labeling three-phase, Design A and B, single-speed, 1–125-hp induction motors with their full load efficiency. In addition, most manufacturers voluntarily label larger motors.

Table 2-7 shows the nameplate efficiency index defined by NEMA Standard MG-1, which requires that either a code letter or, more recently, a nominal efficiency percentage be stamped on the motor nameplate. The efficiency on the nameplate is the motor's actual nominal efficiency rounded down to the nearest preset value shown in Table 2-7. When NEMA first adopted these numerical ratings, there were 22 present values for rounding. NEMA has redefined the list several times resulting in the present 51 preset ranges.

The NEMA labeling method currently defines the minimum full load efficiency of a motor as that level corresponding to 20% higher losses than the listed nominal value. In response to concerns that this definition allows for a sizable discrepancy between the efficiency a motor purchaser thinks he is getting and what he might actually get, NEMA's members have adopted a "suggested standard for future design" that tightens the range so that the minimum efficiency of a motor will correspond to 10% higher losses (instead of 20%) than the nominal efficiency. NEMA has also defined minimum and nominal efficiency levels that a motor must meet to be marketed as an energy-efficient model. Details are covered in chapter 9.

To enable meaningful comparisons to be made it is essential that manufacturers measure efficiency in accordance with IEEE Standard 112, Method B, preferably with the additional restrictions imposed by NEMA Standard MG1-12.53(a) and (b) or CSA standard C(390)-M1985. Due to the range in the efficiencies in the NEMA

Table 2-7. Nameplate efficiencies for motor labeling.

a. NEMA Efficiency Letters for Motor Nameplates ANSI/NEMA MG 1-12.54

Efficiency (%)			Efficiency (%)		
Index Letter	Nominal	Minimum	Index Letter	Nominal	Minimum
A	—	>95.0	M	78.5	75.5
B	95.0	94.1	N	75.5	72.0
C	94.1	93.0	P	72.0	68.0
D	93.0	91.7	R	68.0	64.0
E	91.7	90.2	S	64.0	59.5
F	90.2	88.5	T	59.5	55.0
G	88.5	86.5	U	55.0	50.5
H	86.5	84.0	V	50.5	46.0
K	84.0	81.5	W	—	<46.0
L	81.5	78.5			

b. Nameplate Efficiency Ranges ANSI/NEMA MG 1-12.54

Nominal Efficiency	Minimum Efficiency	Nominal Efficiency	Minimum Efficiency	Nominal Efficiency	Minimum Efficiency
99.0	98.8	95.4	94.5	81.5	78.5
98.9	98.7	95.0	94.1	80.0	77.0
98.8	98.6	94.5	93.6	78.5	75.5
98.7	98.5	94.1	93.0	77.0	74.0
98.6	98.4	93.6	92.4	75.5	72.0
98.5	98.2	93.0	91.7	74.0	70.0
98.4	98.0	92.4	91.0	72.0	68.0
98.2	97.8	91.7	90.2	70.0	66.0
98.0	97.6	91.0	89.5	68.0	64.0
97.8	97.4	90.2	88.5	66.0	62.0
97.6	97.1	89.5	87.5	64.0	59.5
97.4	96.8	88.5	86.5	62.0	57.5
97.1	96.5	87.5	85.5	59.5	55.0
96.8	96.2	86.5	84.0	57.5	52.5
96.5	95.8	85.5	82.5	55.0	50.5
96.2	95.4	84.0	81.5	52.5	48.0
95.8	95.0	82.5	80.0	50.5	46.0

The nominal efficiency listed is the lowest value for each index; the minimum efficiency corresponds to 20% higher losses than the nominal values. The actual nominal motor full load efficiency is used to determine the NEMA efficiency letter (a) and the NEMA efficiency number (b). For example, a motor with a full load nominal efficiency of 93.5% would have a NEMA efficiency label of D, or 93.0.

Source: Lawrie 1987, NEMA 1991a.

labeling system, it is better to obtain the actual nominal or minimum efficiencies for the motor model under consideration, and to determine whether the minimum is guaranteed. The source of this information can be either a motor catalog or the manufacturer.

Depending on one's perspective, it may make more sense to use the nominal or the minimum efficiency for analysis. For example, if many motors are being considered for the same application, the nominal values will give a good indication of what the overall energy usage and savings will be. On the other hand, if the application of a single motor is being analyzed, and it is important for the usage to be no greater than a particular value, then it would make more sense to use the minimum value.

The preceding discussion explains the testing that is used to label motors and how to interpret those labels in making purchase decisions. But labels are of limited value in understanding the performance of motors in the field. A relatively simple field test procedure is described next.

Field Measurements of Motor Efficiency

A user undertaking a serious energy-efficiency program should attempt to characterize the efficiencies of the most important loads in the building or plant. In the case of motors, full load efficiency may not be known, especially for old or rewound units. To solve this problem, the user can approximate motor efficiency under field conditions by measuring load and no-load voltages, currents, power factors, shaft speeds, and stator resistances (Lobodovsky 1989). The testing equipment costs around $4,000, and typical test times are 1 to 1½ hours per motor. At least one manufacturer (Esterline Angus) provides special-purpose equipment and software to analyze the results of these tests and to perform field efficiency tests using the same approach.

The motor test procedure is adapted from IEEE Standard 112, Methods E/F. The mechanical output is estimated indirectly by determining the losses:

$$\text{Efficiency} = \frac{\text{Mechanical Output}}{\text{Electrical Input}} = \frac{\text{Electrical Input} - \text{Losses}}{\text{Electrical Input}}$$

This method, though not as accurate as NEMA MG 1-12.53, provides some useful estimates of the motor load and efficiency, identifying low-efficiency motors, low-efficiency rewinding jobs, and oversized motors, which can be identified by measuring the input for both no-load and normal operating conditions (compared to the full load rating) and the losses. In addition, the speed of the motor compared to the rated full load speed gives an indication of the motor's loading.

The results of the test, together with the motor load profile, allow the user to evaluate whether to keep the motor, replace it with a more efficient version, or replace it with a motor better matched to the load.

A quicker, though less accurate, method for checking motor efficiency in the field is to take advantage of the nearly linear relationship between motor slip and load (see Figure 3-9). Caution should be used in applying this technique, because the slip also varies with motor voltage possibly resulting in errors of over 5% (Nailen 1987); the technique is not reliable for rewound motors. The necessary test equipment includes a watt-meter and a tachometer accurate to plus or minus 1 rpm. The tachometer is used to measure the actual motor speed, which is used to determine the motor load (output power). The watt-meter is used to measure the motor input power.

For example, a 10-hp motor was rated at 1,745 rpm at full load. The measured speed was found to be 1,778 rpm, and the measured power was 3.8 kW. Load is nearly proportional to slip, so the fraction of full load is approximately proportional to the fraction of the full load slip:

$$\frac{1,800 \text{ rpm} - 1,778 \text{ rpm}}{1,800 \text{ rpm} - 1,745 \text{ rpm}} = 40\% \text{ of full load (or 4.0 hp output)}$$

The efficiency is thus approximately:

$$\frac{4.0 \text{ hp} \times .746 \text{ kW/hp}}{3.8 \text{ kW}} = 79\%$$

If the application never runs at a higher load, a 5-hp EEM might make a good retrofit or replacement (from Table A-1, A-2, or 2-3, 5-hp EEM efficiency is about 89%), depending on the operating hours, load profile, and cost of electricity.

Summary

The three-phase squirrel cage induction motor accounts for over 75% of U.S. drivepower input, followed by single-phase induction motors, and all others (synchronous, wound-rotor induction, DC, and so forth). Energy-efficient motors (EEMs) are available for most three-phase and single-phase induction motor applications. These EEMs offer reduced energy and peak-power costs as well as increased life compared to standard-efficiency motors. EEMs are generally cost-effective in new applications or as alternatives to rewinding, and they can be cost-effective as retrofits depending on the specifics of the application. It is important to obtain up-to-date information on the efficiency of the specific motor in question. Motor rewinds can seriously degrade the efficiency of motors; the extent of this degradation, and solutions to it, would be fruitful areas for research and development.

Chapter 3

System Considerations

T his chapter discusses a number of important but often overlooked determinants of motor system efficiency, including
power supply quality, the distribution network that feeds the
motor, the match between the load and the motor, the transmission
and mechanical components, and maintenance practices. Simple
and inexpensive diagnostic techniques for identifying some common motor system problems also are presented. Unfortunately, the
lack of field data makes it difficult to quantify the extent of energy
losses and equipment damage from poor system optimization. In
general, older facilities modified in pieces and loaded closer to
capacity are more likely to have problems than newer facilities.

Power Supply Quality

AC motors, particularly induction motors, perform best when fed
by symmetrical, sinusoidal waveforms of the design voltage value. Deviations from these ideal conditions can reduce the motor's
efficiency and longevity. Such distortions in power quality include
voltage unbalance, undervoltage or overvoltage, and harmonics.

Voltage Unbalance

In a balanced three-phase system, the voltages in the three phases
can be represented by three vectors of equal magnitude, each out
of phase by 120 degrees. A system that is not symmetrical is called
an unbalanced system.

There are several common causes of voltage (or phase) unbalance. The first is a nonsymmetrical distribution of single-phase
loads in the plant. Most plants contain a mixture of three-phase
loads (such as motors) and single-phase loads (such as most lighting

and electrical outlets as well as single-phase motors). Putting a disproportionate share of the single-phase loads on one of the three phases can cause voltage unbalance. A second cause of voltage unbalance is an open circuit in one of the phases, often caused by a blown fuse.

Finally, different size cables carrying the phases of a three-phase load can lead to unbalanced conditions. This can happen in an older plant when a load is converted from single phase to three phase. Different cable sizes produce different voltage drops, which in turn lead to the unbalanced voltages.

Voltage unbalance is problematic for several reasons. First, it wastes energy. As Figure 3-1 illustrates, voltage unbalance leads to high current unbalance, which in turn leads to high losses. A modest phase unbalance of 2% can increase losses by 25%. Second, prolonged operation under unbalanced voltage can damage or destroy a motor. The excess heat generated in a motor running on a 2% unbalance can reduce insulation lifetime by a factor of eight (Andreas 1982). An unbalance of 5% or more can quickly destroy a motor. Because of this, many designers include phase unbalance and phase failure protection in motor starters. Another negative impact of phase unbalance is the reduction in motor torque, particularly during start-up. Figure 3-2 shows the reduction of rated power as a function of the voltage unbalance.

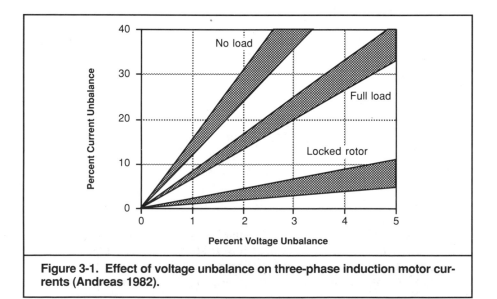

Figure 3-1. Effect of voltage unbalance on three-phase induction motor currents (Andreas 1982).

While severe unbalance (over 5%) causes immediate, obvious problems, small unbalances in the 1–2% range are insidious because they can lead to significant increases in energy use without being detected for a long time, particularly if a motor is oversized. To avoid this situation, the voltages in a plant should be regularly monitored, and an unbalance over 1% should be remedied; unbalance under 1% is generally not a cause for action.

The diagnosis of voltage unbalance is a simple operation requiring the measurement of the voltages in the three phases. The following formula can be used for the approximate calculation of the voltage unbalance:

$$\text{Voltage Unbalance } (\%) = \frac{\text{Maximum difference of the voltages in relation to the average voltage}}{\text{Average voltage}} \times 100$$

Suppose the measurements in the three phases give the following values:

$$V_a = 200V$$
$$V_b = 210V$$
$$V_c = 193V$$

The average voltage is 201V
Maximum difference from the average $= 210V - 201V = 9V$
Voltage unbalance $= (9V / 201V) \times 100 = 4\%$

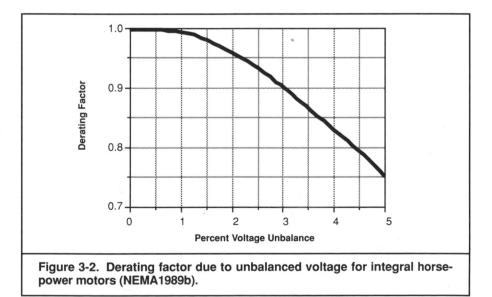

Figure 3-2. Derating factor due to unbalanced voltage for integral horsepower motors (NEMA1989b).

Voltage Level

When an induction motor is operated above or below the rated voltage, its efficiency and power factor change. Motors are designed to operate successfully at full load with a plus or minus 10% voltage fluctuation (NEMA 1991a). Voltage fluctuations exceeding 10% can decrease motor efficiency, power factor, and lifetime if the motor is running at or nearly at full load. Voltage fluctuations normally result from improperly adjusted transformers, undersized cables (leading to large voltage drops due to higher resistance in the small cables), or with poor power factor in the distribution network.

When a motor is underloaded, reducing voltage can improve the power factor and efficiency, mainly by reducing the reactive current. This practice works for both standard and efficient motors, although EEMs are less affected by voltage fluctuations.

Because torque is proportional to the square of the voltage, motors subjected to undervoltage might have a hard time starting or driving a high-torque load. For instance, if the voltage dips to 80% of the rated value, the available starting torque is only about 60% of its rated value.

The diagnosis of voltage level problems requires monitoring and recording voltages, preferably for a whole cycle of the plant operation. Patterns in voltage fluctuation over time sometimes help to reveal the cause(s). Measuring voltage is normally easiest at the motor starter terminals; to estimate the voltage at the motor terminals, the voltage drop in the cable connecting the starter to the motor should be calculated. NEMA-rated voltages for three-phase 60-Hz induction motors appear on the motor nameplate and typically allow for a voltage drop of about 4% in the motor feeding cables.

Harmonics and Transients

Under ideal conditions, utilities supply pure sinusoidal waveforms of one frequency (60 Hz in North America and 50 Hz in Europe), like those shown in Figure 2-3. Resistive loads, such as incandescent lights, use all of the energy in that waveform. Other loads, including ASDs and other power electronic devices, arc furnaces, and overloaded transformers, cannot absorb all of the energy in the cycle. In effect, they use energy from only part of the sine wave and thus distort it (Figure 3-3).

The resulting distorted waveform contains a series of sine waves with frequencies that are multiples of the fundamental 60-

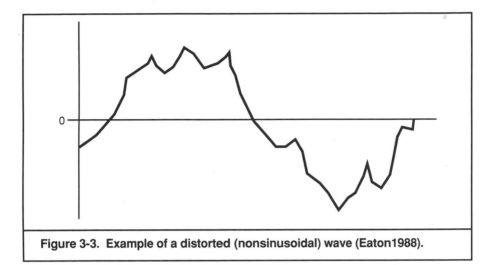

Figure 3-3. Example of a distorted (nonsinusoidal) wave (Eaton1988).

or 50-Hz frequency. These are called harmonics. The 180-Hz (or three times 60 Hz) component is the third harmonic, the 300-Hz (or five times 60 Hz) component is the fifth harmonic, and so on.

Harmonics can increase motor losses, reduce torque, and cause torque pulsation and overheating. Vibration and heat in turn can shorten motor life, by damaging bearings and insulation. Harmonics also can cause malfunctions in electronic equipment, including computers; induce errors in electric meters; produce radio frequency static; and destroy power system components.

Electronic ASDs, discussed further in chapter 4, can both generate and be damaged by harmonics from other sources. It is thus very important that they be installed properly and, in some cases, be isolated from other equipment by separated feeders, transformers, and harmonic filters. Serious harmonics problems from properly installed ASDs are rare. They are most likely with large drives and in situations where ASDs control a large fraction of the total load.

Standard-efficiency motors must sometimes be derated by 10–15% when supplied by an ASD that produces substantial harmonics. Derating is more likely to be needed for constant-horse-power installations than for variable-torque loads. For example, a motor with a nominal rating of 100 hp that was derated by 15% would only be able to drive an 85-hp load. EEMs are better able to cope with harmonics because of their higher thermal margins and lower losses and thus are seldom derated. One notable exception is when the motor runs below about 30–40% of rated speeds. Under these conditions, high losses caused by harmonics combined with the

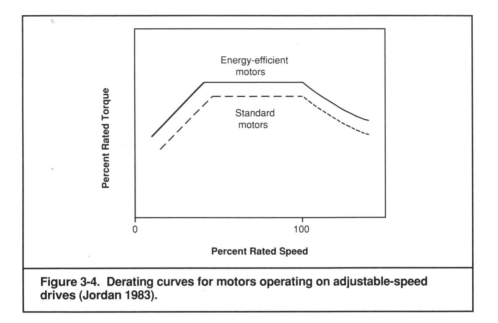

Figure 3-4. Derating curves for motors operating on adjustable-speed drives (Jordan 1983).

lower ventilation available at reduced fan speeds often require even EEMs to be derated. As Figure 3-4 shows, the allowable torque of both kinds of motors falls off sharply at low and very high speeds when powered by an ASD.

Harmonics can substantially disrupt conventional electric meters. One study sponsored by EPRI found errors ranging from +5.9% to −0.8% in meters subjected to harmonics from ASDs (EPRI 1982). During periods when ASDs operate at very light load, errors have exceeded 10% and caused severe waveform distortion. Harmonics generated by ASDs or other equipment may cause metering errors and overbilling, providing customers with another reason to correct or suppress harmonics at the source. Customers with medium or large ASDs on their premises might even consider installing solid-state watt-hour meters that generally give accurate readings even in the presence of harmonics (Peddie 1988).

Besides harmonics, the voltage waveform may also contain another form of undesirable distortion called transients. Transients are brief events, usually microseconds in length, and appear either as voltage spikes or voltage notches in the sinusoid. Fast transients result from the commutation of power electronic devices and circuit-breakers, and from lightning.

Transients, if rare, have little impact on energy consumption, but if they occur repeatedly and at frequent intervals, they can behave like harmonics and thus increase losses in a motor. If they are very large, as in the case of a lightning strike, they can damage or destroy equipment.

Generally, transients present a problem only in facilities where large loads are cycled, producing distortion in the voltage waveforms. For example, a facility with a large induction furnace where power is applied intermittently might have a problem with transients.

Diagnostics and Mitigation of Harmonics and Transients

Equipment to accurately monitor transients and harmonic distortion is available but fairly expensive (see appendix B). Several less expensive tools can roughly assess the level of harmonics and indicate whether more precise measurements are warranted.

For instance, an oscilloscope can be used to generate a picture of the voltage waveform, distortions that signal the presence of meaningful harmonics, which should be measured with a harmonic analyzer. Another technique is to compare the voltage readings from two AC digital voltmeters, one with true root mean square (RMS) capabilities and the other without. The true RMS meter gives the correct voltage even if there is harmonic distortion, whereas a normal meter only gives the correct value if there are no harmonics. If the two meters are calibrated, readings differing by more than a few percent indicate a significant harmonic content.

Harmonics should be reduced to an acceptable level (less than 5% of the fundamental current in medium-voltage systems and less than 1.5% in high-voltage systems) as close as possible to the source (IEEE 1981). Mitigation at the source is normally most effective, as it prevents the losses from harmonics propagation in the network.

Surge suppressors are available and effective for suppressing transients that may interfere with the operation of computing and communications equipment. Claims that these devices save energy, however, are unfounded.

In ASDs, harmonics are most commonly controlled by installing filters at the ASD input circuit to provide a shunt path for the harmonics and to perform power factor compensation. IEEE Standard 519 contains guidelines for harmonic control and reactive power compensation of power converters. The cost of the harmonic filter to meet this standard is typically around 5% of the cost of the ASD.

The Federal Communications Commission (FCC) has produced a set of regulations regarding the electromagnetic interference (EMI) produced by "computing devices." These regulations, which are also becoming widely accepted in the ASD market, set permissible radiation and conduction levels. FCC standards define two classes of products: Class A systems used in commercial and industrial environments and Class B systems used in residences. The Class B standards are more strict, to avoid noticeable interference with radio and television use in the home. Although ASDs are expected to meet only Class A standards, some manufacturers offer ASD equipment that performs within Class B requirements. Radiated EMI can be brought down to FCC standards by proper layout and by shielding the enclosure.

System Oversizing

Motor systems become oversized when designers adopt successive safety factors. Designers do this to allow for growth in a system's peak requirements, and they assume the extra capacity cost is a small premium to pay to insure the system will be able to cope with maximum demand. For example, a designer might choose a pump with a 30% safety margin (some margin for increase in the process requirements and some margin for scale build-up in the pipes) and then round up in choosing among standard motor sizes and specify a motor with 20% extra horsepower. Such oversizing sometimes is warranted and sometimes leads to costly waste.

Motor Oversizing

As Figures 3-5 and 3-6 illustrate, the efficiency and power factor

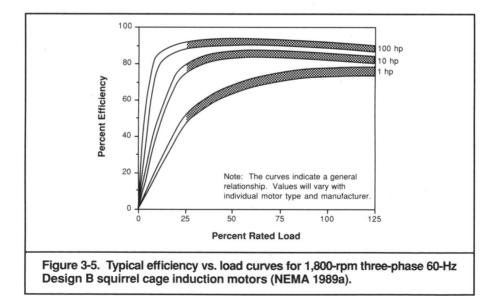

Figure 3-5. Typical efficiency vs. load curves for 1,800-rpm three-phase 60-Hz Design B squirrel cage induction motors (NEMA 1989a).

of a motor vary with the load. The efficiency of most motors peaks at around 75% load and drops off sharply below 40% load, although this range varies among motor designs.

In a recent test of 37 motors at various loads, the standard-efficiency motors peaked near 100% load and the high-efficiency models peaked nearer 75% load (Colby and Flora 1990). Power factor drops steadily with the load. Even at 60% load, the power factor often needs compensation, and it drops sharply below 60% load. Figure 3-7 (next page) shows that low-speed motors have substantially lower power factors than high-speed motors of the same size.

To stay within its optimal operating bounds, a motor should be sized to run at 50% or more of its rated load most of the time. A grossly oversized motor (generally below 40% loading) will run at low efficiency and low power factor, thus increasing energy costs and requiring either costly power-factor compensation or added utility charges to pay for the reactive current. In addition, larger motors cost more to buy and install and require larger and more expensive starters.

Although oversized motors present these drawbacks, they also can accomodate unanticipated high loads and are likely to start and operate more readily with undervoltage conditions. A modest sizing margin, however, can generally provide these advantages.

The question of motor sizing is not limited to new installations.

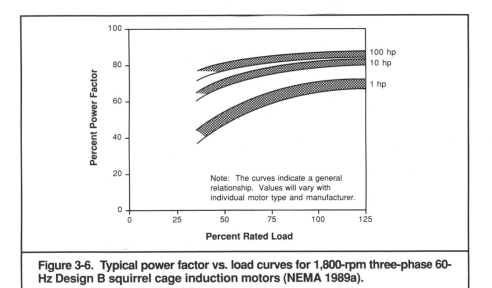

Figure 3-6. Typical power factor vs. load curves for 1,800-rpm three-phase 60-Hz Design B squirrel cage induction motors (NEMA 1989a).

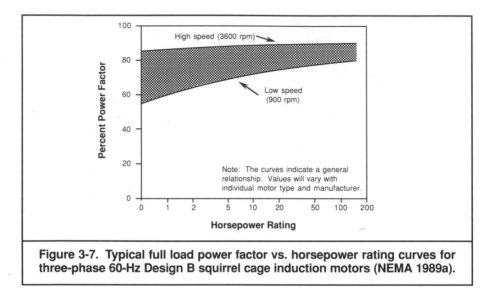

Figure 3-7. Typical full load power factor vs. horsepower rating curves for three-phase 60-Hz Design B squirrel cage induction motors (NEMA 1989a).

It is often economic to replace existing motors with smaller, high-efficiency units. Users must carefully evaluate which motors to downsize. Motors that run many hours per year at light loading are obvious candidates because they are running far below their optimal efficiencies. Motors that operate in the 50–100% load region are not likely candidates, because they are operating close to peak efficiency. For example, a 150-hp motor driving a 120-hp load is normally more efficient than a 125-hp motor driving the same load, because the efficiency of many motors peaks at approximately 75% load, and larger motors generally have higher efficiencies.

Because the relationship between efficiency and load varies among different sizes and types of motors, there is no definitive rule about downsizing. Larger motors generally maintain efficiency at low loading better than smaller motors do. Similarly, high-efficiency motors have a flatter efficiency curve than standard-efficiency models. For example, efficiency might drop rapidly at 48% of full load in a standard motor but not until it reaches 42% of rated load in an EEM. In general, motors that always run below about 40% load are strong candidates for downsizing.

Figure 3-8 shows the change in efficiency over a range of 25% to 150% load for seven new 5-hp motors (three EEMs and four standard units) and eight new 10-hp motors (five EEMs and three standard-efficiency motors) tested by the North Carolina Alternative Energy Corporation (Colby and Flora 1990). These figures represent one of the few published measurements of efficiency below 50% loading. Manufacturers typically list efficiencies at 100%, 75%, and 50% loading only.

Two points are notable. First, the high-efficiency motors better maintained their efficiencies across the full range of loading. Second, the efficiency of all motors fell off sharply as loads fell from 50% to 25%; efficiency fell six to seven points in the standard-efficiency units and about four points in EEMs.

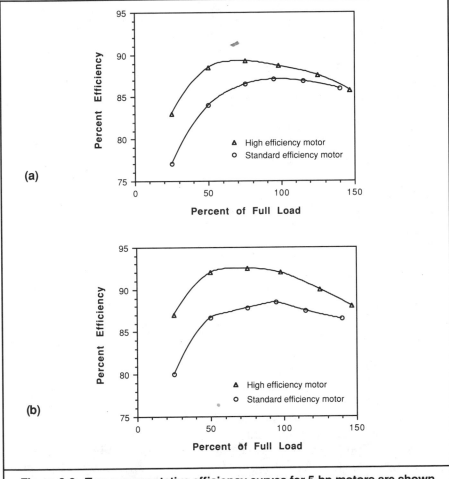

Figure 3-8. Two representative efficiency curves for 5-hp motors are shown in (a). Note that the high-efficiency motor maintains its high efficiency level over a wider load range than does the standard motor. A similar comparison for 10-hp motors is presented in (b) (Colby and Flora 1990).

Evaluating whether to downsize a motor requires knowing both its typical and maximum loads. Consider, for example, two 100-hp motors running different fans in a plant. Both motors are metered, and neither one requires more than 35 hp during the metering period. The first fan's specifications reveal it will never need more than 40 hp. Since the motor will never exceed about 40% of its rated 100-hp load, it is appropriate to replace it with a high-efficiency 50-hp unit. The second fan's specifications indicate that it occasionally must use up to 80 hp, even though it typically draws only 35 hp. Thus, while the motor often runs at only 35% of its rated 100-hp capacity, it should not be down-sized—or at least not by much—unless the system's maximum power requirements are reduced. Such measures are discussed in chapter 5.

Another option in such a situation is to use a smaller motor with a high service factor, so that it can withstand overloading on rare occasions. However, this determination should be made by an engineer familiar with the process.

Diagnostics of Motor Oversizing

The load on a motor can be estimated by comparing a wattmeter reading of the power input with the motor's rated power. If the motor load changes over time, a simple wattmeter will not suffice. Instead, a chart recorder or data logger should be used to monitor the absorbed power under different operating conditions. A clamp-on ammeter does not give a good estimate of the motor load since the power factor drops sharply at low loads, and amperage readings are greatly affected by power factor.

The same technique described on page 63 for estimating motor efficiency in the field can be used to roughly determine the load on a motor. It is based on the fact that the slip speed is almost a linear function of the load (Figure 3-9). The first step is to measure the motor's speed with an inexpensive optical (strobe or reflective-beam) tachometer. Once the operating speed is measured, the load can be calculated by comparison with the nameplate data. Consider a four-pole motor with a full load speed of 1,750 rpm, a synchronous speed of 1,800 rpm, and an operating speed, as measured by the tachometer, of 1,780 rpm. Subtracting the rated full load speed from the synchronous speed yields the full load slip. Subtracting the measured operating speed from the synchronous speed gives the actual slip. The ratio of the actual slip to the full load slip in this case is 40%. Plugging in 40% of full load slip on the y-axis of Figure 3-9 corresponds to about 42% of full load on the x-axis.

$$\text{Full load slip} = 1,800 - 1,750 = 50 \text{ rpm}$$
$$\text{Actual slip} = 1,800 - 1,780 = 20 \text{ rpm}$$
$$\text{Fraction of full load slip} = 20/50 = 40\%$$

Extent of Oversizing

Data presented in chapter 6 suggest that roughly one-fifth of motors 5 hp and larger are running at or below 40% of rated load. The distribution of these loads—the proportion at 35–40% load, 30–35% load, 25–30% load, and so on—is not known. We assume, with no supporting data, that most are clustered in the high end of the range, between, say, 25% and 40% loading. As discussed above, there is very little published information on efficiency loss at loading below 50%. The data shown in Figure 3-8 suggest that loading between 25% and 40% leads to a drop of two to eight percentage points below the performance at 50% load. Loading below 25% will lead to an even larger drop in efficiency. Based on admittedly sketchy data we assume an average efficiency loss of five percentage points in those motors running below 40% loading.

While some oversized motors should be downsized, many should instead be equipped with controls that enable them to operate more efficiently at partial load. Various means of doing this are discussed in chapter 4.

Distribution Network Losses

In-plant distribution losses can be reduced by selecting and properly operating efficient transformers and by correctly sizing distribution cable.

Distribution transformers normally operate above 95% efficiency,

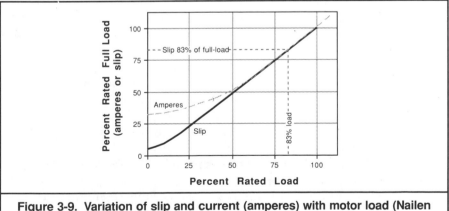

Figure 3-9. Variation of slip and current (amperes) with motor load (Nailen 1987). Note that this relation holds only at the rated motor voltage; slip varies inversely with the square of the actual motor voltage.

unless they are old or very lightly loaded. Transformers over 30 years old should be replaced by new models that are more efficient. It is usually more efficient to run one transformer at moderate to full load than to operate two of them in parallel lightly loaded .

Many large customers are primary metered (on the high voltage side of the distribution transformer) and own their own transformers. Some utilities are encouraging their primary metered customers to install high-efficiency transformers in new facilities or for plant expansions.

Cable Sizing

Losses in distribution wiring are a function of the square of the current times the cable's resistance (I^2R). Most facilities size distribution cable on the basis of the National Electric Code, which is concerned with preventing overheating and fire risk and with providing adequate starting current to motors, not with energy efficiency. Because resistance falls as the diameter of the wire increases, larger wire has lower losses and helps to decrease the voltage drop between the transformer and the motor.

In general, it is cost-effective (in new installations and remodels but not as a retrofit) to install cable larger than that required by code if (1) the larger cable can be installed without increasing the size of the conduit, (2) the motor or set of motors is expected to run at or near full load, and (3) the system operates a large number of hours per year.

Table 3-1 compares losses (W/ft) and costs of using several cable sizes to supply a 30-A load, assuming 8,000 hours of annual operation and an electricity cost of $.06/kWh. The larger cables offer very attractive payback times. Note that the payback is very sensitive to operating hours: at 4,000 hours, the paybacks in Table 3-1 would double.

Small feeders typically use the minimum size conduits an electrician can easily pull the wire through. In general, the size of small feeders can be increased without increasing the size of the conduit, making the use of oversized feeders cost-effective. Conduit for feeders for larger motors is determined by the size of the wire, so the conduit size will often need to be increased if the wire size is increased. As a result, the use of oversized wires for larger circuits needs to be evaluated case by case.

Energy savings are not the only reason to install larger

Table 3-1. Savings from lower-loss distribution wiring.

Marginal Savings and Costs of Lower-Loss Distribution Wiring (Compared to #8 Wire) Assuming 100% Load

Wire Size[1]	Loss W/Ft $3I^2R$/Ft	Loss $/Ft-Yr	Saving $/Ft-Yr	Installed 1989 $/Ft First Cost			Marg. Cost +$/Ft	Simple Payback(Yr)
				Conduit	Wire	Total		
#8	2.01	$0.96		$3.16	$1.21	$4.37		
#6	1.28	$0.61	$0.35	$3.16	$1.59	$4.75	$0.38	1.1
#4	0.81	$0.39	$0.57	$3.86	$2.20	$6.06	$1.69	3.0
#3	0.65	$0.31	$0.65	$3.86	$2.67	$6.53	$2.16	3.3

Marginal Savings and Costs of Lower-Loss Distribution Wiring (Compared to #8 Wire) Assuming 75% Load

Wire Size[1]	Loss W/Ft $3I^2R$/Ft	Loss $/Ft-Yr	Saving $/Ft-Yr	Installed 1989 $/Ft First Cost			Marg. Cost +$/Ft	Simple Payback(Yr)
				Conduit	Wire	Total		
#8	1.18	$0.56		$3.16	$1.21	$4.37		
#6	0.75	$0.36	$0.20	$3.16	$1.59	$4.75	$0.38	1.9
#4	0.47	$0.22	$0.34	$3.86	$2.20	$6.06	$1.69	5.0
#3	0.38	$0.18	$0.38	$3.86	$2.67	$6.53	$2.16	5.7

[1]Sizes in American Wire Gauge. For diameters, see page 294.
Assumptions: 30 amp load, 8,000 hours per year, 6¢/kWh.

Source: Lovins et al. 1990.

The Southwire Company's Wire-Sizing Policy

The Southwire Company, a billion-dollar industrial firm with an aggressive energy management program, wires all new loads under 100 A with conductor one size above code and uses larger than normal wire for larger loads when doing so is cost-effective. Jim Clarkson, corporate energy manager, said that because his staff does not have time to evaluate every new wiring job, uneconomic oversizing may occur in some installations, but that, overall, the policy saves the firm money, energy, and time (Clarkson 1990).

---

distribution cable. The added distribution capacity provides room to expand loads later without having to remove and replace the old wiring. It also provides lower voltage drops, which improve motors' starting and operating performances.

Power Factor Compensation

As discussed in chapter 2, low power factor has undesirable and costly effects often worth mitigating. Figure 3-10 shows the extent to which improving power factor can reduce loss. For example, increasing power factor from 0.75 to 0.90 reduces cable and transformer copper losses by 32%. Improving power factor from 0.60 to 0.90 reduces losses by 57%.

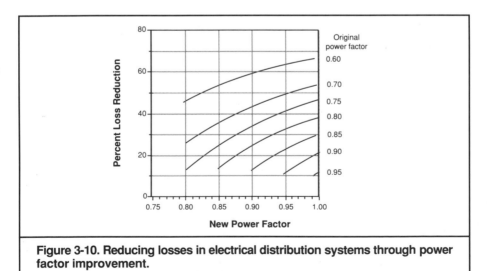

Figure 3-10. Reducing losses in electrical distribution systems through power factor improvement.

In smaller motors power factor generally is lower, and it drops more rapidly as the load drops. As a result, a plant with a large number of small motors but without power factor correction will typically have a lower power factor than a plant with predominantly large motors. Poor power factor can be caused not only by lightly loaded motors but also by other loads such as fluorescent lighting ballasts and certain types of ASDs.

Improving the power factor can save energy and dollars by reducing losses in the customer's distribution system. Greater savings can often be achieved by reducing power factor penalty charges (if these charges are imposed by the utility). Such charges are normally substantial enough to make it cost-effective for the customer to improve his power factor to 0.90. For example, a utility might increase the demand charge by 1% for every 1% the power factor drops below 90%. If a plant has a peak demand of 1,000 kW and a power factor of 81%, the plant will be charged for peak demand as follows:

Adjusted peak = (1,000 kW) $\times$ [1 + (0.90 − 0.81)] = 1,090 kW

If the utility demand charge is $70 per kW-yr, the power factor penalty will be

Power Factor Penalty = (90 kW) $\times$ $70 = $6,300 per year.

In contrast, the energy savings for the same load, assuming it is fed at 480 volts, three-phase through 500 feet of cable, would be about 21,000 kWh, resulting in a cost reduction of $1,300 per year at $.06/kWh.

The consumer can correct power factor either in a distributed manner (capacitors connected to the motor terminals) or in a centralized manner (a capacitor bank at a central location in the plant). Some large plants may have an intermediate scheme with several capacitor banks, each serving several motors. The distributed option reduces the losses between the motors and the central capacitor bank and costs less to install.

The centralized scheme requires controlled switching of the capacitor bank. Switching avoids overcompensation of the power factor when only a few motors are running. Overcompensation causes the same undesirable effects as a low power factor.

The installed cost for capacitors ranges from about $20 to $30 per kVAR of reactive power for dispersed units to $50 to $75 per kVAR for central capacitors. In the above example, 240 kVAR are required (see Figure 3-11, next page). Assuming $25 per kVAR, the cost would be $6,000. The resulting payback is less than one year from penalty reduction alone, five years from energy savings alone, and about nine months from the combined savings.

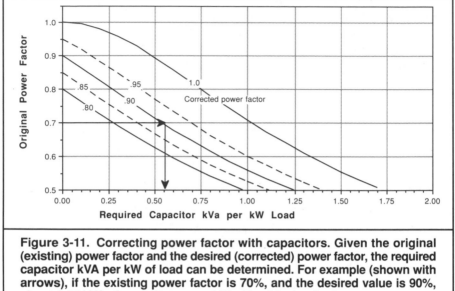

Figure 3-11. Correcting power factor with capacitors. Given the original (existing) power factor and the desired (corrected) power factor, the required capacitor kVA per kW of load can be determined. For example (shown with arrows), if the existing power factor is 70%, and the desired value is 90%, about 0.54 kVA of capacitance per kW of load is required.

Motor manufacturers generally recommend a maximum capacitor size at the motor terminals (see Figure 2-12, page 28). Figure 3-11 shows the capacitor kVAR required to improve the power factor by various amounts. Thus, improving the power factor of a 1,000-kW load from 70% to 90% requires capacitance of just over 500 kVA.

The concern over electrical equipment (primarily transformers and capacitors) containing polychlorinated biphenyls (PCBs) has resulted in the removal and disposal of many power-factor correction capacitors in customers' distribution systems. If these capacitors have not been replaced, power factor has decreased. Since the switchgear and mounting are already in place, new capacitors might be installed cheaply enough to obtain a reasonable payback on energy savings alone, depending on the specifics of each application.

As mentioned in chapter 2, another potentially lucrative benefit of power factor improvement is increased capacity of the distribution system. This advantage is especially relevant in new construction or whenever load approaches the system's capacity (as when plant is expanded or equipment added). If installing compensation eliminates or even postpones the need to replace a transformer, switchgear, feeders, or other equipment, doing so can be very cost-effective, again depending on the specifics of the situation.

Diagnostics of Power Factor Compensation

Medium and large customers generally know if they have low power factor because the utility charges them for the reactive power they draw. An industrial facility where most of the load is for motors that do not have power factor correction will typically have a plant power factor of 70% to 80%. Other equipment, such as rectifiers or arc furnaces, can have power factors as low as 45%. To avoid a large reactive power bill, most such facilities have already installed equipment for power factor correction. Customers such as residential consumers and small commercial buildings below 50–100 kVA, who typically pay small or no power factor charges, generally do not install corrective equipment. Most small consumers have a fairly good power factor because they have a high fraction of resistive loads or compensated loads such as lighting ballasts with internal power factor correction.

The measurement of the power factor requires the use of a wattmeter to measure power, a voltmeter and a clamp-on ammeter. There are also meters that read power factor directly, either as dedicated PF meters or as part of fancier analyzers. Also on the market is equipment that can measure watts, volts, and amps, and can register them continuously on paper or in computer storage (see appendix B). The power factor in a symmetrical three-phase system is given by

Power factor = $P/(3 \ V \times I$), where:
P is the three-phase power,
V is the phase-to-neutral RMS voltage, and
I is the RMS current in each phase.

Load Management and Cycling: General Considerations

Most energy-saving measures described previously reduce demand except in instances of variable loads whose peak coincides with the utility's peak. (For example, installing an ASD on a variable load saves energy at partial load but does not reduce demand at full load.) Consumers, especially those operating large motor systems, should take into account the economic benefits of demand reduction when evaluating the cost-effectiveness of energy conservation investments.

Motor cycling and scheduling can reduce power demand during peak periods. Loads that can be suspended periodically with no serious cost or inconvenience are likely candidates for cycling. Examples include refrigeration equipment, air conditioners, and heat pumps. Loads that frequently idle for extended periods also are good candidates for shut-down or cycling to lower power during idle periods.

Whether equipment cycling lowers energy use as well as demand depends on the application, as described in the following examples:

A large retail store with a constant-volume HVAC system installs an energy management system that turns off one of the four 30-hp fans in a staggered 15-minute rotation every hour. Cycling is an acceptable control method in this building since it creates only minor temperature swings that are well within the limits of comfort for both shoppers and employees. In this case, in addition to demand falling, energy use also declines by 25% for each fan. If the building comfort level can be maintained using only 75% of the HVAC system's ventilation capacity, another option is to slow the fans by means of ASDs or resheaving (changing the pulleys that help connect the motor to the fan).

In another building, several small air conditioning units are each turned off for 15 minutes every hour. However, the setpoints on the thermostats in the building do not change. In this case, while there is a substantial reduction in demand, there is only a small reduction in energy use because the air conditioners have to work harder when they are operating.

Potential Cycling Problems

Starting a motor causes an inrush of current that generates a lot of heat. During start-up, ventilation fans are turning slowly, so they remove only a small portion of this heat. If this heat buildup is excessive, it can reduce the lifetime of the insulation and bearings, and lead perhaps to rapid failure. There are also mechanical stresses from electromagnetic forces associated with large starting currents. In particular, the ends of the windings can suffer fatigue and cracking.

These thermal and mechanical stresses limit the frequency at which a motor can be cycled. Additionally, the electrical equipment that feeds the motor and the mechanical equipment driven by the motor are stressed each time a motor is started. These types of drawbacks can be mitigated with the use of starting controls, which are discussed below.

Allowable Cycling

The heating that occurs when a motor starts is a function of the current and the time used to accelerate the load. The longer the starting time, the greater the motor heating. The time it takes to accelerate a load from start to the rated speed depends on several factors, including the load's inertia, which depends on both the load's mass and its radius. A fan with a large radius has a much larger moment of inertia than a smaller-radius pump of similar shaft power requirement. The higher the load's rated speed, the longer it takes to start. Kinetic energy is proportional to the square of the speed. Torque is also important: the higher the torque required by the load relative to the torque available from the motor, the longer it takes to accelerate to rated speed.

NEMA Standards MG 1-12-50 and MG 10 provide guidance on the number of successive starts that can be made each hour without causing motor damage. Table 3-2 (next page) presents the allowable number of starts per hour and the minimum time between starts, considering the effects of motor horsepower, number of poles (rated speed), and inertia of the load. If a motor operates close to the upper bounds derived from Table 3-2, some reduction in motor lifetime should be expected.

Starting Controls

Three-phase motors use starters that apply all three phases to the motor simultaneously. These starters generally include a motor contactor (a relay to control the flow of electricity to all three phases) as well as devices that protect the motor and the wiring from either a prolonged small overload or from a sudden severe overload.

Because the switching mechanism is a contactor, the conventional three-phase motor starter will apply the full voltage to a motor as soon as the contactor receives power. Since the motor is starting from a dead stop, extra current is required to produce the magnetic field that drives the motor and to supply the initial energy to move the motor and load. As a result, a motor will use between five and seven times the current when starting as it will when operating at full load. This current surge typically lasts for about 30 seconds but may range from a few seconds to several minutes in the case of heavy loads.

These large starting currents may also produce large voltage drops in the feeders, making starting difficult and causing computers to malfunction, lights to dim, and other motors to stall. These problems deserve special attention with large motors and those with long feeders or feeders with small cross sections.

Certain types of electronic controls can ramp up the power during starts instead of forcing the motor to go to full speed from a dead stop. This system, known as a soft start, reduces the inrush of starting current and thus decreases equipment wear.

Table 3-2. Allowable number of starts and minimum time between starts for Design A and Design B motors.

HP	2 Pole			4 Pole			6 Pole		
	A	B	C	A	B	C	A	B	C
1	15.0	1.2	75	30.0	5.8	38	34.0	15	33
1.5	12.9	1.8	76	25.7	8.6	38	29.1	23	34
2	11.5	2.4	77	23.0	11.0	39	26.1	30	35
3	9.9	3.5	80	19.8	17.0	40	22.4	44	36
5	8.1	5.7	83	16.3	27.0	42	18.4	71	37
7.5	7.0	8.3	88	13.9	39.0	44	15.8	104	39
10	6.2	11.0	92	12.5	51.0	46	14.2	137	41
15	5.4	16.0	100	10.7	75.0	50	12.1	200	44
20	4.8	21.0	110	9.6	99.0	55	10.9	262	48
25	4.4	26.0	115	8.8	122.0	58	10.0	324	51
30	4.1	31.0	120	8.2	144.0	60	9.3	384	53
40	3.7	40.0	130	7.4	189.0	65	8.4	503	57
50	3.4	49.0	145	6.8	232.0	72	7.7	620	64
60	3.2	58.0	170	6.3	275.0	85	7.2	735	75
75	2.9	71.0	180	5.8	338.0	90	6.6	904	79
100	2.6	92.0	220	5.2	441.0	110	5.9	1181	97
125	2.4	113.0	275	4.8	542.0	140	5.4	1452	120
150	2.2	133.0	320	4.5	640.0	160	5.1	1719	140
200	2.0	172.0	600	4.0	831.0	300	4.5	2238	265
250	1.8	210.0	1000	3.7	1017.0	500	4.2	2744	440

A=Maximum number of starts per hour
B=Maximum product of starts per hour times load Wk^2
C=Minimum rest or off time in seconds
Allowable starts per hour is the lesser of A, or B divided by the load Wk^2 i.e.
Starts per hour $\leq$ A $\leq$ B/Load Wk^2

Note: Table is based on following conditions:
1. Applied voltage and frequency in accordance with MG1-12.43.
2. During the accelerating period, the connected load torque is equal to or less than a torque that varies as the square of the speed and is equal to 100 percent of rated torque at rated speed.
3. External load Wk^2 equal to or less than the values listed MG1-12.50.

For other conditions, the manufacturer should be consulted.

Source: NEMA 1989a.

Transmission

The transmission subsystem, or drivetrain, transfers the mechanical power from the motor to the driven equipment. The efficiency of drivetrains (output power $\times$ 100/input power) ranges from below 50% to over 95%. As a result, the type of drivetrain used for a given application can have a greater effect on overall system efficiency than the efficiency of the motor itself.

The choice of transmission type depends upon many factors, including the desired speed ratio, horsepower, layout of the shafts, and type of mechanical load. The major varieties include direct shaft couplings, gearboxes, chains, and belts. There is no large-scale survey of the distribution of the different transmission types in the field. Lovins et al. 1989 estimate the distribution for commercial and industrial motors to be as follows:

- 30–50% shaft couplings
- 10–30% gears
- 34% belt drives
- 6% chains

These data were compiled from a small number of sources and may differ from the proportions in any given utility territory.

Shaft Couplings

Shaft couplings have low losses if precisely aligned. Misalignment of the shafts not only increases losses but also accelerates wear on the bearings. The use of couplings is constrained by space and shaft location and is limited to applications where load speed does not vary with respect to motor shaft speed.

Gears

Gears or gear reducers are the primary drive elements for loads that must run slowly (generally below 1,200 rpm) and require high torque that might cause a belt to slip. Gears are also frequently used for loads exceeding 3,600 rpm. The ratings for gear drives depend on the gear ratio (or the ratio of the input shaft speed to the speed of the output shaft) and on the torque required to drive the load. Several types of gears can be used in motor transmissions, including helical, spur, bevel, and worm gears (see Figure 3-12, next page, and the glossary).

The losses in gears result from friction between the gears and in the bearings and seals, from windage, and from lubricant churning. A large number of gear combinations can be used for a given speed ratio.

Helical and bevel gears are most widely used and are quite efficient, reaching 98% efficiency per stage (each step of reduction or increase in shaft speed). With helical gears, the input and output shafts are parallel; with bevel gears they are at right angles. Spur gears are used for the same purpose as helical gears but are less efficient, so they should not be used in new applications.

Worm gears allow a large reduction ratio (5:1–70:1) to be achieved in a single stage. Their efficiency ranges from 55% to

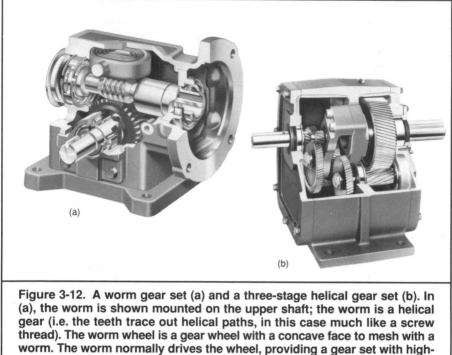

(a)

(b)

Figure 3-12. A worm gear set (a) and a three-stage helical gear set (b). In (a), the worm is shown mounted on the upper shaft; the worm is a helical gear (i.e. the teeth trace out helical paths, in this case much like a screw thread). The worm wheel is a gear wheel with a concave face to mesh with a worm. The worm normally drives the wheel, providing a gear set with high-reduction-ratio connecting shafts with nonintersecting axes at right angles. In (b), gears with helical teeth are used to transmit power between parallel shafts. For bevel and spur gears, see glossary. (Reprinted with permission from Reliance Electric.)

94% and drops quickly as the reduction ratio increases, due to the increase in the friction between the gears. For this reason worm gears should only be used in drives below 10 hp, where operating costs are low. A large reduction ratio is more efficiently achieved by several stages of helical or bevel gears.

Worm gears cost less than helical gears for applications up to 10–15 hp, but helical gears are less expensive above this rating and are becoming the standard for larger drives. The different efficiencies of these two types of gears also affect costs. For example, in low-horsepower ranges, the efficiency of worm gears at full load is typically between 70% and 80% compared to approximately 90% to 96% for a helical gear. The worm gears' lower efficiency will often force the user to increase the size of the motor, and this added cost must be taken into account when comparing gears: a helical gear with a smaller motor may have a lower initial cost than a worm gear with a larger motor, even for applications below 10 hp.

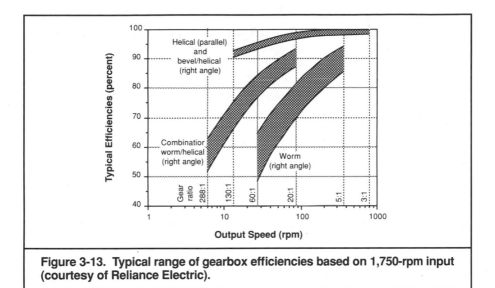

Figure 3-13. Typical range of gearbox efficiencies based on 1,750-rpm input (courtesy of Reliance Electric).

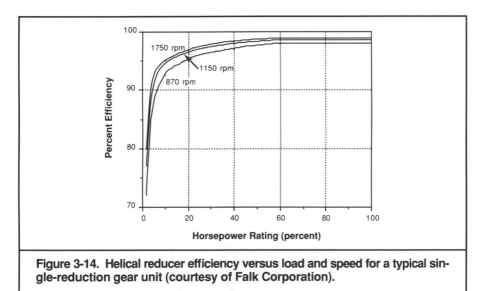

Figure 3-14. Helical reducer efficiency versus load and speed for a typical single-reduction gear unit (courtesy of Falk Corporation).

Figure 3-13 shows the comparative efficiencies of several types of gearboxes, as a function of the speed ratio.

Gear drives are similar to motors in that their efficiency drops markedly below 50% of full load (Figure 3-14) since some of the losses are direct functions of load. For a large gearbox, these fixed losses represent about half the total losses at full load.

In large gearboxes, reducing losses is even more important because the temperature and lifetime are diminished by high temperatures. Using low-friction bearings, gears with a high-quality finish, and improved lubricants can bring the efficiency of a single-stage helical gear to over 99%.

Because gear reducers come in an assortment of in-line and right-angle configurations and sizes, more efficient reducers are difficult to retrofit without major changes to the equipment, as the new reducers are likely to have different dimensions, configurations, or both.

Belt Drives

About one-third of motor transmissions use belts (Lovins et al. 1989). Belts allow flexibility in the positioning of the motor relative to the load, and, using pulleys (sheaves) of suitable diameters, belts can increase or decrease speeds. There are several types of belts: V-belts, cogged V-belts, synchronous belts, and flat belts (Figure 3-15).

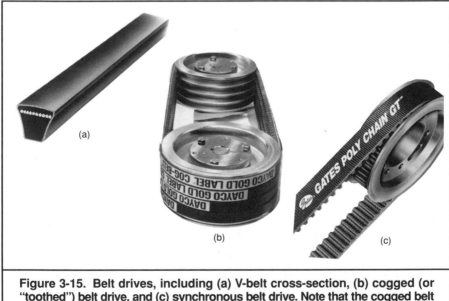

Figure 3-15. Belt drives, including (a) V-belt cross-section, (b) cogged (or "toothed") belt drive, and (c) synchronous belt drive. Note that the cogged belt drive uses the conventional V-belt (smooth) pulleys, while the synchronous belt has meshing teeth on the belt and pulleys (or "sprockets"), preventing slip. Flat belts are similar to synchronous belts (wide and thin) but are smooth on both sides and ride on smooth flat pulleys (courtesy of Gates Rubber Company [a] and [c], and Dayco Products [b]).

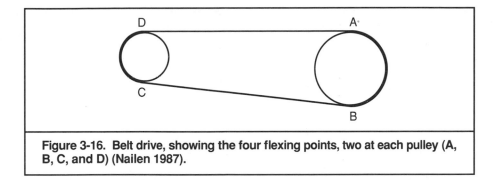

Figure 3-16. Belt drive, showing the four flexing points, two at each pulley (A, B, C, and D) (Nailen 1987).

V-belts are the most common type and have an efficiency in the 90–96% range. The V-belt losses stem from flexing (Figure 3-16), slippage, and, to a lesser extent, windage. Flexing losses are caused by the bending and unbending of the belt material when the belt enters and leaves the pulley.

Belt tension critically determines belt performance. Too much tension can stress the belts, bearings, and shafts; too little tension causes slip, high losses, and premature failure of the belt. With wear, V-belts stretch and need retensioning. They also smooth with wear, and thus become more vulnerable to slip, which, if the V-belt is not properly maintained, will increase and lower efficiency, possibly to below 90%.

Cogged V-belts have lower flexing losses, since less stress is required to bend the belt, and so they deliver 1–3% better efficiency than standard V-belts. Cogged belts can easily be retrofitted on the same pulleys when the V-belts wear out. Cogged V-belts cost 20-30% more than V-belts, but their extra cost is recovered over a few thousand operating hours. One industrial management expert maintains that cogged belts offer a vast potential for energy and dollar savings because they are almost universally substitutable for V-belts, do not require new pulleys, and last twice as long (Johnston 1990).

The most efficient belt is the synchronous design, which can be 98-99% efficient since it has low flexing losses and no slippage. Figure 3-17 (next page) shows the relative performance of synchronous belts in comparison with conventional V-belts. Synchronous belts have no slip because their teeth engage in the teeth of the sprocket pulleys. V-belts rely on friction between the belt and the pulley grooves to transmit the torque, and that friction can be affected by liquids, dust, wear, and other factors. Synchronous belts are designed for minimum friction between the belt and the pulley

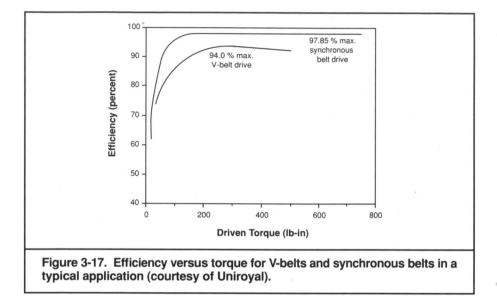

Figure 3-17. Efficiency versus torque for V-belts and synchronous belts in a typical application (courtesy of Uniroyal).

and can withstand much harsher conditions. The efficiency curve of synchronous belts is not only higher but also flatter than that of V-belts, with larger percentage savings as the load decreases.

Because of their construction, synchronous belts stretch very little and do not require periodic retensioning. Synchronous belts typically last over four times longer than V-belts, and these savings in labor and materials for replacements in most cases more than offset the extra cost of the belts.

Retrofitting synchronous belts requires installing sprocket

Case Study: More Efficient Belt System

The Inland Rome Lumber Company mill in Rome, Georgia, used eight V-belts in parallel with a 200-hp, 1,800-rpm motor. These belts lasted only about 1,000 hours and their combined cost was $260. To reduce maintenance requirements, the firm decided to replace the belts and sprockets with one synchronous belt and a pair of sprockets at a cost of $480 for the belt and $1,500 for the sprockets. The new belts last over four times longer than V-belts and improve transmission efficiency by 3-4%. The combination of energy savings and reduced maintenance and replacement costs has resulted in a simple payback of under two years (Thompson 1990).

pulleys that cost several times the price of the belt. In cases where pulley replacement is not practical or cost-effective, cogged V-belts should be considered.

Synchronous belts offer the most efficient and maintenance-free performance. They are available in sizes from fractional-horsepower applications to over 1,000 hp. Due to their positive transmission, they are suitable for applications requiring accurate speed control. They are not, however, suited for shock loads, where abrupt torque changes can shear sprocket teeth. Some manufacturers have doubled the belt's resistance to shock loads by using polyurethane compounds instead of neoprene rubber. Another drawback of synchronous belts is that they do not slip if a machine jams, and they can thus pose a possible safety problem. Possible solutions, other than using a different belt type, include using a clutch or a shear pin that breaks and disengages the equipment in the case of a jam.

High-performance, thin, flat belts using aramid fibers and high-friction surface compounds feature low stretching and low flexing losses. This type of belt is common in Europe. Thin, flat belts can accommodate some slip when there is a surge in the torque, yet maintain an efficiency level close to that of synchronous belts under normal conditions.

Higher-efficiency belt operation leads to lower belt temperature. As with motor insulation, belt life is reduced by half if operating temperature increases by 10°C.

The meshing of the belt teeth in the sprocket makes synchronous belts noisier than V-belts, but a sound-reducing shield can mitigate this problem. It can also protect personnel from the belt and other moving parts, and protect the equipment from debris.

Chains

Chains, like synchronous belts, do not slip. Traditionally, belts have been applied in relatively high-speed, low-torque applications, whereas chains have been used in low-speed, high-torque applications.

Chains also feature high load capacity, the ability to withstand high temperatures and shock loads, long life if properly lubricated, and virtually unlimited length. Chain drives of several thousand horsepower have been built. The efficiency of a well- maintained chain-and-sprocket combination can reach 98%, but wear lowers efficiency a few percentage points.

There are several types of chains, including standard roller chain (both single strand and double strand), double pitch, and "silent chains."

With the exception of silent chains, chains are noisier than belts. Silent chains also offer slightly higher efficiency (up to 99%) but are 50% more expensive in the low-horsepower range and 25% more expensive in the high-horsepower range than roller chains.

Although the steel in the chain stretches only minimally when tensed, the chain sags and needs readjustments as links and sprockets wear. Inadequate lubrication increases wear. Keeping high-speed chains well lubricated is difficult because centrifugal forces eject the lubricant; enclosing the chains and providing constant relubrication, however, as is done in camshaft drives in many auto engines, can solve this lubrication problem.

The lubricant can also quickly lose its effectiveness in environments contaminated with dust or liquids. Under these conditions the use of synchronous belts may prove more attractive. When the chain wears, the sprockets also normally need to be replaced, which increases maintenance costs.

Maintenance

Regular maintenance of the motor system, including inspection, cleaning, and lubrication, is essential for peak performance of the mechanical parts and to extend their operating lifetime.

Lubrication

Lubrication is required to reduce the friction and rapid wear of metal parts moving against one another. Both under- and over-lubrication can cause higher friction losses in the bearings and shorten the bearing lifetime. Under-lubrication may occur either because an insufficient amount of lubricant was applied during routine maintenance or because routine maintenance has not been done frequently enough. In either case, the friction of the bearings will increase, and the energy used by the motor will increase to overcome the increased resistance. As a result, the motor will run hotter, further decreasing its efficiency, and the increased temperature will decrease the lubricity and lifetime of the lubricant (Figure 3-18).

Most maintenance staff people will try to avoid under-greasing by applying "plenty of grease," which, unfortunately, often leads to over-greasing of motor bearings.

Bearing grease must be highly viscous to properly lubricate the moving parts when the motor gets hot. If applied in excess, grease develops internal friction that impedes the bearings and increases the force necessary to turn the shafts. Tests have shown that over-greasing can increase bearing losses by up to 25%, thereby

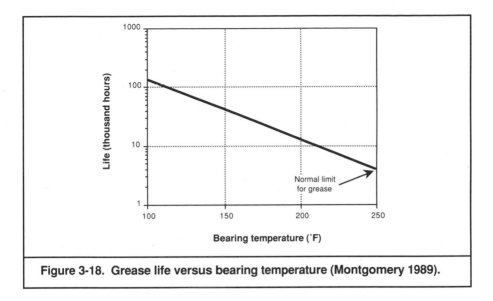

Figure 3-18. Grease life versus bearing temperature (Montgomery 1989).

dropping the overall motor efficiency by perhaps 0.2 to 0.5 percentage point (Katz 1990). In addition, over-greasing may damage the seals and increase churning losses, leading to overheating and early failure of the bearing. Over-greasing also can cause accumulation of grease and dirt on the motor windings, leading to overheating and premature failure.

Old grease should be removed during greasing, and the bearing chamber should generally be filled not more than one-third full of grease. Contamination of the lubricant, especially with water, can also substantially degrade the lubricant performance and lifetime.

Oils and grease are available in a variety of special formulations, with additives to decrease friction and wear and increase lubricant life. Additives may be put in natural (usually mineral) oils, or the oil may be entirely synthetic to meet specific lubrication needs. The most common friction-reducing additives include molybdenum disulfide (MoS_2) and polytetrafluoroethylene (Teflon™). The energy-saving potentials of such lubricants in gearboxes and motors is discussed in chapter 7. Other additives and synthetic formulations are used to improve the lubricants' ability to resist degradation due to high temperatures. While the primary benefit of this improved high-temperature stability is longer lubricant life, indirect energy savings can result from the resulting constancy of desired lubricant properties, especially important where lubricant maintenance is neglected (Lovins et al. 1989).

Periodic Checks

The temperature (the first and quickest indicator of trouble) and the electrical and mechanical condition of a motor should be checked periodically. In general, most facilities with a good maintenance program will grease and inspect a motor every six months.

Bearing wear may be signalled by overheating and increasing noise and vibration; a cracked rotor cage can produce the same effects. The condition of the motor windings should be checked by measuring the resistance of the windings and of the insulation between the windings and the ground. The motor drive train should also be checked so that belt tension can be adjusted or worn belts replaced. Gear reducers should be checked to see if they are properly lubricated.

If a motor is not operated for a large number of hours and is located in a humid place, a heating resistor should be placed inside the motor to avoid condensation. Moisture will decrease the insulation resistance between the windings and ground. Motors with abnormal conditions should be repaired or replaced.

Cleaning and Ambient Conditions

As noted in chapter 2, the cooler a motor operates, the higher its efficiency and the longer its lifetime. Higher temperature increases

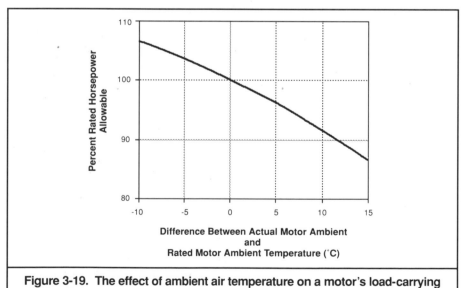

Figure 3-19. The effect of ambient air temperature on a motor's load-carrying ability (Nailen 1987).

the windings' resistivity and, therefore, their losses. Cleaning the motor casing and the ventilation filters and openings of open drip-proof motors is important because the operating temperature increases as dust and dirt accumulate. In extreme cases, failure may occur if thick layers of dust accumulate. Adding paint to the case is not recommended since the extra paint acts as insulation and decreases the ability of the motor to dissipate heat. Figure 3-19 shows the effect of ambient temperature on the allowable horsepower.

Summary

We have discussed many factors that determine the reliability, longevity, and efficiency of a motor-driven system. High-quality power supply; proper equipment sizing; careful attention to harmonics, transients, power factor, and distribution loss; good load management practice; optimized transmission systems; and careful maintenance of the entire drivepower system are all key goals. In chapter 4 we turn to one particularly important set of system components, motor controls.

Motor Control Technologies

M otor speed control offers the single largest opportunity for energy savings in drivepower systems. Most motors are fixed-speed, AC models. However, adjusting the speed to match the requirements of the loads, which generally vary over time, can enhance the efficiency of motor-driven equipment. The potential benefits of speed variation include increased productivity and product quality, less wear in mechanical equipment, and energy savings of 50% or more for some types of applications.

Speed controls can save the most energy in centrifugal machines, which include most pumps, fans, blowers, and some compressors. Speed control is also effective in chillers, mills (such as the rolling mills that produce sheet metal in a steel plant), traction drives (such as subway cars), conveyors, machine tools, and robotics.

The available options for motor speed control include multi-speed and DC motors, shaft-applied drives (including mechanical drives, hydraulic couplings, and eddy-current drives), and electronic adjustable-speed drives (ASDs). These are discussed and compared in light of their typical applications, advantages and limitations, and costs. Electronic ASDs, because they are increasingly important, are covered in detail further on in this chapter, although much of this information is summarized in Table 4-1 (next page). (Readers not concerned with the technical details of how electronic ASDs work should skip the section titled Characteristics of Electronic Adjustable-Speed Drives.)

Speed-control technology should match the characteristics of the load, including the load profile (number of hours per year at each level of load from minimum to maximum), horsepower range, speed range, price of energy, overall energy efficiency of the motor and control systems, reliability and maintenance requirements, physical size

Table 4-1. Adjustable-speed motor drive technologies.

	Technology	Applicability (R = Retrofit; N = New)	Cost	Comments
Motors	Multispeed (incl. PAM[1]) Motors	Fractional-500 hp PAM: fractional-2,000+ hp R,N	1.5 to 2 times the price of single-speed motors	Larger and less efficient than 1-speed motors. PAM more promising than multi-winding. Limited number of available speeds.
	Direct-Current Motors	Fractional-10,000 hp N	Higher than AC induction motors	Easy speed control. More maintenance required.
Shaft-Applied Drives (on Motor Output)	Mechanical — Variable-Ratio Belts	5–125 hp N	$350–$50[2]/hp (for 5–125 hp)	High efficiency at part load. 3:1 speed range limitation. Requires good maintenance for long life.
	Mechanical — Friction Dry Disks	Up to 5 hp N	$500–$300/hp	10:1 speed range. Maintenance required.
	Eddy-Current Drive	Fractional-2,000+ hp N	$900–$63/hp (for 1 to 150 hp)	Reliable in clean areas. Relatively long life. Low efficiency below 50% speed.
	Hydraulic Drive	5–10,000 hp N	Large variation	5:1 speed range. Low efficiency below 50% speed.
Wiring-Applied Drives (on Motor Input) — Electronic Adjustable Speed Drives	Voltage-Source Inverter	Fractional-1,000 hp R,N	$1500–$80/hp (for 1 to 300 hp)	Multi-motor capability. Can generally use existing motor. PWM[3] appears most promising.
	Current-Source Inverter	100–100,000 hp R,N	$200–$30/hp (for 100 to 20,000 hp)	Larger and heavier than VSI. Industrial applications, including large synchronous motors.
	Others	Fractional-100,000 R,N	Large variation	Includes cycloconverters, wound rotor, and variable voltage. Generally for special industrial applications.

[1]PAM means Pole Amplitude Modulated. [2]The prices are listed from high to low to correspond with the power rating, which is listed from low to high. Thus, the lower the power rating, the higher the cost per horsepower. [3]PWM means Pulse Width Modulation.

limitations, control and protection requirements, equipment lifetime, and first cost of the drive system. The ideal drive for any given application should be capable of varying both speed and torque to match the requirements of the load. Adjustable-speed loads can be classified into three groups according to relationship between torque and speed:

Variable-Torque Loads

Here, the torque increases with the square of the speed, as in centrifugal pumps; fans; and compressors common in heating, ventilating, and large air conditioning systems. Because these machines move a very small volume of fluid when they start and when they are operating at slow speeds, the required torque under these conditions is very low.

Constant-Torque Loads

The classic example is a conveyor belt. The torque required to move a conveyor depends on the load on the belt, not its speed. Since the load is independent of the speed, the drive may need to produce maximum torque at any speed.

Constant-Power Loads

With such loads, torque decreases with increasing speed, but power, the product of speed times torque, remains constant. The most familiar example is a vehicle transmission, in which the power can remain constant across the full range of gears.

Figure 4-1 shows typical torque to speed characteristics for the three classes of loads. Variable torque loads are by far the most common, accounting for 50% to 60% of the total motor energy use in the commercial and industrial sectors (see Table 6-11, page 164).

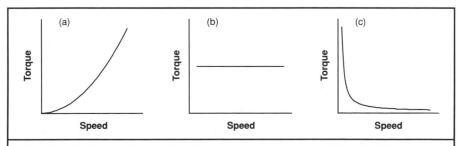

Figure 4-1. Types of motor loads: (a) torque increasing with speed (for example, in centrifugal fans and pumps); (b) constant torque (for example, in positive displacement pumps and compressors); (c) constant power (for example, in vehicle drives).

Speed-Control Technologies Other than Electronic ASDs

Multispeed Motors

Some motors are designed to operate at two, three, or four speeds. (Two-speed motors are the most common.) As explained in chapter 2, the speed of an induction motor depends upon the number of pole pairs of the motor. Multispeed motors are available up to 500 hp, and are very reliable but have several drawbacks:

- The stator slots have to be bigger than those of single-speed motors in order to accommodate two or more windings. As a result, the motors are bulkier and cannot be easily retrofitted.
- The current-carrying capacity of the copper is poorly used since only one set of windings is active at one time.
- Fundamental aspects of their design lead to lower efficiency than comparably sized single-speed motors.
- The available speed ratios are limited.
- The motor starters typically cost up to twice as much as single-speed motor starters.
- Multispeed motors cost 50–100% more than single-speed motors.

Two-speed motors can be used to save energy in such applications as air volume control in facilities that have large differences in day-to-night or weekday-to-weekend air flow requirements. An 1,800/1,200-rpm motor, for instance, can reduce fan energy requirements at night and on weekends by 70%. All that is required is the two-speed motor with starter, a timer, and a relay.

The pole-amplitude-modulation (PAM) motor is a single-winding, two-speed, squirrel cage induction motor that avoids some of the drawbacks of conventional two-speed designs. PAM motors are available in a wider range of speed ratios than standard multispeed motors, but they are limited to ratios based on synchronous speeds. They include 900/720, 1,200/720, 1,200/900, 1,800/720, 1,800/1,200, 3,600/720, and 3,600/900 rpm versions. PAM motors are more compact than other multispeed motors. In fact, they have the same frame size as single-speed designs.

The lower speed can be used for soft-starting, resulting in lower inrush current and less heating. PAM motors are especially well suited for driving large fans or pumps with ratings from a few horsepower to thousands of horsepower in applications for which a two-speed duty cycle is appropriate. In the case of a retrofit, using an existing throttling device (valve or damper) allows the fine tuning of the flow once the main adjustment is made through selection of the speed, while reducing the heavy losses of the throttle-only control.

Like multispeed motors, PAM motors are available for variable-torque, constant-torque or constant-horsepower applications. PAM motors and their starters cost about the same as standard multispeed motors and have similar efficiencies.

Pony Motors

An increasingly popular technique for motor drives with two distinct operating conditions is to use two motors for a single application. The second, smaller motor is called a pony motor. For shaft-driven equipment, two motors can drive pulleys for the same shaft with controls so that only one motor can operate at a time. In pumping applications, two pumps with different capacities and speeds will often be installed in parallel.

Pony motors are becoming a common option on cooling towers and air handlers. They produce energy savings with the use of standard motors and starters, and are easy to maintain and repair. In addition, because they use two different drive belts, they offer greater flexibility in speed selection than the limited ratios available in multispeed and PAM motors.

Direct-Current Drives

Although expensive and of limited reliability (see chapter 2), DC motors can produce high starting torques, and their speed can be controlled with great precision, down to 1% of the nominal speed of the motor, typically by varying the voltage level. They are used in applications up to about 10,000 hp. Figure 4-2 (next page) shows the torque-to-horsepower characteristics for DC motors.

The basic operating theory of DC motors is covered in chapter 2. The commutation subsystem of the DC motor is its weakness. Both the brushes and the copper blade terminals wear as a result of friction and arcing. For this reason, DC motors require periodic maintenance and are not suitable for use in explosive or corrosive environments. In addition, because of the complexity of the rotor, DC motors are substantially more expensive than induction motors, are bulkier due to the dead volume required by the commutator and brushes, and are less efficient than AC motors. High-horsepower DC motors also have lower speed limits than their AC counterparts because of the centrifugal stresses on their larger, heavier rotor. Large AC drives can make use of higher voltages than DC motors, which are subject to a voltage limit because of arcing in the

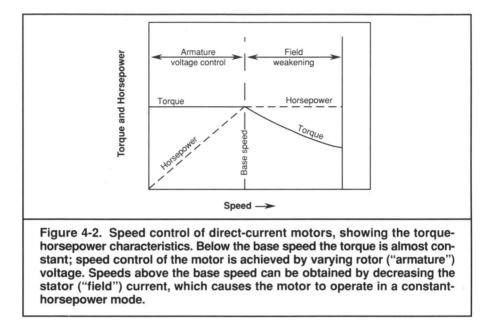

Figure 4-2. Speed control of direct-current motors, showing the torque-horsepower characteristics. Below the base speed the torque is almost constant; speed control of the motor is achieved by varying rotor ("armature") voltage. Speeds above the base speed can be obtained by decreasing the stator ("field") current, which causes the motor to operate in a constant-horsepower mode.

commutator. Higher voltages are desirable because they result in proportionately lower currents for the same power consumption. Lower currents lead to lower power losses in the electrical distribution system and allow smaller, cheaper wire to be used.

DC motors have traditionally been used in applications where high starting torque is required (such as traction devices and cranes) or where accurate speed control is needed (such as rolling mills, lathes, paper machines, and winders).

Using DC motors for speed control requires converting the available AC power to DC. Historically, this was done using either a motor-generator set or a rectifier. In the motor-generator set, an AC motor is used to operate a DC generator, which in turn powers the DC motor. Because each piece of equipment has losses, the overall efficiency of this system can be 50% or lower. For example, it might require 100 kW of electrical input at the AC motor to drive a machine with a DC motor that needs 67 hp (or 50 kW) of power applied to the shaft.

The conversion efficiency from AC to DC can be higher (up to 98%) with solid-state rectifiers than with motor-generator sets or mercury rectifiers. However, many facilities that use rectifiers were initially designed with a central DC power supply and one large rectifier. As new DC tools with their own rectifiers are added to such facilities, the load on the central rectifier goes down and so does the overall efficiency of the central system.

In the past five years, more efficient solid-state controllers for DC motors have appeared. These units, which have some features in common with AC adjustable speed drives, provide DC power at relatively high efficiencies for many existing speed control applications.

However, due to the drawbacks with DC motors mentioned above and the availability of better alternatives discussed later in this chapter, DC motors are now seldom used in new applications and their production is dwindling fast. As discussed in chapter 5, in some applications DC drives should be replaced with AC motors and ASDs. Examples include high-performance drives in steel and paper mills and electric transportation.

Shaft-Applied Speed Control: Mechanical, Hydraulic, and Eddy-Current Drives

Mechanical, hydraulic, and eddy-current (induction clutch) drives are grouped together because they all are installed between the constant-speed motor shaft and the driven equipment. Generally, these drives are bulky, are not very efficient, and require regular maintenance.

Shaft-applied drives are not normally used in retrofits due to their space requirements. In new applications, they are generally installed only in low-horsepower applications where they may be less expensive per horsepower than electronic ASDs. However, when ongoing maintenance and energy costs are included in the analysis, it is often more cost-effective to use an electronic ASD.

In addition, because of the relatively low efficiency of many of these drives, particularly when operating at low loads, it is some-times cost-effective to retrofit a shaft-applied drive with an elec-tronic ASD based on the value of the energy savings.

Mechanical Drives

Mechanical devices for controlling speed include variable gear-boxes, adjustable pulleys (sheaves), and friction dry discs. Vari-able gearboxes usually employ conical drums and can be applied only to small- and medium-size drives, generally under 100 hp. Belt-slipping problems and maintenance requirements are making them less and less attractive relative to other drive options.

Adjustable pulleys (Figure 4-3, next page) are simple devices that allow speed to be varied typically over a 3:1 range by adjusting the gap between flanges of the pulley sheaves. This adjustment can be performed either pneumatically or by a small servomotor. These

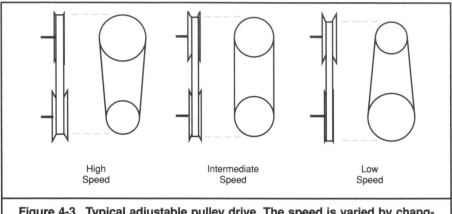

High
Speed

Intermediate
Speed

Low
Speed

Figure 4-3. Typical adjustable pulley drive. The speed is varied by changing the gap between the flanges of the pulley sheaves, thus changing the effective pulley diameters. The top shaft is connected to the motor, the bottom to the driven load.

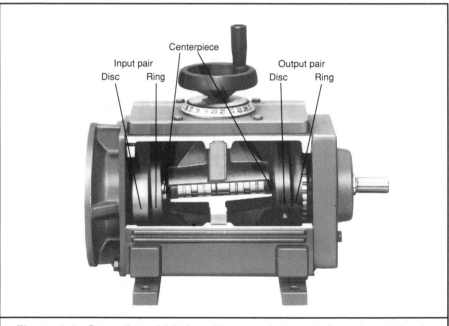

Centerpiece

Input pair

Disc   Ring

Output pair

Disc   Ring

Figure 4-4. Operation of friction disc speed control. Speed variation is achieved by manually turning a crank, which moves the point of contact between one or more friction disc/ring pairs. Moving the contact point in turn changes the effective diameter of the friction discs, thus changing the transmission ratio (courtesy of Reliance Electric).

devices are very efficient (in the 95% range) and fairly inexpensive (prices range from $50/hp for a 100-hp drive to $300/hp for a 5-hp drive). Because of belt-slipping problems, they are not suitable for shock loads and are available only below 125 hp. Adjustable pulleys have been used to control the speed of small- and medium-size fans.

Friction dry discs (Figure 4-4) allow a wide range of speed ratios (up to 10:1), but they are limited to small loads (up to a few hp) and they are expensive ($300 to $500/hp). They are expensive because they require precision parts, and are only used with small motors (most drives are less expensive per horsepower when used with a large motor). Speed is varied by manually turning a crank, which changes the transmission ratio (Payton 1988). These drives are typically 95% efficient. However, the high level of maintenance they require, their inability to be automatically controlled, and their low power-handling capability make friction dry discs inappropriate for many applications.

In general, mechanical drives have a limited horsepower range. Because they have movable parts, some of which rely on friction for transmitting power, they require regular maintenance. New developments in electronic ASDs provide more reliable, flexible, less bulky, and increasingly cost-effective alternatives.

Hydraulic Couplings

The output speed of a hydraulic (or fluid) coupling is controlled by the amount of slip between the input and output shafts. Thus, the output shaft speed cannot exceed the input shaft speed while the motor is driving the load. The torque converter in automobiles with automatic transmissions is a type of hydraulic coupling. In the fluid coupling, the input shaft drives a vaned impeller, and a vaned runner drives the load.

Figure 4-5 (next page) shows the structure of a hydraulic drive. Speed is controlled by varying the amount of oil in the working circuit, achieving a typical speed range of 5:1. This speed ratio is effected by deliberately introducing losses in the system. As a result, the bigger the speed reduction, the lower the efficiency of the system. For an output speed of 50% the overall efficiency of the hydraulic coupling is typically 40%.

Although hydraulic couplings can be used in applications from a few horsepower to tens of thousands of horsepower, their use is acceptable only when most of the duty cycle is in the upper speed range; at lower speeds the losses are too high. In addition, because the couplings are bulky, retrofits, which generally require

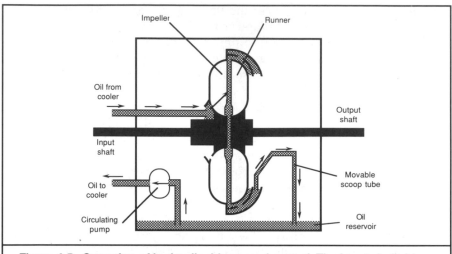

Figure 4-5. Operation of hydraulic drive speed control. The input shaft drives a vaned impeller, and a vaned "runner" drives the load through the output shaft. Note that the input and output shafts are not connected except through the hydraulic circuit. Speed variation is achieved by changing the amount of oil in the working circuit (between the impeller and the runner) through a movable scoop tube. Since the output speed is controlled by the amount of slip between the impeller and the runner, the output shaft speed cannot exceed the input shaft speed. The available speed range is typically 5:1 (Andreas 1982).

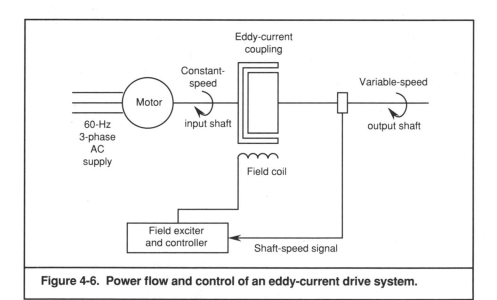

Figure 4-6. Power flow and control of an eddy-current drive system.

repositioning of heavy equipment and construction of new foundations, can be prohibitively expensive.

Eddy-Current Drives

The eddy-current drive couples an eddy-current clutch to an AC induction motor (Figure 4-6). A rotating drum connected to the induction motor surrounds a cylinder attached to the output shaft. The concentric cylinder and drum are coupled by a magnetic field whose strength determines the amount of slip. Speed is controlled by a low-power solid-state controller that varies the current in the winding that produces the magnetic field. This field excitation typically consumes 2 percent of the drive's rated power (Magnusson 1984).

The eddy-current drive is a slip device like the hydraulic coupling, albeit with slightly better efficiency. Waste heat, generated by the motion of the drum and cylinder relative to the magnetic field, is the main source of power loss, and is removed either by air or water cooling. Air-cooled drives are available with ratings from 1/4 hp through 200 hp. Water cooling is also used for some drives ranging from 200 hp to over 2,000 hp.

Eddy couplings operate reliably in a clean environment. They are bulky, typically occupying twice the space of the induction motor itself. Typical prices range from $200/hp for a 5-hp drive to $150/hp for a 15-hp drive to less than $100/hp for a 100-hp drive. Prior to recent decreases in the cost of electronic adjustable speed drives, eddy-current drives were often specified for speed control in HVAC systems and wastewater treatment plants. Although today's electronic ASDs have higher efficiencies and are competitive in cost, eddy-current drives have the advantage of not producing significant harmonics or voltage transients. They still may be an acceptable choice in installations where the load operates at 70% or more of the rated speed most of the time.

Characteristics of Electronic Adjustable-Speed Drives

Solid-state electronic ASDs were developed about 20 years ago. Early versions were complex, expensive, and only moderately reliable. Advances in semiconductor technology for power devices and especially for microelectronics have been dramatic in the last two decades. Their costs have decreased substantially, and their performance and reliability have improved. As a result, electronic ASDs are becoming more and more attractive to potential users.

This section provides the reader with a technical overview adequate for understanding application issues and costs. Readers interested in a somewhat more technical discussion of ASD design and operation should consult Ryan 1988 or Greenberg et al. 1988. For more in-depth information, see Bose 1986 or Dewan 1984.

Because the speed of AC motors is proportional to the frequency of the power supply, most electronic ASDs control motor speed by synthesizing electrical power of the desired frequency. In this manner it is possible to control the speed over a wide range—from 0% to 300% of rated speed.

Because ASDs are more compact than mechanical or hydraulic adjustable-speed controls, and because they do not have to be mechanically coupled to the motor, they can be easily retrofitted. The main ASD components do not have moving parts, and therefore require little periodic maintenance. When properly applied, ASDs can be extremely reliable. They are available in a power range that covers fractional horsepower motors (typical of home appliances) to a few hundred horsepower (as in commercial building HVAC systems) to the tens of thousands of horsepower used by the pumps and fans of large electric power plants.

Types of ASDs

Electronic ASDs are classified by the type of electronic input they require and the way they control a motor's speed. There are four basic types of ASDs: inverter-based, cycloconverters, wound-rotor slip recovery, and voltage-level controls.

Inverter-based ASDs

Inverter-based ASDs are the most common systems for induction

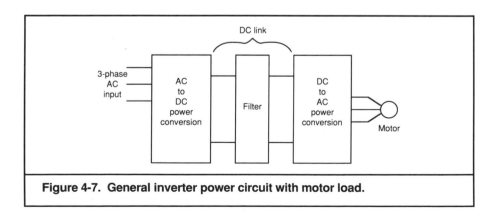

Figure 4-7. General inverter power circuit with motor load.

motors and can be used with synchronous motors as well. They account for well over 90% of the ASDs currently sold (PEAC 1987).

The general diagram for an inverter-based ASD is shown in Figure 4-7. Some ASDs operate on single-phase power (which is found in most residences and many small commercial buildings) and drive single-phase motors; others operate three-phase motors.

Figure 4-7 shows that, in the first stage, the input AC power supply is converted to DC using a solid-state rectifier. The DC link, which carries the DC power from the first stage to the second stage, includes a filter to smooth the electrical waveform.

In the second stage, the inverter uses this DC supply to synthesize an adjustable-frequency, adjustable-voltage AC waveform by releasing short steps or pulses of power. The speed of the motor will then change in proportion to the frequency. Usually, the output voltage waveforms can be synthesized over the frequency range of 0–120 Hz, but they are available up to 180 Hz.

There are three main types of inverter-based ASDs: voltage-source inverters (VSIs), pulse-width modulation (PWM) inverters, and current-source inverters (CSIs). Each has its own advantages and disadvantages as well as its own niche in the market.

VSIs and PWM inverters each generate a variable-frequency, variable-voltage waveform. The former synthesizes a square wave: the latter creates a pulse-width-modulated output made of a series of short duration pulses, as shown in Figure 4-8. In both cases, the

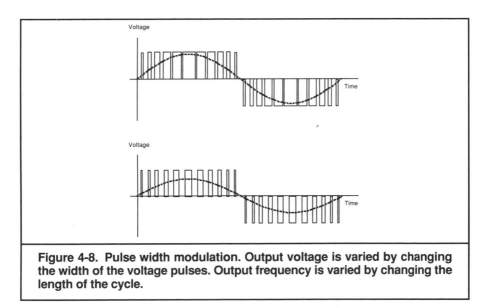

Figure 4-8. Pulse width modulation. Output voltage is varied by changing the width of the voltage pulses. Output frequency is varied by changing the length of the cycle.

output has the frequency that will produce the desired speed, but the shape of the output is not as smooth as the sinusoidal AC waveform on a good power distribution system.

VSIs (also known as square-wave inverters) are used in low- to medium-power applications, typically up to several hundred horsepower, and can operate several motors at once. Multimotor operation is desirable when several motors are operated at the same variable speed, as is often the case in the textile industry. Moreover, it is much cheaper to use one 200-hp ASD to drive ten 20-hp motors than to buy 10 ASDs to drive the same motors. One drawback to multimotor operation is that external overload protection must be provided to each motor.

PWM inverters have fewer problems than square-wave inverters although their efficiency is a bit lower due to the higher switching losses. Because of their better performance at low speed, lower harmonics, and their ability to maintain good efficiency in the high-frequency range (by switching from pulse-width modulated to square-wave output), PWM converters have gained ground during the past few years. In fact, PWM ASDs are becoming predominant in applications below 200 hp and are available up to about 500 hp.

Neither square-wave nor PWM inverters have regeneration capabilities. Regeneration is the capability to return energy to the supply system when a motor is slowing down—essentially operating the motor as a generator. This feature saves energy in drives with a high start-stop duty cycle (such as electric traction in urban rapid transit systems) as the braking energy is pumped back into the AC supply.

CSIs, also called current-fed inverters, behave like a constant current generator, producing an almost square-wave of current. Current-source inverters are used instead of VSIs for large drives (above 200 hp) because of their simplicity, regeneration capabilities, reliability, and lower cost. Although more rugged and reliable than VSIs, CSIs have a poor power factor at low speeds and are not suitable for multimotor operation. A special type of CSI, the load-commutated inverter, can be used with synchronous motors, typically in applications above 1,000 hp.

Cycloconverters

Cycloconverters convert AC power of one frequency to AC power of a different frequency without using an intermediate DC link. The output frequency can range from 0% to 50% of the input frequency. Cycloconverters feature regeneration capabilities and are used in large drives (above a few hundred horsepower), for low-speed applications. There are no common applications for cycloconverter ASDs in residential or commercial buildings, but typical industrial applications include ball mills and rotary kiln drives in the cement industry, where their low-speed capability eliminates the need for gears.

Wound-Rotor Slip Recovery ASDs

As noted in chapter 2, the speed of a wound-rotor induction motor can be altered by inserting an external resistor in the circuit. However, controlling the speed in this manner is very inefficient. A wound-rotor slip recovery ASD recovers and reuses some of the power wasted when the speed of a wound rotor motor is controlled by an external resistor. Its use is limited to very large motors (typically over 500 hp).

Two types of wound-rotor ASDs are available. The static Kramer drive is commonly used in applications requiring 50–100% of the synchronous speed, such as large pumps and compressors (Bose 1986). The more expensive static Scherbius drive is used in large pumps and fans, where higher-than-synchronous speeds or regenerative braking are important (Leonard 1984).

Voltage-Level Controls

Unlike other electronic ASDs, variable-voltage controls do not vary the frequency of power supplied to the motor. Instead, the effective AC supply voltage applied to the stator windings is varied. When the applied voltage level decreases, the motor slows down. Although simple, this control method is not widely used due to its low efficiency and the high level of harmonics generated (Mohan 1981). Essentially the same technology is used in the so-called power factor controllers, described below.

Application of ASDs

ASDs are used for two basic reasons: to provide accurate process control or to match the speed of a motor-driven device to varying load requirements. The most dramatic energy savings from speed control occur with loads whose losses fall at reduced speed. This is true of centrifugal machinery, including most pumps, fans, and some compressors. Their energy use is often proportional to the cube of the flow rate, so small reductions in flow can yield disproportionately large energy savings. For instance, a 20% reduction in flow can, under conditions spelled out in chapter 5, reduce energy requirements by nearly 50%.

ASDs are ideally suited to modifying the speed of centrifugal machines to provide the exact flow required by the system. Contrast this with the conventional practice in fan and pump systems of running the motor at full speed and controlling flow via throttling devices, like inlet vanes or outlet dampers on fans, and valves on pumps. Such flow constriction is analogous to controlling the speed of a car with the brake while the accelerator is pushed to the floor—a very wasteful practice. Fans and pumps represent such a large proportion of drivepower energy use, and are such attractive candidates for speed control with ASDs, that we devote much of chapter 5 specifically to these applications.

In other equipment, lowering the speed produces less dramatic savings. For example, conveyors are sometimes equipped with speed controls for process reasons. The energy needed to drive a conveyor depends primarily on the load on the belt and secondarily on its speed. As a result, the energy savings that can be achieved with speed controls depend on the load profile of the conveyor.

In addition to the energy-efficiency advantages of speed control, other features may be even more important to the user, particularly in industry. For instance, response time and process control are improved. Moreover, by eliminating control valves, ASDs reduce the number of parts exposed to the fluid, which may be important in some applications to reduce contamination problems.

General Considerations for Selecting ASDs

Although pumps and fans provide the best applications for ASD retrofits, speed controls are not necessarily cost-effective for all pumps and fans. The load profile (time variation of the pressure and flow requirements) is very important for determining the cost-effectiveness of an application. For example, if a system must operate at full flow at all times, then none of the flow control schemes is a good choice. However, the typical system will have flow requirements that vary considerably over time. The greater the amount of operation at relatively low flows, the more cost-effective it is to efficiently provide flow control.

The best way to determine the cost-effectiveness of a proposed ASD installation is to look at the power needed at each operating condition with and without an ASD. The energy savings can then be calculated by taking the reduction in power at each condition and estimating the savings based on the actual (or expected) operating time at that condition. Sample calculations appear in appendix A.

In general, good applications for variable-speed flow control, and in particular ASD control, are those which:

- Are fixed at a flow rate higher than that required by the load.

- Are variable-flow, where the variation is provided by throttling (by valves or dampers) and where the majority of the operation is below the design flow.

- Use flow diversion or bypassing (typically via a pressure reducing valve).

- Are greatly oversized for the flow required. This situation can occur where successive safety factors were added to the design, where a process changed so that the equipment now serves a load less than the original design, and where a system was overdesigned for possible future expansion.

- Have long distribution networks.

- Have flow control by on-off cycling. Such systems are usually less cost-effective retrofit candidates than those that use a throttling control.

- Have a single large pump or fan rather than a series of staged pumps or fans that come on sequentially as the process needs increase.

- Can reduce the pressure at the outlet of the fan or pump at lower flows. For example, a pump that pumps water into a long pipeline that can move the water at a lower pressure when the flows are low (due to the decreased frictional losses in the pipes) would be a good candidate for an ASD. A pump supplying a fire protection system where the piping is oversized and a constant pressure is needed regardless of flow would probably not save enough for an ASD to be cost-effective.

Once a good ASD application is identified, the question arises of what kind of drive to use. This choice involves selecting among the major categories of ASDs discussed above as well as choosing a version suited for variable-torque loads (including pumps, fans, and compressors) or constant-torque loads (such as conveyors, some machine tools, and winders). Most manufacturers make two lines of ASDs, each suited to either variable- or constant-torque loads.

Constant-torque ASDs are typically 10% to 20% more expensive than variable-torque equipment because the electronics must be designed and built to withstand the high currents that occur when a constant-torque machine starts with a full load. Figure 4-9 (next page)

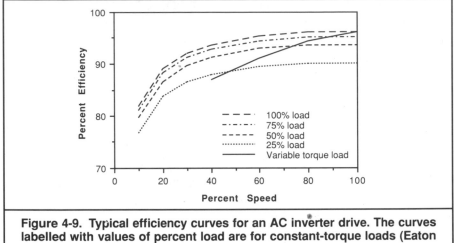

Figure 4-9. Typical efficiency curves for an AC inverter drive. The curves labelled with values of percent load are for constant-torque loads (Eaton 1988).

shows the efficiency of variable torque and constant torque inverters as a function of both speed and load.

Process Controls and the Integration of ASDs

To work effectively, reduce the motor speed, and save energy, an ASD must be integrated into some type of control system. Most ASDs have a provision for controlling the speed by adjusting a setting on the local ASD panel. This type of relatively unsophisticated control is used where there are no major changes in the process that correspond to a desired speed of the equipment. For example, an ASD is often used to control the speed of a conveyor that carries a product through a freezing tunnel or a drying oven. The slower the belt, the longer the product is held in the tunnel. In these situations, the operator sets the belt speed based on some characteristics of the product and only changes the speed when the product changes. This is called open-loop control, as the system output is not monitored to regulate the performance. Typical applications, in addition to conveyors, include some types of ventilation equipment and pumps and fans that run at constant speed where relatively large flow fluctuations due to external disturbances can be tolerated.

In more demanding applications, a control system, of which the motor and the ASD are a subsystem, has to be designed to satisfy the process requirements. Generally, the control system will have one or more sensors to monitor the state of the process variables. The sensor(s) provide data into a controller or computer programmed with the control strategy. The controller compares the actual level of the process variable with a preset desired level. Based on this comparison, the controller or computer will send a signal to change the system operation so that the actual value of the variable correlates with the desired value. A common example is a building HVAC control system, where the speed of a chilled water pump might be controlled by the pressure of the water in the circulating loop of the building.

ASDs feature several types of control inputs that allow them to be easily controlled by an external signal. The value applied at the control inputs determines the speed of the motor. Most ASDs are equipped with low-voltage or low-current control inputs, or a low-pressure pneumatic control input. Additionally, modern ASDs also feature a computer interface that allows the plant supervisory computer to communicate directly with the ASD. This situation is common in the process industries such as pulp-and-paper, chemicals, and refineries.

Potential Drawbacks of ASDs

The electronic ASD holds great promise, but improper selection and use of the technology can lead to a number of problems.

ASDs can often be retrofitted on an existing motor. In rare cases, however, it is necessary to replace the motor to provide an adequate system to meet the application. For example, because the electrical wave generated by an ASD is slightly irregular, a motor will heat up slightly more when used with an ASD than when run off the standard line power. As a result, it is occasionally necessary to install either a more efficient motor or a larger motor with an ASD if the present motor, without an ASD, is operating at or above its rated horsepower. Similarly, while most ASDs can be ordered with an option to drive a motor at up to twice its rated speed, it is sometimes necessary to replace a motor if the system requires this speed range, because the existing motor's rotor and bearings cannot handle the higher speed, or because the load demands more power at higher speed.

An ASD can save a great deal of energy by slowing a motor to match light loads. Care must be taken, however, because motors run at small fractions of their rated speed can overheat or suffer irregular rotation known as cogging. Most manufacturers do not recommend running a motor at less than 10% to 15% of rated speed.

ASDs have some losses in the circuitry. As with most electric equipment, these losses result in increased heating of the drive. Most drives are equipped with cooling fins so that the heat does not build up and trip the electronic circuitry; many are also equipped with fans. In general, the ASD cooling system will work adequately when the drive is located in a room with normal temperature conditions. Some large drives and those located near hot spots in a plant may need to be placed in an air-conditioned room.

A standard ASD houses the circuitry in a box that closes but does not tightly seal. Most manufacturers can package ASDs in housings that are impervious to dust, water, or explosive vapors. These special enclosures can add 10% to the cost of small ASDs and 5% to larger ones.

There are many other add-on features that can increase the cost of an ASD installation. These can arise when special control interfaces are needed; when the machine has special requirements for acceleration, deceleration, or reversal of direction; when manual or automatic bypass of the ASD is needed; or when equipment is required to protect against overload, voltage fluctuation, short circuits, loss of phase, harmonics, and electromagnetic interference.

The early generation of ASDs often had poor power factors. However, most small ASDs (below 300 hp) today use an input circuit with a high power factor over the entire speed range. In fact, these types of ASDs have a better power factor over the entire operating range than motors connected directly to the line power (which have lower power factors at low loads). These units are generally identified in catalogs as having high input power factor (0.95 or above) over the entire speed range or as a PWM-type input circuit. For small drive units for general applications, it is almost always possible to find an ASD with good power factor.

Larger ASDs may have poor power factors at low speed (Table 4-2). The effect of poor power factor is partially mitigated by the

Table 4-2. Displacement power factor versus speed for typical large electronic adjustable-speed drives.

Percent Speed	Power Factor
100	.94
90	.95
80	.76
70	.67
60	.58
50	.50
40	.41
30	.32
20	.23
10	.14

Source: Eaton 1988.

fact that less power is used at low speeds than at full speed, especially in the case of variable-torque loads. With constant-torque loads, the reactive power consumed at low speeds can reach unacceptably high values. It is possible to correct the low power factor caused by the ASDs with capacitors and filters (for displacement power factor and harmonics, respectively). However, an ASD differs from other motor loads in that the displacement power factor should be corrected at a central location with a switched capacitor bank instead of at the individual piece of equipment.

ASDs larger than 200 hp—such as cycloconverters and current-source inverters—can produce harmonics and electromagnetic interference (EMI) that can disrupt power line signals, computers, and other electronics and communications equipment. Smaller pulse-width-modulated ASDs damp harmonics better and generally do not pose problems unless there are many of them in the plant. The problems caused by harmonics and EMI, and their diagnosis and mitigation, were discussed in chapter 3.

Trends and Developments in ASD Technology

There has been continuous progress over the past two decades in the technologies used for both major components of electronic ASDs: the microelectronics used in control circuits and the power electronics used to condition the input current to the motor. These technology trends and the associated cost reductions are helping to accelerate ASD market penetration.

Increasingly powerful microprocessors and large-scale-integration devices have allowed complex control functions and algorithms to be incorporated in compact and inexpensive ASD packages. The advances in power electronics technology, although substantial, have not been so dramatic, either in performance or cost.

The integration of power electronics devices and microelectronics in single packages known as power integrated circuits (PICs), or smart power devices, will lead to further miniaturization. PICs have the potential to slash the number of ASD components, reduce costs, and improve reliability.

Another interesting development is the integration of sensors, in a silicon chip, for all kinds of variables, including temperature, pressure, light, force, acceleration, and vibration. Near-term integration of PICs with sensors will help to decrease the cost not only of the ASD but also of the overall control system of which the ASD is a part. The capability already exists to package electronic ASDs routinely with motors, especially in low-horsepower devices such as home air conditioners, heat pumps, washing machines, and other appliances. The Japanese are doing this in many heat pumps and air conditioners, and a few U.S. manufacturers, including Trane and Carrier, have integrated ASDs into several of their furnaces, air conditioners, and heat pumps. (See box on Carrier furnace in chapter 2.)

The general trends in electronic ASDs point to increased compactness, efficiency, and reliability, as well as more flexibility (added control and protection features), less power-line pollution, and decreasing cost per horsepower. With these trends, a staggering growth in the electronic ASD market can be expected, particularly as the cost per horsepower decreases.

Economics of Speed Controls

The economics of motor-speed-control technologies are briefly discussed below. For further details, see appendix A.

Mechanical ASDs, such as adjustable pulleys, are fairly inexpensive, with prices ranging from $50/horsepower (equipment only) for a 125-hp drive to $350/hp for a 5-hp drive. Typical equipment prices for eddy-current drives range from $200/hp for a 5-hp drive to $150/hp for a 15-hp drive and less than $100/hp for a 100-hp drive.

While the equipment prices of these mechanical ASDs are typically low, they are seldom used for retrofits because of the difficulty of repositioning the motor or the driven load. Electronic ASDs are generally much better suited to retrofit applications because they are connected into the motorized system only through the wiring. Their price, in terms of dollars/horsepower, is a function of the horsepower range, the type of AC motor used, and the additional control and protection facilities offered by the electronic ASD. Typical capital costs for ASDs for induction motors are shown in Table 4-3. These data are based on 1990 trade prices from several manufacturers of ASDs and assume variable-torque equipment. The costs are slightly lower for most commercial applications since the industrial models are built for heavier duty and harsher environments.

Drives for synchronous motors above 1,000 hp are about twice as expensive per horsepower as drives for induction motors of the same size. As noted earlier, the integration of ASDs into mass-produced appliances lowers costs dramatically. ASDs built into Japanese variable-speed heat pumps reportedly add only $25/hp to the manufacturer's cost (Abbate 1988).

The total installed cost for the ASD depends on the cost of

	Commercial Application ASD Cost ($)	Industrial Application ASD Cost ($)
Table 4-3. Summary of ASD capital costs for commercial and industrial applications (1990).		
Size (hp)		
5	2,100	2,700
7.5	2,300	2,900
10	2,800	3,600
15	3,500	4,400
20	3,800	4,800
25	4,400	5,600
30	4,800	6,100
40	5,800	7,100
50	7,200	8,800
60	8,000	10,000
75	9,600	12,000
100	11,000	14,000
125	13,000	16,000
150	16,000	20,000
200		22,000
250		25,000
300		29,000
350		33,000
400		38,000
500		45,000

labor; the type of control system; and the costs of filters and other devices for mitigating harmonics, low power factor, or other problems. Typical installation costs range from $250 per drive where all of the feeder wires and controls are already present (or would need to be installed in the alternative for new construction) to over $5,000 where special controls and sensors are needed to integrate the ASD in the control system and where abatement of harmonics is necessary. The most recent summary of installed costs for ASDs is based on 1987 data (PEAC 1987).

Typical energy savings from ASDs range from 15% to 50% and simple paybacks of one to eight years are common, based on energy savings alone. The payback is of course sensitive to the price of electricity, labor costs, the size of drive, the load profile, whether the application is new or retrofit, and other factors.

In addition, there are costs and benefits that are difficult to quantify, including maintenance requirements, reliability, reduced wear on equipment, less operating noise, regeneration capability, improved control, soft-start, and automatic protection features. In many instances, these difficult-to-quantify factors, like improved process control, motivate a user to install ASDs.

ASD Case Studies

The following profiles of documented ASD applications include engineering background, cost, energy savings, and other benefits. Only projects that have submetered data are used. Unfortunately, such case studies are not very common, and it is difficult to know how typical they are. Further research is needed to provide more information about the motor stock in terms of possible applications for flow control (including well-documented case studies and surveys to indicate how many "typical" applications are potentially available).

Boiler Feed Pump at Fort Churchill Power Plant

Like many small- to medium-size oil- and gas-fired power plants, Sierra Pacific Power Company's 110-MW Fort Churchill plant acts as spinning reserve, operating at minimum power until needed. Its minimum load was 16 MW, provided at relatively high cost due primarily to high fuel costs. An added problem, and one that limited the extent to which the plant's output could be reduced, was the large pressure drops in the throttling valves, which increased maintenance costs as well. The boiler feed pump provided over 2,700 psi of pressure, and all but 250 psi were wasted through restrictive operation of the feedwater control and turbine stop valves (EPRI 1985). In other words, most of the energy delivered to the water was dissipated by the throttling valves.

An analysis of the plant's operation, including the load profile (number of hours per year at each fraction of full load) and the heat rate (amount of fuel required per kWh output at each fraction of full load), showed that the plant could be turned to an even lower load, and valve wear reduced, if the induction motor drive of the boiler feed pump was retrofitted with an adjustable-speed drive. The retrofit on the 2,000-hp pump was performed in 1984 as an EPRI demonstration project (Oliver and Samotyj 1989).

The following results were achieved:

• Minimum power was lowered to 12 MW, resulting in large fuel savings.

- Pump input power at minimum plant power was reduced from 815 kW to 293 kW.
- Annual savings from fuel and pump electricity totalled $1,600,000, which paid off the $480,000 installation in about four months.
- Other benefits (including reduced maintenance and reduced stack emissions) were significant but have not been quantified.

Coolant Pumps at Ford's Dearborn Engine Plant

The Dearborn Engine Plant cooling system includes five 75-hp pumps to circulate cooling fluid to cutting tools. The pre-retrofit operation was for three pumps to operate in parallel at 64 psi and 1,325 gallons per minute each, for 5,700 hours per year (the other two pumps were maintained as spares).

An analysis of the system showed that the pumps were operating most of the time at excessively high pressure in order to meet occasional peak loads (the required pressure is 50 psi). The solution was to install an ASD on one of the pumps, along with a control system not only to maintain pressure levels but also to shut off coolant flow to machines not operating. Thus, the pump staging plus ASD control was able to exactly meet the system requirements.

The results of the retrofit were as follows:

- Reducing the required flow and meeting that requirement more efficiently through reduced pressure reduced energy use by 48%.
- The $75,000 cost of the retrofit was paid back in 1.4 years from the annual $55,000 savings.
- Other benefits (not quantified) included: reduced misting (from the coolant nozzles); improved indoor air quality; reduced pump wear; improved directional control of coolant (due to constant pressure), thereby improving machining quality; and improved coolant filter performance and decreased filter maintenance due to reduced flows. The control system also allows pump wear to be monitored, assisting in planned maintenance (Strohs 1987).

Boiler Fans at Ford's Lorain Assembly Plant

The assembly plant at Lorain, Ohio, is served with 125-psi steam from three coal-fired boilers installed in 1957. In 1976, environmental regulations required replacing the induced draft fans with fans equipped with outlet dampers in order to better control combustion so as to lower emission levels. With this change, particulate emissions were reduced to acceptable levels at high boiler loads but not at loads below about 40% of the maximum 80,000 lb/hr of steam for each boiler. In the summer, steam load dropped as low as 15,000 lb/hr, and large amounts of steam were vented to the atmosphere to prevent excessive emission levels.

After the damper-equipped fans were installed, tightening environmental requirements forced even further modification. In 1986 forced and induced draft fans were equipped with ASDs, which improved control of the boilers and met the particulate emission requirements even at boiler outputs below 25% of maximum.

In addition to avoiding fines for excessive emissions, the controls:

- Saved $53,000 in coal costs by greatly reducing steam venting.
- Saved $41,000 in electricity costs for the six fans.
- Paid back the $90,000 retrofit cost in slightly less than one year.

On the downside, the controls created a low power factor and potential electromagnetic interference, which may require ASD modification or additional power factor correction equipment and filters (Futryk and Kaman 1987).

Ventilation Fans in a New Jersey Office Building

ASDs were installed on two supply fan and two return fan motors in variable-air-volume (VAV) ventilation systems in a 130,000-square-foot commercial office building. The volume had previously been controlled by inlet vanes on the fans.

As a test, two control schemes were compared. The first used the pre-retrofit duct static pressure control with the existing setting of 2.5 inches of water. The second reset the duct pressure to 1.5 inches when system loads were reduced.

The retrofit had the following results:

- With the same control strategy as before the retrofit, energy savings were 35%, amounting to $5,200 annually, which would pay off the $40,000 installation cost in 7.7 years.
- With the modified (pressure reset) strategy, savings increased to 52% or $8,700 annually, with a simple payback period of 4.6 years.
- Due to the inefficiencies of the inlet vanes, resetting the duct pressure with inlet vane control did not result in significant savings compared to the base case (Englander and Norford 1988).

Other Controls

The electronic ASD is only one of the new control technologies to have emerged from the electronics revolution during the past ten years. Electronics process controls, sensors, fast controllers for compressors, power factor controllers, and energy management systems (EMS) for controlling mechanical and lighting systems in buildings are all becoming more sophisticated and effective.

In many cases, the new control systems are used to improve the product being processed, to improve the yield of the product, or to improve the comfort in a building. Energy conservation is sometimes a primary goal, and sometimes a side benefit of installing control systems for process reasons.

Improved controls and sensors can save energy by monitoring more system variables than was previously possible and by responding more rapidly and accurately than ever before. Because of their slow response time, process controllers historically have used setpoints with built-in safety margins to ensure that the process meets minimum performance requirements. A new generation of intelligent controllers is currently available that looks at both the historic pattern of control and the recent system changes and automatically retunes the process control parameters to optimize response time for each situation.

These controllers yield energy savings for two reasons. First, the rapid response ensures the process is operating within desired limits most of the time. Energy wasted when the process is slightly out of spec is now conserved. In addition, the increased accuracy allows the operator to set the process control variable to the exact setpoint without incorporating a safety factor.

Building and industrial process control loops in the past were typically designed to control one or perhaps two variables. With the advent of the microprocessor, control systems can monitor and respond to many parameters. For example, the temperature of heated or chilled supply air or water in building HVAC systems has traditionally been fixed at one preset level, or perhaps was varied with outside air temperature, which is only one indication of cooling or heating requirements in a large building. With this approach, the HVAC system tends to provide excessive cooling or heating much of the time. Newer energy management systems allow the temperatures of the cooling and heating medium to be reset based on the actual demand for heating and cooling in different zones of the building. Significant energy savings can result from this type of control strategy since only the minimum amount of heating and cooling are used to meet the needs of the building.

Historically, industrial process controllers were limited in application because sensors were not available for specific applications, or they tended to drift and need frequent recalibration. As a result, many processes were controlled manually with fairly crude adjustments. The electronics revolution has produced sensors that can detect small concentrations of specific ions, can sense humidity without fouling and drifting, and can measure other process variables. These new sensors, combined with the advent of central control (which reduces the cost for each control point), have expanded the range of processes that can be automated. For example, humidity sensors can now be purchased that survive in the harsh environment of a lumber kiln and remain in calibration for long periods of time, allowing humidity to drive the operation of equipment.

Vector Control

DC motors historically have been used in high-performance applications such as servodrives, rolling mills, robotics, and web winders, where accurate torque and speed control is required. The development of inexpensive microprocessors and ASDs now allow the more reliable AC induction motor to be used in such tasks, in a process called vector control. In vector control, the motor current, voltage, and position are continuously monitored. These values are then plugged into mathematical formulas called algorithms, which then precisely control torque, speed, position, and other critical parameters. Vector control can be coupled with the main types of ASDs and has been successfully used in a wide range of applications. The general approach to closed loop, microprocessor-based vector control is shown in Figure 4-10. For more on vector control, see Leonard 1986 or Bose 1986.

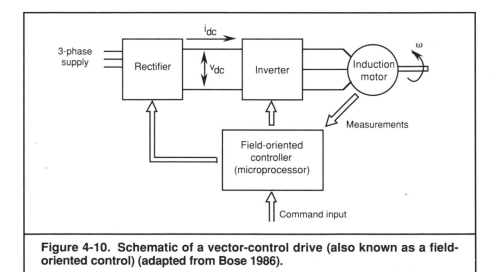

Figure 4-10. Schematic of a vector-control drive (also known as a field-oriented control) (adapted from Bose 1986).

Fast Compressor Controllers

Centrifugal and axial compressors are used to provide compressed air for multiple applications in many industries. To protect against surge—a damaging back-up and reversal of flow in compressors—traditional controls recirculate the compressed fluid or bleed off pressure from the discharge pipe. To compensate for the resulting pressure drop during periods of high demand, the pressure setpoints are set substantially above the minimum pressure required by the loads. Thus, traditional controller characteristics lead to energy waste and are too slow to prevent surges; rather, they dissipate the surges to prevent system damage.

The purpose of fast controllers is to match the output of compressors to their loads and to prevent surge from occurring in the first place. These high-speed microprocessor-based controls* can achieve 5–25% electricity savings through a combination of reduced pressure blow-off, better matching of pressure setpoints to load, and more effective load sharing among multiple compressors in parallel systems. The fast controller monitors variables such as discharge pressure, mass flow, power, and efficiency, and continuously adjusts the compressor's operation accordingly. Benefits in addition to energy savings include longer equipment life, less maintenance and downtime, and reduced pressure fluctuation.

The cost of fast controllers is independent of the compressor size, which makes their application attractive for large compressors (over 500 hp). The typical payback time is around one year for a 1,000-hp compressor, assuming 10% savings. Fast controllers save energy additional to that saved by the ASDs controlled by them. The fast controller computes the optimal speed of the compressor and sends a signal to the control input of the ASD in order to produce the desired motor speed. Thus, the combined use of the fast controller with the ASD maximizes savings in compressor operations.

Power-Factor Controllers

Many motors run a varying load that needs a constant speed. For example, a motor driving a saw blade must maintain constant speed regardless of whether the blade is cutting a 1/2" board or a 2" board, a hardwood or a softwood. As a result, the motor runs with a light load when cutting wood below its maximum capacity or when idling between cuts.

---

* Developed by Compressor Controls Corporation, Des Moines, Iowa.

For applications below 15 hp in which the motor runs much of the time at full speed but at light or zero loading, electronic variable-voltage controls, also known as power-factor controllers (PFCs), can increase both power factor and efficiency. These devices eliminate part of the sine wave fed to the motor. As a result, the average current (represented by the area under the sine wave) is lower. Since the power is the product of the voltage times the current, PFCs reduce the power used at low loads and thus improve motor efficiency by reducing magnetic and I^2R losses.

Energy savings range from 10–50% at light loading to zero at full load. Overall energy savings of 10% typify the limited number of attractive installations, including saws, grinders, granulators, escalators, punch presses, lathes, drills, and other machine tools that idle for extended periods. The potential savings are greater for single-phase motors than for three-phase motors, because the former have much larger no-load losses. Similarly, the higher and flatter efficiency curves of large motors make them unattractive candidates for power factor controllers. PFCs also can incorporate soft-start capabilities at little extra cost. In fact, many controllers marketed as soft-start devices include the power-factor control capability.

Because of the way power consumption is sensed and controlled, PFCs also improve power factor. In general, as the load on a motor decreases, the power factor deteriorates as a result of reactive current, which shifts the voltage sine wave out of phase with the current sine wave. When the controller eliminates part of the sine wave, it not only reduces the average magnitude of that waveform but also shifts the center of the waveform so that the voltage and current are closer to being in phase, thus improving the power factor.

PFCs can generate significant harmonics, which need to be suppressed. They also have internal losses typically equal to a few percent of rated power.

List prices of 10-hp PFCs are $30–$60/hp. Their cost-effectiveness can be evaluated by estimating the load profile and the net efficiency gains at each point of part-load operation.

Summary

Speed control for motors, particularly when applied to fans and pumps, is an extremely effective way to produce energy savings in motor drives. While speed can be controlled using multiple motors, multispeed motors, and an assortment of mechanical devices, most retrofits and many new installations use electronic adjustable-speed drives because these devices are easy to install on existing equipment, and their costs and reliability are making them increasingly attractive to the user.

An ASD produces energy savings most effectively when integrated into a larger control system. Modern control systems not only control motor speed for adjustable-speed applications but also reduce energy use by improving process control.

Chapter 5

Motor Applications

Previous chapters have discussed components of the drive-power system—motors, wiring, controls, and transmission hardware—upstream from the load. In this chapter we take a different perspective, focusing on several principal loads, most notably fans, pumps, and compressors. There are two reasons for this emphasis. First, these key loads account for more than half of drivepower energy use. Second, it is necessary to understand the theory of fan, pump, and compressor applications if one is to design efficient, reliable systems to drive them.

Fans and Pumps

The fans, pumps, and compressors that move and compress air, water, and other gases and liquids (collectively known as fluids) in industry and commercial buildings use approximately 50% of the electricity used by U.S. motor-driven systems. (See chapter 6 for further discussion of end-use estimates.) Most of the electricity used by this group is for fans and pumps, the focus of this chapter. Fans and pumps are used in a wide variety of applications and range in size from fractional-horsepower units in residential appliances to tens of thousands of horsepower required in utility power plants. Despite the range of size and usage, nearly all fan and pump applications have time-varying flow requirements, and most of the flow variation is done inefficiently, if at all.

Fluid Flow Fundamentals

All fan and pump applications share certain characteristics, one of which is the nature of fluid flow.

To produce a flow of fluid through a pipe, duct, damper, or

valve, a pressure difference must be created across the component. A common example is the garden hose. In order to force water through it, the pressure must be greater at the faucet than at the far end of the hose. The greater the pressure difference, the greater the flow.

For a given system, a curve can be drawn to show the pressure difference required at any given flow. An example is shown in Figure 5-1, where the pressure difference (in pump jargon, "head") is shown as a function of flow (gallons per minute). As the figure shows, the relation is quadratic. That is, the pressure difference is proportional to the square of the flow rate. Or, expressed in mathematical terms:

$$\Delta P \propto Q^2, \text{ where:}$$

ΔP is the pressure difference in pounds per square inch (psi) or inches or feet of water,

Q is the flow rate in cubic feet per minute (cfm) or gallons per minute (gpm), and

$\propto$ means "is proportional to."

Thus, if the flow doubles, the pressure drop quadruples. This squared relation holds for systems of all fluid types. The slope of the curve is determined by the system components' resistance to flow. For example, a water distribution system constructed from 2-inch-diameter pipe will have much larger resistance than one constructed

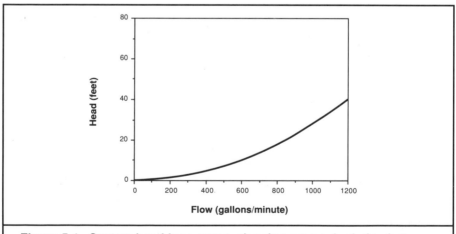

Figure 5-1. System head loss curve, showing squared relation between pressure ("head") and flow. This system has no minimum pressure requirement.

from 4-inch pipe, and would thus result in a greater pressure drop than the 4-inch system for any given flow. In other words, the more restrictive the system, the steeper the system curve.

In the United States, units for measuring pressure and flow differ between fan and pump applications. For fans, the pressure difference is given in inches of water column (one inch corresponds to 0.0361 psi; thus, 27.7 inches of water equals one psi). For pumps, the pressure (or head) is given in feet of water column (one foot corresponds to 0.434 psi; 2.31 feet of water equals one psi). For fans, the flow rate is given in cubic feet per minute (cfm); for pumps, in gallons per minute (gpm). (To convert U.S. units to Systeme Internationale units of pascal, multiply inches of water by 249, or multiply psi by 6,894.)

The power required to create a given flow relates directly to the shaft power required by the fan or pump from the drive motor (which in turn relates to the required electrical input power). A basic relation that follows from the physics of fluid flow is that the theoretical power required to create the pressure difference needed to produce a given flow is proportional to the product of the pressure and flow. That is:

$$\text{Power} \propto \Delta P \times Q.$$

Thus, there is a unique theoretical power required for any given combination of pressure and flow. The terms water-horsepower (for water pumps) and air-horsepower (for air fans) are often used to denote the theoretical power required in these systems. The relationship between the theoretical and the actual power requirements is discussed in the following section on fan and pump characteristics.

For fans and pumps, there is a set of relations known as affinity laws. One of the affinity laws states that, for a given fan or pump, installed in a given (unchanging) system, the flow rate is directly proportional to the speed of the fan or pump:

$$Q \propto N, \text{ where } N \text{ is speed.}$$

For example, if the speed of a fan is doubled, the flow through the fan and system attached to it is also doubled.

Another affinity law is that the power required by a fan or pump increases with the cube of its speed:

$$\text{Power} \propto N^3.$$

For example, when a fan's speed is doubled, the power requirement grows eightfold (two to the third power). This cubic relation follows directly from the concepts previously discussed. Since the power required is proportional to the product of the pressure and the flow, and the pressure in a given system is proportional to the square of the flow, the power required is proportional to the cube of the flow. Since the flow is proportional to the speed, the power is proportional to the cube of the speed.

The "cube law" has great significance in the energy used by motors in fluid-flow applications. For example, reducing the flow (by reducing the speed of the fan) in an oversized ventilation system by only 20% halves the power required by the fan. In many systems, the flow can be varied on a continuous basis to meet a constantly-fluctuating demand. Methods of flow variation are discussed below under "System Control and Optimization Techniques."

The affinity laws can be very useful in fan and pump applications, but the user must be sure that the laws hold in a specific case. The laws only hold if all of the other variables are held constant. For the cube law to apply, the system curve must be of the form in Figure 5-1: at zero flow, the pressure difference must be zero. Examples of such systems include residential and other ventilation systems that were originally designed to operate at constant air flows, and water circulation systems that do not use pressure controls or other means to create flow-independent pressure differences. The affinity laws also assume that the efficiency of the fan or pump remains constant at varying speeds.

Another fairly common type of system has a fixed pressure requirement, even at zero flow, as shown in Figure 5-2. Examples include pumping between two reservoirs, where there is an elevation increase, and most variable-air-volume building ventilation systems,

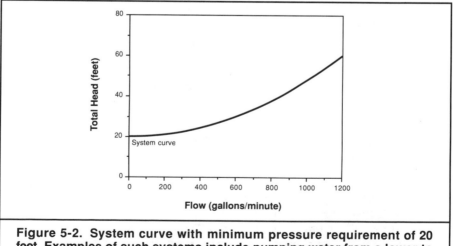

Figure 5-2. System curve with minimum pressure requirement of 20 feet. Examples of such systems include pumping water from a lower to a higher reservoir and variable-air-volume building ventilation.

which are designed to maintain constant pressure in the ductwork upstream from the dampers that serve each area. Since the cube law does not hold where the pressure does not drop to zero at zero flow, a more tedious analysis must be performed to determine the operating conditions of such systems.

Fan and Pump Characteristics

Just as each system has a characteristic curve for the pressure differences required by different rates of flow, so does each fan and pump have a performance curve. More precisely, each fan and pump has a family of curves that, like system curves, are plotted on a graph with pressure on the vertical axis and flow on the horizontal axis. These curves describe where the energy goes (to some combination of pressure and flow) when a certain amount of energy is added to the fluid. Figure 5-3 (page 136) shows a typical fan curve, and Figure 5-4 (page 137), a typical pump curve.

Centrifugal fans and pumps are the only types dealt with in detail here because they collectively consume more energy than other types (see Table 6-11). However, most of the analytical methods apply to some other types as well, especially propeller-type (axial) fans. Centrifugal fans are used in many air-moving applications—residential furnaces, commercial and industrial HVAC equipment, and large blowers in utility power plants. Similarly, centrifugal pumps are used in applications ranging from fractional-horsepower residential units to industrial pumps of thousands of horsepower.

The curves for centrifugal fans and pumps have about the same shape: starting at some pressure at zero flow, with constant or slightly increasing pressure as flow increases, then decreasing in pressure as flow increases further. In Figure 5-3, the different curves are for the same fan operated at different speeds, which are shown on the solid curves. In Figure 5-4, the different curves are for the same pump with different diameter impellers (the bladed wheel-shaped devices attached to the rotating shaft). The reason that different speeds are shown for fans is that fan speed is usually easy to set at any given point through the use of different belts and pulleys that connect the fan to the motor. For pumps, the most common arrangement is for the motor to be directly coupled to the pump, forcing both to operate at the same speed. Different performance characteristics for the same pump are provided by machining the pump impeller to a smaller diameter. Of course, it is possible to operate pumps at different speeds, as discussed in the following section. With fans, on the other hand, it is generally not possible to reduce the impeller diameter to obtain different characteristics, so flow restriction or speed control must be used.

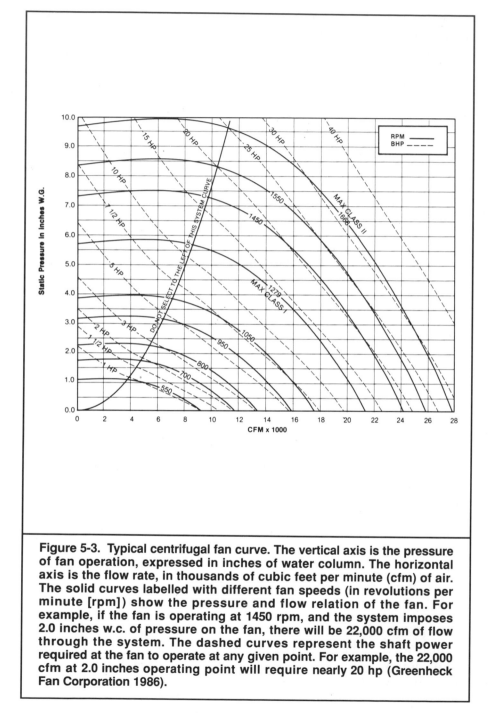

Figure 5-3. Typical centrifugal fan curve. The vertical axis is the pressure of fan operation, expressed in inches of water column. The horizontal axis is the flow rate, in thousands of cubic feet per minute (cfm) of air. The solid curves labelled with different fan speeds (in revolutions per minute [rpm]) show the pressure and flow relation of the fan. For example, if the fan is operating at 1450 rpm, and the system imposes 2.0 inches w.c. of pressure on the fan, there will be 22,000 cfm of flow through the system. The dashed curves represent the shaft power required at the fan to operate at any given point. For example, the 22,000 cfm at 2.0 inches operating point will require nearly 20 hp (Greenheck Fan Corporation 1986).

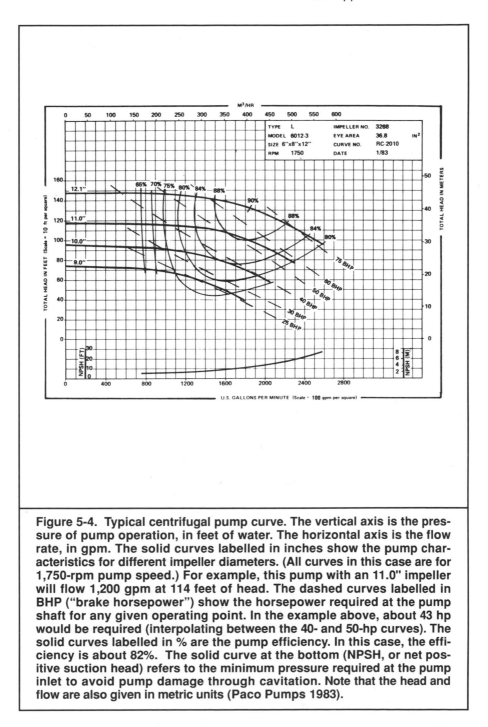

Figure 5-4. Typical centrifugal pump curve. The vertical axis is the pressure of pump operation, in feet of water. The horizontal axis is the flow rate, in gpm. The solid curves labelled in inches show the pump characteristics for different impeller diameters. (All curves in this case are for 1,750-rpm pump speed.) For example, this pump with an 11.0" impeller will flow 1,200 gpm at 114 feet of head. The dashed curves labelled in BHP ("brake horsepower") show the horsepower required at the pump shaft for any given operating point. In the example above, about 43 hp would be required (interpolating between the 40- and 50-hp curves). The solid curves labelled in % are the pump efficiency. In this case, the efficiency is about 82%. The solid curve at the bottom (NPSH, or net positive suction head) refers to the minimum pressure required at the pump inlet to avoid pump damage through cavitation. Note that the head and flow are also given in metric units (Paco Pumps 1983).

The cubic relation between speed and power requirement in certain fan and pump loads can affect the efficiencies gained by EEMs. High-efficiency motors often have lower slip, hence up to 1% higher speed than their standard-efficiency counterparts. A 1% increase in motor speed can increase power draw by 3%, negating much of the gain from switching to an EEM. To insure such gains are not lost, pump impellers can be trimmed and fan sheave and belt systems modified to slow the fan down slightly. When sheaves and belts are changed for this reason, the efficiency of the transmission system can also be improved by substituting cogged V-belts or synchronous belts for conventional V-belts.

The curves also depict efficiency (generally only for pumps, in curves of constant efficiency in percent) and the shaft horse-power required to operate at any given combination of flow and pressure (shown in the figures as dashed lines sloping downward from left to right). The efficiency of a pump or fan is the ratio (usually expressed in percent) of the theoretical power required to the actual power required and can range from well below 50% to above 90%. The information for fans is often provided not as a graph (as in Figure 5-3) but as a table containing the same information. For example, Table 5-1 lists flow rates in the first column; thus, the rows of data are at constant flow. Pressures are listed across the top, so the columns of data are at constant pressure. The intersection of flow and pressure shows the necessary speed and power required.

A fundamental concept for analyzing fluid-flow applications is the so-called operating point, the combination of pressure and flow at which a given system and fan (or pump) operates. The operating point is determined by plotting the system curve and the fan (or pump) curve on the same graph of pressure versus flow. The operating point is simply the intersection of the two curves; it represents the equilibrium flow point where the pressure drop through the system equals the pressure added to the fluid. For example, in Figure 5-3, assume the curve labeled "do not select to the left of this system curve" is the applicable system curve. If the flow requirement is 10,000 cfm, then the pressure required will be about 7.7 inches of water, the fan speed must be about 1490 rpm, and the shaft power required is about 18 hp.

Once a desired operating point is determined, the system designer must choose a fan or pump to meet this condition. This choice is an important stage in the process of determining the energy and power requirements of the system. Unfortunately, choosing the best equipment for the application is difficult, even with pumps, where the efficiency is explicitly given once the pump and operating

STATIC PRESSURE IN INCHES W.G.

CFM	OV	¼" RPM	¼" BHP	½" RPM	½" BHP	¾" RPM	¾" BHP	1" RPM	1" BHP	1¼" RPM	1¼" BHP	1½" RPM	1½" BHP	1¾" RPM	1¾" BHP	2" RPM	2" BHP	2¼" RPM	2¼" BHP	2½" RPM	2½" BHP	2¾" RPM	2¾" BHP	3" RPM	3" BHP
4500	870	353	0.30	426	0.51	492	0.74	551	0.98	609	1.26	661	1.55												
5000	967	375	0.36	444	0.58	505	0.82	564	1.09	616	1.37	668	1.68	716	2.00										
5500	1064	398	0.42	463	0.67	521	0.93	576	1.20	628	1.50	675	1.80	723	2.14	767	2.49	809	2.85						
6000	1161	423	0.50	483	0.76	539	1.04	590	1.33	640	1.64	687	1.96	730	2.29	774	2.65	816	3.03	856	3.42	893	3.82		
6500	1257	448	0.59	505	0.87	558	1.16	607	1.47	653	1.79	700	2.13	743	2.48	783	2.83	823	3.22	863	3.62	900	4.04	936	4.46
7000	1354	474	0.69	527	0.99	577	1.30	625	1.62	670	1.96	712	2.30	755	2.67	795	3.05	833	3.43	870	3.83	907	4.26	943	4.70
7500	1451	500	0.81	550	1.12	598	1.45	644	1.79	687	2.14	728	2.50	768	2.88	808	3.27	845	3.67	881	4.08	915	4.50	950	4.95
8000	1547	527	0.94	574	1.26	620	1.61	663	1.97	705	2.34	745	2.72	783	3.10	820	3.51	858	3.93	893	4.36	927	4.79	960	5.23
8500	1644	554	1.09	599	1.42	642	1.79	683	2.16	724	2.55	762	2.95	800	3.35	835	3.76	871	4.19	906	4.64	940	5.09	972	5.55
9000	1741	581	1.25	624	1.60	665	1.98	705	2.37	744	2.78	781	3.19	817	3.62	852	4.05	886	4.48	919	4.94	952	5.41	985	5.88
9500	1838	609	1.43	649	1.79	689	2.19	727	2.60	764	3.02	801	3.46	836	3.89	869	4.34	902	4.80	934	5.26	965	5.73	997	6.23
10000	1934	636	1.63	675	2.01	713	2.41	750	2.84	786	3.28	820	3.74	855	4.19	887	4.66	919	5.13	951	5.61	981	6.09	1010	6.59
10500	2031	664	1.84	702	2.24	738	2.66	773	3.10	808	3.56	841	4.03	874	4.51	906	4.99	937	5.48	968	5.98	997	6.48	1026	6.99
11000	2128	692	2.08	728	2.49	763	2.92	797	3.38	830	3.86	863	4.35	894	4.84	926	5.34	956	5.85	985	6.36	1014	6.88	1043	7.41
11500	2224	720	2.34	755	2.76	788	3.21	821	3.69	853	4.18	885	4.68	915	5.19	945	5.71	975	6.24	1004	6.77	1032	7.31	1060	7.85
12000	2321	749	2.62	782	3.06	814	3.52	846	4.01	877	4.52	907	5.04	937	5.57	966	6.11	995	6.65	1023	7.20	1051	7.76	1077	8.32
12500	2418	777	2.92	809	3.38	840	3.86	870	4.35	901	4.88	930	5.41	959	5.96	987	6.52	1015	7.08	1043	7.65	1070	8.22	1096	8.80
13000	2515	806	3.25	837	3.72	867	4.21	896	4.73	925	5.26	953	5.81	981	6.38	1009	6.95	1036	7.54	1063	8.12	1089	8.71	1116	9.31
13500	2611	835	3.61	864	4.09	893	4.59	922	5.13	950	5.67	977	6.24	1004	6.82	1031	7.41	1058	8.01	1083	8.62	1109	9.23	1135	9.84
14000	2708	864	3.99	892	4.49	920	5.00	948	5.55	974	6.11	1002	6.69	1028	7.29	1054	7.89	1080	8.51	1105	9.14	1129	9.77	1154	10.40
14500	2805	893	4.40	920	4.91	947	5.44	974	6.00	1000	6.57	1026	7.17	1052	7.78	1077	8.40	1102	9.04	1127	9.68	1151	10.33	1174	10.98
15000	2901	922	4.84	948	5.36	974	5.90	1000	6.48	1026	7.07	1051	7.67	1076	8.30	1100	8.94	1124	9.59	1149	10.25	1173	10.91	1196	11.59
15500	2998	951	5.31	976	5.83	1001	6.40	1027	6.98	1052	7.59	1076	8.21	1100	8.85	1124	9.51	1148	10.17	1171	10.84	1195	11.53	1217	12.22
16000	3095	981	5.81	1004	6.34	1029	6.92	1053	7.52	1078	8.14	1101	8.77	1125	9.43	1149	10.10	1171	10.78	1194	11.47	1217	12.17	1239	12.88
16500	3191	1010	6.34	1032	6.88	1057	7.48	1080	8.08	1104	8.72	1127	9.37	1150	10.03	1173	10.72	1195	11.42	1217	12.13	1239	12.84	1262	13.57

Table 5-1. Typical centrifugal fan table. This is an alternative to the fan curve method (e.g. Figure 5-3) of showing the fan characteristics at various pressures and flows. The flow (in cubic feet per minute) is given in the first column; the second column gives the outlet velocity in feet per minute. The column headings across the top are pressure, expressed in inches of static pressure (i.e. inches water column). For each pressure, the speed and shaft power required are shown for each flow. For example, if 15,000 cfm are required in a system that imposes 2 1/2 inches of static pressure on the fan, a fan speed of about 1149 rpm is required, and 10.3 hp will be required at the fan. Intermediate values are determined through interpolation. This table is for the same fan as in Figure 5-3 (Greenheck Fan Corporation 1986).

point are known. Given an operating point, a designer wishing to get the most efficient pump from a specific manufacturer uses an equipment selection chart such as the one in Figure 5-5.

Then the specific pump curve is inspected (Figure 5-4) to establish the efficiency at the operating point. These last two steps are repeated for pumps available from all manufacturers until the best pump is found. For fans, where the efficiency is not directly available, the process is analogous, except that the designer would search for the lowest shaft power requirement for a given operating point. This difficult and time-consuming process is seldom performed exhaustively, even in new systems. In existing systems that are modified (and thus have a different system curve), seldom is the fan or pump analyzed in an effort to reoptimize the system. The result is that many systems suffer from unnecessarily large energy consumption due to pumps and fans operating far from their points of maximum efficiency.

Another common cause of energy waste is oversizing of the driven equipment, the motor, or both. As discussed in chapter 3, systems are usually designed to have excess capacity. Pumps or fans are often oversized because exact determination of the system curve is difficult. In addition, the system may change over time. For example, pumps will require more capacity as scale and corrosion builds up inside pipes, and fans may require additional capacity as the leakage and dirt buildup in the ductwork increases. To prevent excessive flows in oversized systems, some means of flow control are often necessary, as discussed next.

System Control and Optimization Techniques

In order to minimize the energy and peak power requirements of a fluid-flow system, a designer must "get the big picture" of why the flow is needed in the first place. For example, if a flow of chilled water is needed to cool a building, consider options to reduce the required flow: reducing the cooling load, increasing the size of the cooling coils, reducing the temperature of the chilled water, increasing the air flow through the cooling coils, or some combination of all these measures. Of course, some of these options involve trade-offs between pumping energy and energy to chill water or move air, and most of them involve trading first cost and operating cost, so interdependent sets of optimization variables are at work. For further discussion of economics, see appendix A.

Once the optimum flow is ascertained, the system should be designed to achieve this flow with the minimum possible pressure drop (since the required power is proportional to the product of

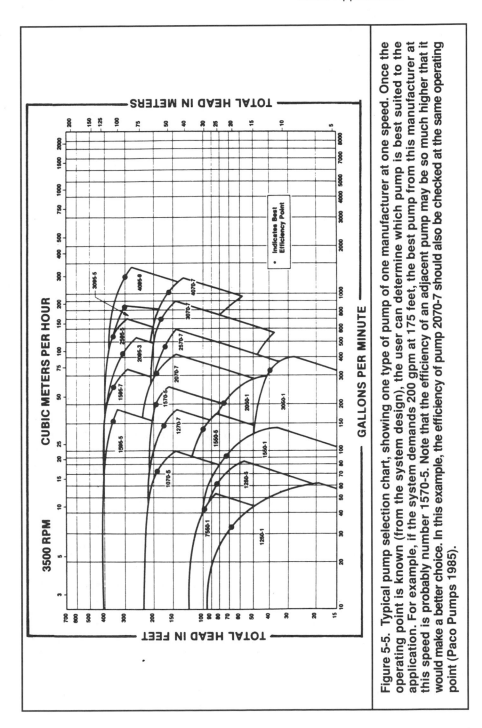

Figure 5-5. Typical pump selection chart, showing one type of pump of one manufacturer at one speed. Once the operating point is known (from the system design), the user can determine which pump is best suited to the application. For example, if the system demands 200 gpm at 175 feet, the best pump from this manufacturer at this speed is probably number 1570-5. Note that the efficiency of an adjacent pump may be so much higher that it would make a better choice. In this example, the efficiency of pump 2070-7 should also be checked at the same operating point (Paco Pumps 1985).

flow and pressure). Low pressure drops are achieved by using large-diameter system components, smooth surfaces, and gradual bends and transitions in the elbows, tees, and so on. Again, there are trade-offs between first cost and operating costs.

Once the flow and pressure requirements are reduced to the lowest practical values (thus minimizing the theoretical power requirement), the overall efficiency of the pump (or fan) package (including the pump, the transmission between pump and motor, and the motor) should be scrutinized. The object here is to meet the required operating point with the minimum amount of electrical power. In the industry jargon, the overall efficiencies of pumps and fans are known respectively as wire-to-water efficiency and wire-to-air efficiency.

Once the cost-effective minimum electrical power for meeting the design operating point is determined, a flow control scheme should be worked out for efficiently meeting system requirements at flows below design flows (since most systems operate below design load most of the time). For example, in the case of chilled water for cooling, the flow used for design is that necessary to meet the peak cooling load on the hottest day of the year. The rest of the time, the system could easily operate at lower flow rates.

The best opportunities for cost-effective overall system optimization exist in new systems designed from the beginning to work with variable flow. However, many existing constant-flow systems can be converted to variable flow, with large potential savings. For example, constant-volume building ventilation systems can be converted to variable-air-volume (VAV). While most medium to large buildings are now being built with VAV systems, most buildings in the existing stock are likely to have constant-volume systems.

While optimization opportunities are specific to each case, some general opportunities for energy savings should be pursued for fan and pump systems, both new (including renovations) and retrofit. For fans and pumps in new applications, the following principles apply:

- Reduce restrictions in ductwork and fittings or in pipes and fittings (use larger sizes, gradual bends, and so on).

- Reduce flow in variable-volume heating or cooling systems by, respectively, increasing or reducing the temperatures of the supply air, water, or both. This step will often involve a trade-off in energy use between the fan and the boiler or chiller.

- Regulate pressure in variable-volume systems with a reset control, based on the actual needs of the worst-case zone of the system.

For example, in a VAV system that provides cooling, the warmest zone would dictate the system supply pressure.

- Use an ASD control to vary the fan speed in VAV applications and the pump speed in variable-volume pumping applications (see following discussion on ASD versus other control schemes).

For fans and pumps in retrofit applications, the following measures are recommended:

- Use an ASD control to vary the fan or pump speed either to convert constant-volume to variable-volume systems, or to replace inlet vanes, discharge dampers, or throttling valves.
- Reduce pressure in variable-volume systems using worst-zone reset.

Variable-flow systems can be controlled directly or indirectly. That is, the flow control components in the system may respond to feedback from a flow sensor or from another sensor responding to other parameters, including pressure, temperature, and velocity. A few basic techniques, and variations of them, can vary flow: throttling devices, including discharge dampers on fans and throttling valves on pumps; multiple fans or pumps; and speed controls.

Throttling devices, which are essentially adjustable restrictions, operate by changing the system curve. Figure 5-6 shows the

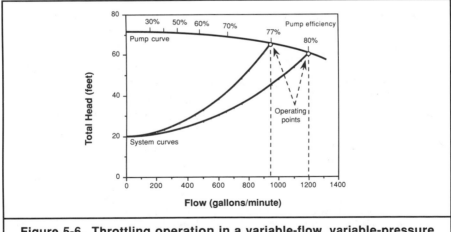

Figure 5-6. Throttling operation in a variable-flow, variable-pressure pumping system at 80% flow. Throttling makes the system curve more restrictive (i.e. steeper), which causes the intersection of the system and pump curves (i.e. the operating point) to occur at a lower flow. Throttling generally decreases the power requirement slightly relative to full-flow operation (adapted from Baldwin 1989).

effect of throttle control on flow; by steepening the system curve, throttle control causes system and pump curves to intersect at a lower flow (the operating point is shifted to the left along the pump curve). The power required for throttled flow is generally somewhat less than for full flow, since the flow reduction is a greater percentage than the pressure increase. The extent of this power reduction depends on the shape of the pump or fan curve. While throttling devices are relatively inexpensive and can give fairly precise flow control, using them is the least energy-efficient flow-control technique. This inefficiency is due to the fact that throttles dissipate flow energy that was provided by the fan or pump.

Another type of throttling device for fans, the inlet vane (also called the variable-inlet-vane or VIV), works by changing the fan curve. While these devices are more efficient than outlet dampers, they still control flow by dissipating energy across the control device.

Multiple fans or pumps can be used, in series or parallel (or both) to adjust the flow rate. This scheme works by changing the effective fan or pump curve. That is, the fluid-moving machine seen by the system is the combination of two or more fans or pumps. Thus, the operating point is changed, again resulting in flow control. Because this control scheme works in steps, it is not as precise or efficient as might be desired.

Speed control works by changing the fan or pump curve. For fans, the usual fan curve or table can be used to see the effect of speed control. For pumps, special variable-speed (rather than variable impeller diameter) curves can be obtained for certain impeller sizes (see Figure 5-7 for an example). For other impeller sizes, or where variable-speed curves cannot be obtained, variable-speed curves can be calculated by several methods, including interpolation between several available curves for the same pump at different fixed speeds (for example, 3,500, 1,750, 1,150 rpm for 60-Hz motors and 2,900, 1,450, and so forth for 50-Hz); use of the affinity laws (though the assumption of constant efficiency may result in significant error); or approximating the speed curves using the impeller diameter curves. The details of constructing such custom curves are beyond the scope of this book; see Garay 1990 for further information.

Speed control can be used with single- or multiple-fan or pump combinations, or as a partial or complete replacement for throttling control. Figure 5-8 (page 146) shows the comparison between a throttling control and a speed control application in terms of power requirements.

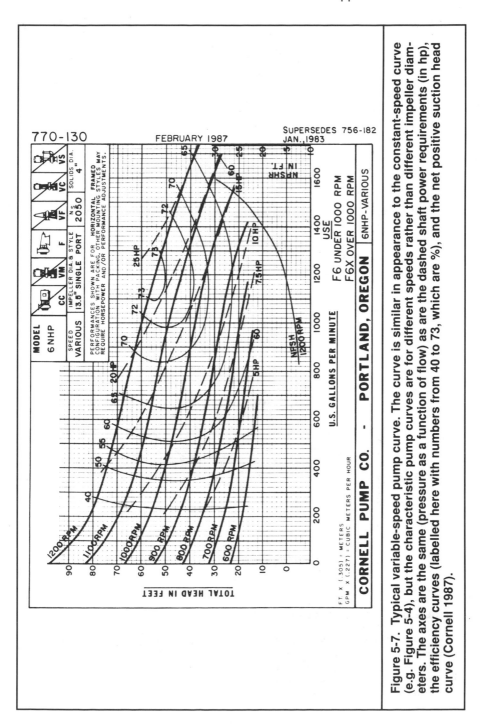

Figure 5-7. Typical variable-speed pump curve. The curve is similar in appearance to the constant-speed curve (e.g. Figure 5-4), but the characteristic pump curves are for different speeds rather than different impeller diameters. The axes are the same (pressure as a function of flow) as are the dashed shaft power requirements (in hp), the efficiency curves (labelled here with numbers from 40 to 73, which are %), and the net positive suction head curve (Cornell 1987).

Speed control is generally the most energy-efficient flow control technique, since it supplies only the amount of flow energy required. In addition, it is the most suitable for retrofit applications where the fan or pump is already in place and varying flow with multiple staged pumps is impractical. The equipment necessary to provide speed control, ranging from mechanical friction disks and adjustable pulleys to electronic adjustable-speed drives, is covered in chapter 4.

System Optimization Case Studies

A number of ASD case studies were presented in chapter 4, documenting the potential of this technology to save energy and money in fan and pump applications. In many cases, however, energy savings can be achieved in pump and fan systems using other techniques. Two such examples are given below.

Farm Irrigation

An irrigation system at a large farm consisted of a series of pumps that supplied water to irrigated fields. The pumps were arranged so that water for fields close to the water source was pumped directly

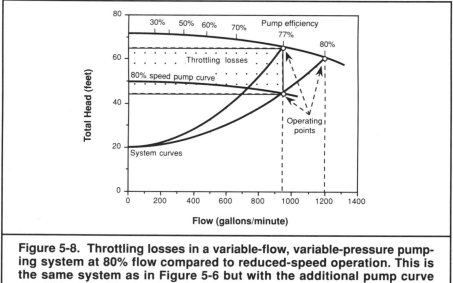

Figure 5-8. Throttling losses in a variable-flow, variable-pressure pumping system at 80% flow compared to reduced-speed operation. This is the same system as in Figure 5-6 but with the additional pump curve at reduced speed showing the alternative operating point at the same flow. The theoretical power loss due to throttling is proportional to the shaded area (since theoretical power is proportional to the product of pressure and flow) (adapted from Baldwin 1989).

from the source while water for distant fields passed through a series of booster pumps.

The conservation strategy consisted of the following system changes:

• The sprinkler heads on the irrigation system were converted from high-pressure nozzles to low-pressure nozzles, halving the pressure needed to supply water to the fields.

• The nozzles at the ends of each pipe run were equipped with small booster pumps. These pumps increased the pressure for the 5% of the water that went to the nozzles on the end of the system. As a result, the pressure in the main system could be lower.

• The decreased pressure requirements allowed the owner to trim pump impellers, thus reducing the brake horsepower needed for pumping a given volume. (Note that significantly reducing the impeller diameter may decrease the pump efficiency to the point where it could be cost-effectively replaced with a pump better matched to the application.)

• Trimming the impellers caused many of the motors to be oversized for the application. These motors were marked and replaced, where physically possible, with smaller energy-efficient motors.

• Traditionally, the fields were irrigated on a fixed schedule designed to provide enough water under worst-case hot weather conditions. Soil moisture sensors installed as part of the conservation package allowed the farmer to water fields only when they needed irrigation. As a result, less water was needed to maintain the quality and quantity of the crop.

• In the past, enough pumps had been run to supply more than an adequate amount of water to the system. After system upgrading, pumps were scheduled to meet minimum flow requirements without excess water.

The results of the retrofit included annual energy savings of 34% of the base energy consumption for the system, a simple payback for the installation of 2.3 years, and reduced wear on the pumps due to the lower operating pressure and reduced use.

Pony Pump and Motor Addition

A school had a hot-water system to meet the space heating needs of the building. The system flow was designed to meet the worst-case needs of the building (the heating needs under the coldest expected condition of 5°F ambient temperature). However, the temperature during most of the year was far above the design criteria.

The system had six hot-water circulation pumps with a connected load of 95 hp.

The conservation retrofit entailed installing a second set of pumps in parallel with the existing set. These pumps provided approximately 60% of the full-rated flow and 40% of the full-rated pressure of the system. Each set of pumps was equipped with a set of controls and valves that operated the main pumps if the ambient temperature was below 35°F, operated the pony motor pumps if the ambient temperature was above 35°F, and turned off all the pumps if the ambient temperature was above 62°F. The total connected load for the pony motors and pumps was 20 hp.

Data from the building energy management system following the retrofit were as follows:

- With the original system, the hot water circulation pumps ran for 5,800 hours per year. After the retrofit, the main pumps ran for 750 hours per year, the pony motors and pumps ran for 4,050 hours per year, and the pumps were off for the remaining period.
- Energy savings totalled 225,000 kWh, amounting to $14,700, per year, offering a 1.6-year simple payback on the investment.
- The system provided added reliability because both sets of pumps were unlikely to break at the same time.

Other Motor Applications

Fans and pumps, the largest users of motor energy, are the main targets for energy-saving motor controls and system optimization measures. There are, however, many other motor-driven loads in commercial and industrial applications. While this book cannot discuss in detail all end-use systems, several additional important systems are discussed to give the reader a sense of the conservation possibilities.

Most of these applications require custom engineering, due to the special requirements of individual systems. The costs and savings are typically site-specific.

Air Compressors

Compressed air systems are used in many industries and can be a significant end use in some facilities. Air compressors are either centrifugal or positive displacement (such as reciprocating). System optimization opportunities on compressed air systems generally include the following:

- Reduced air leakage.

- Reduced air use through the use of low-flow nozzles, alternate valves and actuators, and elimination of air blow-off for pressure regulation.
- Reduced pressure requirements from increased pipe sizes (reduced friction) or alternative end-use devices.
- Improved efficiency of compressors through staging of parallel units, some of which can be shut down during low load conditions.
- The use of higher efficiency compressors for new installations.

Centrifugal Compressors and Chillers

Centrifugal compressors and chillers can often benefit from "cube law" savings via speed control in much the same way that pumps and fans can. Wasteful throttling devices and frequent on-off cycling of the equipment can be largely avoided with precise speed control, leading to both energy savings and extended equipment lifetime. Accurate control of centrifugal chillers is especially beneficial in buildings where space conditioning systems run regularly at partial load. The capability of motors with ASDs to operate at high speeds can eliminate the need for speed-increasing gearboxes, with corresponding savings in initial investment and in energy and maintenance costs.

In addition, centrifugal compressors on compressed air systems are subject to the same types of system optimization strategies used for other types of air compressors.

Substituting AC Motors and ASDs for DC Drives

AC motors used in tandem with ASDs are supplanting DC drives for many applications in industry. As noted in chapter 2, AC motors are favored by many industries in new applications because they require less maintenance, and the maintenance that is required costs less: it is less expensive to rewind or replace an AC motor than to rewind or replace a DC motor. However, there are still a large number of applications that use DC motors.

Energy savings from these conversions accrue for several reasons:

- The greatest energy savings occur when the old DC system produces current using a motor generator set (see chapter 4). In these applications, the overall system efficiency can be improved from approximately 55% to approximately 85% when an ASD and AC motor are installed.
- Substantial energy savings can also occur when rectifiers and DC motors are changed to ASDs because of the losses in the rectifier as well as the higher efficiency of the AC motor.

- AC motors with ASDs are currently being installed on systems where a constant mechanical tension is required, as in winders. With the older DC systems, the tension adjustment was often done by imposing friction on the system, thus decreasing the overall system efficiency. Eliminating the need for such energy-wasting devices improves the system efficiency.

- DC drives were the standard for traction devices such as subway cars and ship propulsion. Such loads can be served instead by AC drives with ASDs with regenerative capabilities (which put power back into the grid when braking). As a result, the overall energy use for the system decreases.

Examples of areas where AC drives can replace DC drives with resultant energy savings include mills and kilns (steel, paper, cement, and mining industries), traction drives (transportation), winders (paper machines and steel rolling mills), and machine tools and robotics.

Conveyors

The power required by a conveyor consists of the power used to move the material on the conveyor and the power used to move the conveyor belt.

The power required to move the material is set by the rate of flow of the material. For example, a conveyor operating at its design capacity, measured in, say, pounds per minute, uses 100 hp to move the material. When only half as much material is needed, it will require 50 hp to move the material, regardless of whether the belt is running at half load (in units, say, of pounds per foot of belt) and full speed or full load and half speed.

However, if the belt speed is slowed at partial capacity with an ASD, the power required to move the belt will go down. For most short conveyors, the power needed to move the belt when it is fully loaded is less than 10% of the power needed to move the material. However, in long conveyors, such as those found at power plants and mines, the power to move the belt can be substantial.

As an example of energy savings at reduced speed, if moving the material on a fully loaded conveyor needs 150 hp, a belt needing 20 hp uses 12% of the power at full load. Running the belt at half load and full speed will use 95 hp (75 hp for the material and 20 hp for the belt). Running the belt at half speed and full load will use 85 hp (75 hp for the load and 10 hp for the belt) for a savings of 11%. Likewise, the savings at 25% of full load will be 26%.

Based on these calculations, there is some potential for energy savings from the use of ASDs on conveyors. However, the savings are not as dramatic as the savings that are available when ASDs are used on pumps and fans.

Summary

Pumps and fans provide a fertile area for energy conservation, both because of the characteristics of fluid flow and because of the relative importance of these machines as a percentage of the total motor population. Pumps and fans are good candidates for both system optimization techniques that reduce either the pressure or flow requirements of a system and for speed control technology to match the pump or fan output to the system requirements.

Other motor systems presenting conservation opportunities include air compressors, refrigeration compressors, systems currently using DC drives for speed control, and conveyors.

The next two chapters estimate the aggregate conservation potential based on the technologies described in chapters 2 through 5.

Chapter 6

A Profile of the Motor
Population and Its Use

M ore than half of all electricity in the United States and
most other nations, and about two-thirds of all industrial
electricity, flows through motors. (Motors actually consume only a small portion of their energy input; the rest is used by
the driven equipment.) By way of comparison, primary energy input
to motors exceeds fuel use in all U.S. highway vehicles. One would
think that a class of devices responsible for such a sizable portion of
world energy flows would be thoroughly profiled. Remarkably, however, less is known about the stock, performance, and usage of
motors than about any other major category of energy-using
equipment.

Data Sources and Limitations

Most of the literature published to date on the United States motor
stock contains little or no reliable field data on how many motors of
what types and sizes are used for which purposes for how many
hours per year. Rather, most reports cite previous works, which in
turn cite earlier projects, all eventually leading back to two major
data sources. The first is a study prepared by the consulting firm
Arthur D. Little, Inc., for the U.S. Federal Energy Administration in
1976 using 1972 data. Entitled *Energy Efficiency and Electric
Motors* and hereinafter referred to as A.D. Little 1976, this report
was largely superseded by a follow-up project in 1980 by the same
firm, under contract to Argonne National Laboratory, for the U.S.
Department of Energy. Two versions of the resulting document were
published under the same title, *Classification and Evaluation of
Electric Motors and Pumps*. The draft version, written in the closing days of the Carter Administration and released in February

1980, is cited herein as A.D. Little 1980. The final version, released in September 1980 (Argonne National Laboratory 1980), contained a number of changes in assumptions about the costs and likely penetration rates of high-efficiency motors, the need for standards, and other policy matters. The nature of these changes is discussed further in chapter 9.

The 1976 and 1980 studies represent the most comprehensive available source of information on the population, distribution, and usage patterns of electric motors and pumps in the United States. Unfortunately, the most current of these is based on 1977 data, and no more recent, comprehensive survey exists.

Even these reports, however, were based on scattered and sometimes inconsistent statistical sources and disturbingly little measured field data. To their credit, the authors did a remarkable job given the limited information at their disposal and did not profess to any greater accuracy than was warranted. Unfortunately, much of the more recent literature on motors tends to rely heavily on these reports without acknowledging that the underlying information, while plausible, has never been systematically verified.

In the intervening years, various local or regional motor surveys have been conducted by utilities, motor distributors, and government entities. To the extent that they gathered field data, most of these researchers focused on nameplate information like motor size, frame type, and speed (rpm). Some logged hours of operation as well. But very few investigations measured actual operating efficiency at full and partial load, judged whether motors were sized properly for their tasks, or monitored how motor loads varied over time. In this chapter we attempt to glean as much useful field data as possible from recent surveys and use it to verify or modify, where appropriate, the data from A.D. Little 1980, thus arriving at an updated profile of the current motor stock and how it is used.

Clearly, even this effort is not sufficient. Given the enormity of motor energy use and the lack of data, governments, utilities, and other groups should place a high research priority on broadly based, well-designed field surveys of the motor stock. Policies and programs would be more easily directed to the largest and most lucrative savings opportunities if those opportunities could be identified and accurately quantified. Of course, there are limits to the value and applicability of general field data. Unlike that other gold mine of electricity savings, commercial lighting—where there are millions of virtually identical fixtures operating under similar conditions—drivepower savings are far more application-specific. With these caveats in mind, let us now examine the data.

Motor Population, Distribution, and Use by Size and Type

In 1977, an estimated 720 million electric motors 1/6 hp and larger were operating in the United States, and the motor population was growing by about 3% per year (A.D. Little 1980). Steady growth at this rate would put the 1990 U.S. motor population at slightly over 1 billion. One might well ask how an energy-efficiency program can practically be applied to a billion motors. Luckily, the task is not quite this daunting, as a relatively few large motors offer the lion's share of the motor savings potential. Less than 10% of all motors (those larger than 1 hp) account for at least 80% of all motor energy input. A.D. Little (1980) goes even further than this, asserting that less than 2% of the motor population (those units above 50 hp) account for some 78% of all motor input energy. Although we believe that A.D. Little slightly exaggerates the usage by large motors and underestimates motor energy use by fractional-horsepower motors, the report's general thesis that most of the savings potential lies in larger motors remains sound.* The relationship of motor population and energy use by size class is depicted in Figure 6-1, adapted from a similar figure in A.D. Little 1980.

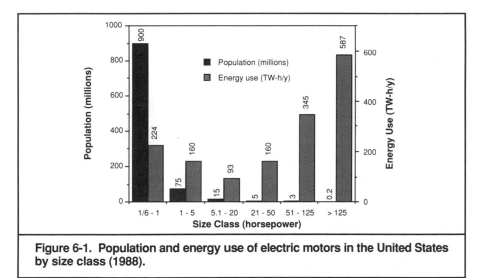

Figure 6-1. Population and energy use of electric motors in the United States by size class (1988).

---

* As summarized in Table 6-11, motors in the residential sector—most of which are under 1 hp—account for 21% of total U.S. motor input. A.D. Little 1980 states without explanation that only 30 TWh of a total 720 TWh motor input in 1977 (4%) went to fractional-horsepower units. We believe our detailed end-use analysis yields a more reliable estimate.

A.D. Little 1980 estimated that only 1.3% of drivepower energy in 1977 went to DC motors and that nearly all the rest was used by AC induction motors. Motor sales data in Table 6-1 suggest that DC units might have a somewhat larger share. Lacking solid usage data, we estimate that about 3% of integral-horsepower drivepower energy goes to DC motors, less than 1% to synchronous units, and 96% to AC induction motors. Thus a program focusing on integral-horsepower induction motors has the potential to capture most of the available savings.

Table 6-1. United States motor sales 1989.

	Quantity
Fractional-horsepower	
AC	
Single-phase	
Shaded pole	68,265,514
Permanent split-capacitor	21,764,620
Other	27,685,392
Polyphase	828,852
DC	10,958,947
Universal (AC/DC)	17,969,864
Total	147,473,189
Integral-horsepower	
AC	
Single-phase	1,798,343
Polyphase, induction	
1–5 hp	986,679
5.1–20 hp	493,016
21–50 hp	145,826
51–100 hp	58,769
101–125 hp	19,910
126–200 hp	17,923
201–500 hp	8,642
Over 500 hp	2,605
Synchronous motors	243
Universal (AC/DC)	2,386,549
DC (includes motors and generators)	
Mechanically commutated	102,623
Electrically commutated	163,677
Total	6,184,805

Source: Bureau of the Census 1990.

Note: Fractional hp motor data exclude hermetics and other rotating equipment, and motors used in automobile accessories, aircraft, spacecraft, toys, and clock-type timing. Universal motors can operate on either AC or DC power. While a significant number of universal motors are sold, these motors account for a very small share of integral horsepower motor energy use because the maximum size of universal motors is approximately 3 hp.

Although the greatest savings potential in a systems sense lies with the largest motors, the opportunities with fractional-horsepower motors should not be ignored. Most of the energy use by fractional-horsepower motors occurs in major home appliances—notably refrigerators, freezers, furnaces, and air conditioners. Fractional-horsepower units receive an estimated 14% of all motor input energy and are considerably less efficient than larger motors. Moreover, as discussed in chapter 2, the difference in efficiency between standard- and high-efficiency motors is greatest in the small size ranges, so the opportunity for improvement is great. As Table 7-1 shows, replacing standard-efficiency motors with EEMs would save more energy in fractional-horsepower units than in any other size class.

Distribution by Speed and Type of Frame and Enclosure

As discussed in chapter 2, high-efficiency motors are readily available in T-frame, NEMA Design B motors with open drip-proof (ODP) and totally enclosed fan-cooled (TEFC) enclosures, and for the most common speeds—1,200, 1,800, and 3,600 rpm. Price and performance vary among these categories; for this reason, as discussed in chapter 9, several utilities offer different rebates for different speeds and enclosure types.

The availability of high-efficiency motors outside these standard categories varies. Some manufacturers make high-efficiency, explosion-proof, C-face, and 900-rpm T-frame motors. High-efficiency replacements for some non-T-frame motors can be obtained with special modifications to the mounting hardware or other special order. In other instances, high-efficiency models are simply not available.

Analyzing the aggregate potential for EEMs and designing programs to promote them requires some estimate of the composition of the existing motor stock by speed, frame, and enclosure type. We were unable to locate any national breakdown by these criteria, but several regional motor surveys offer some insights. These include an unpublished field survey of 2,641 motors in commercial and industrial facilities in Rhode Island (New England Power Service Company [NEPSCO] 1989); a proprietary survey of 405 motors in six Pacific Northwest factories (Seton, Johnson & Odell [SJO] 1983); an inventory of 106 facilities containing 128,000 motors, representative of the industrial sector in Wisconsin Electric Company's service territory (Xenergy 1989); a survey of about 160 motors in southeastern pulp and paper mills (Carolina Power and Light 1986); and measured data on 97 HVAC motors in Stanford University buildings (Wilke and Ikuenobe 1987). Two of the surveys have data on enclosure types. Summarized in Table 6-2, they show 56% to 73% of installed motors with open drip-proof housings, 21% to 30% with totally enclosed housings, and under 10% with explosion-proof housings.

Three of the surveys sort motors according to speed. As shown in Table 6-3, these data indicate that 60% to 96% of the motors operate at 1,800 rpm, 2% to 16% operate at 1,200 rpm, 1% to 8% operate at 3,600 rpm, and 1% to 13% run at various other speeds.

Table 6-4 presents the distribution of frame types. T-frames

Table 6-2. Distribution of enclosure types.

Survey	Open Drip-Proof #	Open Drip-Proof % of Total	Totally Enclosed #	Totally Enclosed % of Total	Explosion-Proof #	Explosion-Proof % of Total
SJO 1983	226	55.9	122	30.2	39	9.7
NEPSCO 1989	814	72.9	236	21.1	66	5.9

Table 6-3. Distribution of motor speeds.

	3600 #	3600 %	1800 #	1800 %	Nominal RPM 1200 #	1200 %	900 #	900 %	720 #	720 %	Other #	Other %
Survey												
SJO 1983	33	8	243	60	64	16	14	3	4	1	47	12
NEPSCO 1989	129	7	1475	79	216	12	37	2	0	0	2	0.1
Wilke and Ikuenobe 1987	1	1	91	96	2	2	0	0	0	0	1	1

Table 6-4. Distribution of frame types.

Survey	T-Frame #	T-Frame %	U-Frame #	U-Frame %	Other #	Other %
SJO 1983	223	55	160	40	22	5
NEPSCO 1989	1359	87	159	10	44	3
Wilke and Ikuenobe 1987	52	56	41	44	0	0

clearly predominate, with a range of 55% to 87% in these three data sets. Nearly all remaining motors are U-frames.

Tables 6-2 to 6-4 suggest that most operating motors are types for which high-efficiency versions are readily available. The potential for improvement depends upon how many EEMs are already installed.

Saturation of High-Efficiency Motors

The saturation of high-efficiency motors throughout the United States has not been measured. Table 6-5 shows the EEM proportion of 1988 motor sales by NEMA member companies.

Roughly fifteen years since their introduction, EEMs now constitute about 20% of motor sales in sizes above 20 hp and a somewhat smaller percentage of smaller, integral-horsepower sizes. These averages mask variation by region and end use. Darryl Van Son, the source of the data cited in Table 6-5, maintains that pulp and paper mills and other industries in Maine, where electric rates are relatively high, are now buying virtually all high-efficiency motors, while sales are much lower in areas where electricity is cheap.

EEMs' share of current sales must be much higher than their share of total stock, given the large population of motors and typical operating lives of 10 to 30 years. This observation is supported by data on 128,000 motors in a cross section of the industrial sector of Wisconsin Electric Company's service territory. The overall horsepower-weighted stock saturation of EEMs in the survey was 3% (Xenergy 1989). Table 6-6 (next page) shows the percentage of EEMs for each of the 17 industrial sectors in the Wisconsin survey.

Table 6-5. EEMs as percent of 1988 motor sales.

Horsepower	% EEMs of total sales
1–5	8
6–20	13.7
21–50	19.4
51–125	20.7
126–200	20.5
201–500	20.0

Source: Van Son 1989.

Note: These data are from NEMA member companies, which account for only 52% of U.S. motor sales.

Table 6-6. Stock saturation of high-efficiency motors in Wisconsin industries in 1989.

Sector	HP-weighted % high-efficiency
Paper Mills	7.9
Motor Vehicles	0.6
Iron/Steel Foundries	0.0
Beverages	0.0
Engines and Turbines	1.3
Commercial Printing	3.1
Constr./Mining Equipment	1.0
Elec. Ind. Apparatus	0.2
Misc. Plastic Products	10.1
Nonferrous Foundries	17.4
Dairy Products	8.5
Converted Paperboard	0.1
Gen. Ind. Mach./Equip.	3.3
Ind. Inorganic Chem.	0.0
Metal Forging/Stamping	0.0
Nondurable Goods	0.0
Durable Goods	0.0
Average	3.0

Source: Xenergy 1989.

One very knowledgeable motor distributor in New England believes EEMs constitute about 3% of installed motor capacity in the United States (Gilmore 1989). Based on admittedly sketchy data, and the views of motor professionals like Gilmore and Van Son, it thus appears that well over 90% of the potential savings from converting the U.S. motor stock to high-efficiency models remains to be tapped.

Duty Factor

One important consideration in determining the savings potential for a given motor is its duty factor, or how many hours per year the motor operates. Motors with high duty factors have the greatest savings potential. Unfortunately, the data on duty factors are limited. A.D. Little 1980 estimated the U.S. distribution of duty factors by size but offered no supporting data. A few of the recent field surveys cited above suggest that the A.D. Little estimates were low across all sizes. In other words, motors run far more hours (and the savings potential from improving them is larger) than the A.D. Little values suggest. Table 6-7 (page 162) compares A.D. Little estimates of 1977 average duty factors by size class with estimates from three recent field surveys. Note that the Rhode Island survey (NEPSCO) was limited to commercial and industrial motors, the Wisconsin survey (Xenergy) to industrial facilities only, and the Stanford data (Wilke and Ikuenobe) to 97 HVAC motors in university buildings.

A survey of 161 motors in Southeastern paper mills also shows far higher duty factors than the A.D. Little estimates suggest. There, 76%, 87%, 80%, and 94% of motors in the respective size classes of 0–5 hp, 5.1–20 hp, 20.1–50 hp, and >50 hp operated between 7,000 hrs/yr and 8,000 hrs/yr. Most of the rest run more than 5,000 hrs/yr (Carolina Power and Light 1986). Pulp and paper mills, however, tend to run their motors more than many other industries.

The wide differences in duty factors shown in Table 6-7 reveal an important gap in the data and challenge the conventional wisdom that the largest motors run the most hours. Even more greatly needed than average values by size class, however, is an accurate profile of the distribution of duty factors within a size class—what percentage of motors of a given size are used 1,000 hrs/yr, 2,000 hrs/yr, and so on. This information would be useful to utility program planners, among others, for determining eligibility criteria in motor rebate programs. New England Electric, for example, offers generous rebates for high-efficiency replacements for 100-hp or smaller motors operating at least 1,250 hrs/yr and for motors larger than 125 hp operating 2,500 hrs/yr. This rebate program is discussed further in chapter 9.

Tables 6-8 and 6-9 (page 162) show distributions of motors by size and duty factor, as drawn from the Rhode Island survey and as estimated by A.D. Little. We include them here principally to illustrate the kind of information needed. As the wide differences between these tables demonstrate, the data are too sparse to lead to any firm conclusions about typical duty factors.

Table 6-7. Motor duty factors by size class.

| HP | Average Hours of Operation/Year | | | |
	A.D. Little[a]	Xenergy	NEPSCO	Wilke-Ikuenobe
0–5	250	2979	6195	——
5.1–20	1500	4132	6161	5156
20.1–50	2300	4132	5041	6190
>50	3500	5539	4670	4500

[a]Weighted averages rounded to nearest 50 hours, based on population and usage estimates by end-use category in Table 3-11 of A.D. Little 1980.

Load Factor and Motor Sizing

In addition to duty factor, the size of a motor relative to its typical load helps determine its operating efficiency and the potential for savings. Load factor is important because as a motor operates further and further below about 50% of its rated load, its efficiency drops. This drop occurs more rapidly for standard-efficiency than for high-efficiency units, as was illustrated in Figure 3-8. As discussed in chapter 3, downsizing should be considered for any motors running below 40% loading.

So endemic is the problem of motor oversizing that Pacific Gas and Electric engineer Konstantin Lobodovsky has concluded after conducting field measurements on about 1,000 (mostly industrial) motors that "Half the motors in the real world are operating at less than 60% of their rated load. And a third are operating at less than half their rated load" (Lobodovsky 1989).

Edward Cowern, an engineer with the motor manufacturer Baldor, concurs with Lobodovsky. He maintains that when users select their own motors, they generally oversize, and that consultant-

---

Table 6-8. Percentages of commercial and industrial motors, sorted by size and duty factor, in Rhode Island, 1989.

Hp	Duty factors (thousands of hours per year)									
	0<1	1<1.5	1.5<2	2<3	3<4	4<5	5<6	6<7	7<8	8<8.8
0–5	1.8	0.2	0.4	9.2	7.5	18.4	5.2	9.4	4.3	43.4
5.1–20	1.3	0.7	0.5	8.7	9.3	18.2	3.5	9.6	4.3	43.4
20.1–50	1.4	1.9	0.1	3.9	2.9	14.7	13.2	7.8	3.9	31.3
>50	0	1.9	0.9	15.6	18.6	17.6	11.7	13.7	8.8	10.7

Source: NEPSCO 1989.

Table 6-9. ADL estimate of (1977) U.S. motor duty factors by percentage and size class.

Hp factors	Percent of motors by size class with various duty (thousands of hours per year)									
	0–1	1–1.5	1.5–2	2–3	3–4	4–5	5–6	6–7	7–8	8–8.7
5.1–20	21.8	29.8	29.8	7.7	5.0	2.0	1.6	0.9	0.8	0.7
21–50	3.6	13.3	13.3	29.7	14.3	11.2	7.4	3.8	1.6	0.8
51–125	9.1	16.1	16.1	21.9	14.3	9.6	6.7	3.6	1.7	0.6

Source: A.D. Little 1980.

designed ventilating systems are also almost always oversized and often run at below 35% of rated load (Cowern 1989). Cowern notes, however, that motors supplied as part of packaged equipment are usually sized well.

Cowern's opinion appears to be confirmed by the survey cited above of 97 standard-efficiency HVAC motors at Stanford University. The motors, which were presumably specified and installed using standard practice, ranged from 7.5 to 75 hp. The survey found that 20% of the motors were operating at or below 40% of rated load, 29% of the motors operated at or below 50% of rated load, and 43% fell at or below 60% of rated load. Weighted by horsepower, these three ranges consumed 21%, 33%, and 45%, respectively, of motor input (Wilke and Ikuenobe 1987). In a separate test of 44 (mainly industrial) motors, 50% were loaded at or below 60%, 34% at or below 50%, and 23% at or below 40% (Lobodovsky, Ganeriwal, and Gupta 1983).

For comparison, a survey of 22,300 motors in small- and medium-size industries in Brazil yielded the breakdown of motor loading shown in Table 6-10. This survey suggests somewhat better sizing practices than those reflected in U.S. sources just cited, as it indicates that only 22% of the motors operate below 50% load (Geller 1990).

As wasteful as motor oversizing can be in energy and dollars, it is sometimes justified in terms of higher reliability, reduced downtime, and the flexibility to accommodate expanded process needs. Individual cases should be evaluated with an eye to the various trade-offs involved in the sizing decision. In general, it is probably not cost-effective from an energy savings point of view to downsize motors running above 40% of their rated load.

Table 6-10. Motor load factors in Brazilian industry.

% Load	% of Motors
0–10	4.4
10–20	0.5
20–30	2.2
30–40	6.0
40–50	10.1
50–60	11.5
60–70	13.7
70–80	14.9
80–90	12.6
90–100	24.5

Source: Howard Geller 1990.

The data on oversizing are too sparse to allow firm conclusions about the extent of the problem. If the numbers cited above are typical, however, they suggest that about one-fifth of motors 5 hp and larger are running at or below 40% of rated load.

Motor Energy Input by Sector, Industry, and End Use

While analyzing motor usage by size class is informative, most efficiency programs are designed and implemented by sector, industry, or specific end use. It is thus useful to break out motor consumption by such categories. As Table 6-11 illustrates, in the United States, about 37% of residential electricity, 43% of commercial electricity, and 74% of industrial electricity power motors. The largest residential uses are refrigeration and space cooling. Space cooling and ventilation dominate commercial-sector motor use. Pumps, blowers, fans, and compressors account for more than half of industrial motor use.

As Table 6-11 shows, the industrial sector dominates motor use. This is to be expected given the observation made earlier that

Table 6-11. 1988 U.S. motor energy input by end use.

End Use	Energy Input (TWhrs/yr)
Residential sector[a]	
Refrigerators	134
Space cooling	113
Freezers	32
Furnace fans	22
Heat pumps (heating mode)	16
Clothes dryer motors	3
Clothes washer motors	3
Miscellaneous	10
Residential electricity used in motors	333
Total residential energy consumption	890
% of residential electricity in motors	37%
Commercial sector[b]	
Space cooling	205
Ventilation	60
Refrigeration	35
Heat pumps (heating)	~4
Commercial electricity used in motors	304
Total commercial electricity consumption	710
% of commercial electricity in motors	43%

most motor energy use is accounted for by a relatively few large motors, which are concentrated in the industrial sector.

Figure 6-2 (next page) shows where, within the large and diverse industrial sector, motor use is concentrated. Using mid-1980s data, the figure illustrates how much electricity various industries use and how much of that use passes through motors.

Table 6-11 *(continued).*

End Use		Energy Input (TWhrs/yr)
Industrial sector[c] (device allocations highly uncertain)		
	Pumps	227
	Blowers and fans	132
	Compressors	116
	Machine tools	63
	Other integral hp applications	82
	DC drives	74
	Fractional hp applications	32
	HVAC	16
	Industrial electricity used in motors	742
	Total industrial electricity consumption	955
	% of industrial electricity in motors	78%
Other sectors[d]		
	Utilities' own use	146
	Other public authorities	45
	Railways	4
	Total	195

Summary of motor input energy			% of total
	Residential	333	21%
	Commercial	304	19%
	Industrial	742	47%
	Other	195	13%
			100%
	Total motor input	1574	
	Total U.S. electrical consumption	2783	
	Motor % of electrical consumption	57%	

Notes:

[a]Values based on saturation and typical usage values from half a dozen sources.

[b]Values adapted from *Commend* data, Lann et al. 1986.

[c]Sectoral use based on sales data from DOE 1989a and industrial self-generation not sold to the grid as estimated by staff of Edison Electric Institute (Monroe and Moscarillo 1989). End-use values are based on percentages estimated in A.D. Little 1976.

[d]Assumes 6% of utility net output is own use and 90% of that is motors; assumes 73% of other public authorities and 90% of railway use (both as estimated by Edison Electric Institute 1987) are motors.

Pumps

Pumps deserve special mention because, by one estimate, they account for about 45% of motor electricity input (A.D. Little 1980). (This is a higher value than Table 6-11 would suggest, but whatever the actual value, pumps are a significant end use.) Of total pump energy use, more than 90% is used by the less than 2% of the pump population larger than 20 hp. Efficiency programs targeted at electrically driven pumps larger than 20 hp thus have the potential to capture most of the savings potential available in this important class of technology.

Where are the big pumps used? A.D. Little compiled a list of major pump users in order of descending shares of total pump input energy. Lacking any more recent data, we present these percentage allocations as the best available (see Table 6-12).

It is interesting to note that electric utilities are the single largest users of electric motors and pumps. Utility facilities thus offer a prime test bed for efficiency improvements, as the EPRI case studies in chapters 4 and 9 describe.

Having examined where motor energy use occurs, we now review the national savings potential represented by drivepower systems.

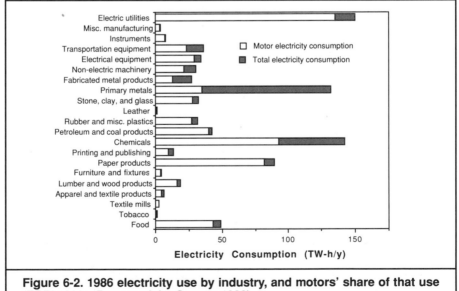

Figure 6-2. 1986 electricity use by industry, and motors' share of that use (EPRI 1988a, Bureau of the Census 1988).

Table 6-12. Percent allocations of pump energy use by user category.

User	Percent of U.S. total pump energy use
Electric utilities	18.2
Petroleum pipelines	14.5
Nonresidential air conditioning and building services	13.7
Chemical manufacturing	8.7
Pulp and paper manufacturing	6.1
Primary metals manufacturing	4.3
Mining and oilfields	4.0
National defense	3.9
Farm and agriculture	3.7
Petroleum manufacturing	3.5
Water and sewage	2.9
Irrigation and water distribution	2.2
Household	1.9
Other	12.4

Source: A.D. Little 1980.

Chapter 7

Estimating the National Savings Potential in Motor Systems

M ost analyses of the potential for improved efficiency in motor systems have focused on only two technologies: high-efficiency motors and adjustable-speed drives. These are indeed very important technologies, but as the preceding chapters made clear, many other measures deserve attention as well:

• Optimal motor sizing.

• New and improved types of motors.

• Better motor repair practices.

• Improved controls in addition to ASDs.

• Electrical tune-ups, including phase balancing, power-factor and voltage correction, and reduction of in-plant distribution losses.

• Mechanical design improvements, including optimal selection and sizing of gears, chains, belts, and bearings.

• Better maintenance and monitoring practices.

Only by paying close attention to all of these parameters and to the synergism among them can designers and users of motor systems truly optimize efficiency and reliability. We now examine the major efficiency measures in turn and evaluate their savings potential.

Induction-Motor Improvements

As discussed in chapter 6, induction motors draw an estimated 96% of all U.S. motor input, or 1,511 TWh/yr. How much of this energy could be saved if readily available high-efficiency induction motors were substituted for standard units? Estimated savings are nearly 60 TWh/yr, equivalent to the output of twelve 1,000-MW power stations (assuming typical 65% capacity factor and 8% grid loss).

This estimate is based on the differences in efficiency between the existing motor stock and new EEMs, as shown in Table 7-1.

The energy inputs to each size class are highly uncertain; they are drawn from Figure 6-1, which in turn is adapted from a similar figure in A.D. Little 1980. If these allocations and the efficiency spreads used here are roughly correct, however, they reveal a surprising fact: the largest savings opportunity from motor replacement lies in the smallest motor sizes, where the spread between standard- and high-efficiency equipment is greatest. Because the smallest motors tend to be used fewer hours per year, however, the savings from replacing them are undoubtedly more expensive than the savings from replacing larger, more heavily used motors. Moreover, most small motors are purchased by original equipment manufacturers, whose increased use of more efficient equipment will require special incentives or standards or both. Federal appliance-efficiency standards may force manufacturers to incorporate high-efficiency motors in some classes of appliances. The potential size

Table 7-1. Savings estimates by size range.

Size range potential (hp)	Est. input-weighted avg. size (hp)[a]	Effic. spread btw. EEMs & stock[b] (% points)	Adjusted input to size class (TWh)[c]	U.S. Savings (TWh)
1/6–<1	1/4	15	161	24.2
1–5	2.1	7.1	115	8.2
5.1–20	11.9	6.1	67	4.1
21–50	32.5	4.5	115	5.2
51–125	86.7	3.6	248	8.9
>125	212	1.9	423	8.0
Total				58.6

[a] Input-weighted average sizes from A.D. Little 1980.
[b] Efficiency spread between today's stock and new EEMs. To determine stock efficiency we compared 1977 standard-efficiency motors with 1990 standard-efficiency models and estimated that the current stock has improved beyond 1977 offerings by one-third of the efficiency spread between 1977 and 1990 models. Lacking data on efficiencies of 1977 fractional-horsepower motors, we calculated the spread for this size class based on 1990 standard- and high-efficiency single-phase models. The actual spread between the stock and 1990 offerings is probably larger. Loss from oversizing and rewind damage are ignored here.
[c] Values from Figure 6-1, first subtracting 4% allotted to DC and synchronous motors, and then subtracting 25% from the remaining input to allow for units that are already high efficiency or do not have readily available high-efficiency replacements.

of the savings in this area suggests the value of gathering accurate field data on energy input and duty factors for all motor sizes.

The Cost of Savings from Induction Motor Improvements

Assuming that a typical motor lasts 10–20 years before failing, the entire motor stock faces either replacement or rewinding about every 15 years (Gordon et al. 1988). When motors under 10 hp fail, they are typically replaced; larger motors are typically rewound. As Figures 2-23, 2-24, 2-26, and 2-27 show, high-efficiency motors in typical (4,000 hrs/yr) operating regimes have simple paybacks of two years or less in new installations (including replacement of failed units) and against rewinds in sizes up to around 200 hp for open drip-proof enclosures and 50 hp for totally enclosed fan-cooled designs.

Clearly, some motors will not fail in the next 15 years, and some might be replaced before they fail. Savings from motors with low duty factors will cost more than this aggregate analysis suggests, and those with very high duty factors will yield cheaper savings. The general conclusion remains, however, that full upgrading of the U.S. motor stock over a 15-year period would provide on the order of 60 TWh/yr in savings (reducing consumer bills by $3–4 billion per year and displacing the need for $12–36 billion worth of generating capacity) with a typical simple payback of two years. Assuming a typical motor life of 15 years and a real discount rate of 6%, a two-year payback corresponds to a cost of saved energy of about $.025/kWh. This analysis ignores the sizable savings from correcting for past rewind damage and oversizing, which are covered next.

Correcting for Past Rewind Damage

As discussed in chapter 2, standard rewind techniques that bake motor cores in high-temperature (>600°F) ovens can damage the motor core and reduce the motor's efficiency. In chapter 2 we estimate that this damage equals 1% of total motor input or 15 TWh/yr. At current retail rates these losses cost $1 billion per year and equal the output of 3 GW of installed capacity. Full replacement of the existing motor stock with new, high-efficiency motors would thus not only save energy equal to the difference in nameplate efficiencies but also eliminate the losses associated with operating a "wounded" motor fleet, essentially for free. Such rewind damage could be avoided in the future by adopting improved motor repair practices as discussed in chapter 2.

Optimal Motor Sizing

When replacing the current motor stock, care should be taken to correct for past oversizing, with the caveat that oversizing in some cases is justified by reliability and flexibility benefits (Baldwin 1989). A precise calculation is impossible, but the limited data discussed in chapter 6 suggest that one-fifth of motors 5 hp and larger are running at or below 40% of rated load. Assuming, as noted in chapter 3, that this oversizing is causing, on average, a 5% efficiency penalty in the affected motors, which have an input of 240 TWh/yr (one-fifth of the input to motors 5 hp or larger shown in Figure 6-1), the savings from correcting oversizing would be on the order of 12 TWh. For various reasons discussed in chapter 3, not all motors running lightly loaded should be downsized, so we reduce the 12 TWh by one-third to account for installations where downsizing is seemingly attractive but in fact not practical or desirable. The remaining savings potential is thus 8 TWh.

Since the oversized motors will be replaced eventually, and since smaller, high-efficiency replacements will often cost less than standard-efficiency replacements of the original size, the marginal cost of the new, downsized motors in some cases may be negative.

Electrical Tune-ups

As discussed in chapter 3, phase unbalance, voltage variations, poor power factor, and poor supply waveforms can reduce motor efficiency and damage equipment. No precise calculations of the aggregate efficiency loss caused by such problems exist, but one study has estimated the range of potential savings from their mitigation at 15–179 TWh (corresponding to a 1–15% savings), with a cost of well under $.01/kWh (Lovins et al. 1989). The wide range of uncertainty in this estimate underscores the need for far more field data. To be conservative we adopt a 1–5% savings range for our analysis, recognizing that the actual value may well be higher.

ASDs

As discussed in chapter 4, the range of potential cost-effective applications for electronic adjustable-speed drives is vast. ASDs are becoming very popular for flow control in new industrial installations. The Electric Power Research Institute (EPRI) estimates that the portion of motor input controlled by ASDs rose from 4% in 1980 to 11% in 1985, yielding savings of 15 TWh/yr (EPRI 1988a).

One industry expert believes that only 10% of the units eligible for ASDs had been reached by 1988 (Zacheral 1988).

Table 7-2 (next page), based on published reports and conversations with several experts, scopes the technical potential for energy savings from ASDs. These values are uncertain, but even the low values, if accurate, represent about $4 billion per year in reduced energy bills. With stakes that large, field surveys to accurately measure the potential would be money well spent.

The cost-effectiveness of ASD applications varies widely. Numerous case studies cite simple paybacks of two years or less, corresponding to costs of saved energy of $.01–.025 (Greenberg et al. 1988, PEAC 1987). These studies are probably skewed toward the most attractive installations, however, with many other installations having paybacks from two to five years. A five-year payback corresponds (assuming $.07/kWh electricity and ten-year equipment life) to a cost of saved energy of almost $.05/kWh. We adopt this value as the upper bound of the cost of savings from ASD installations.

Power-Factor Controllers

As discussed in chapter 4, power-factor controllers (PFCs) can offer up to 10% overall energy savings in certain applications driven by small (15 hp or less) motors that run most of the time at zero or near-zero loading. This 10% savings figure applies to perhaps 20 TWh of input to about 3.5 million hp of lightly loaded equipment across the U.S. economy (15 TWh for the 300,000 mostly 10-hp grinders and granulators in the plastics industry that run at zero load 80+% of the time, 1 TWh for the 25,000 10- to 15-hp escalators, and 4 TWh for all other end uses) (Lovins et al. 1989). The resulting savings potential is thus about 2 TWh/yr.

List prices of 10-hp PFCs are $30–60/hp; prices per horsepower go down with larger models. Assuming a ten-year equipment life, the cost of saved energy for the estimated 2 TWh savings potential is $.016/kWh.

Fast Controllers for Compressors

Fast controllers for centrifugal and axial compressors (discussed in chapter 4) typically save 15–25% of electricity use by eliminating blowoff and fluid recirculation. Manufacturers' estimates suggest that 6,500 centrifugal air compressors with an average size of 500 hp and another 500 large (2,000 hp average) electrically driven centrifugal

Table 7-2. Technical potential for U.S. ASD savings.

Sector	1988 drive input (TWh/yr)	% suited to ASDs	% savings per installation	Potential ASD savings (TWh/yr)
Residential[a]	356	80–90	15–30	43–96
Commercial[b]	304	40–70	15–30	18–64
Industrial[c]	742	20–40	15–40	22–119
Electric utilities[d]	150	70–90	15–40	16–54
Public authorities[e]	45	30–60	15–30	2–8
Subtotal				101–341
Minus savings from ASDs already operating in 1990				−25[f]
Total				76–316

[a] ASDs are already being incorporated into heat pumps, washing machines, furnaces, and air conditioners. They are suited as well to refrigerators, freezers, evaporative coolers, and heat pump water heaters (Greenberg et al. 1988). As Table 6-11 shows, these end uses represent well over 90% of the motor load in the residential sector.

[b] As with the residential sector, over 90% of the large commercial-sector motor uses are potential candidates for ASDs. These include HVAC chillers and pumps, fans, and refrigeration systems. Because some systems are not suited to cycling, and others, such as supermarket refrigeration systems, may serve variable loads more effectively by using parallel unequal compressors, our estimate assumes considerably fewer than 90% of commercial systems will actually be cost-effective ASD candidates. Greenberg at al. 1988 estimate the ASD savings potential in commercial sector air conditioning, ventilation, and refrigeration at 41 TWh/yr. Lovins et al. (1989) estimate a range of 44–93 TWh/yr.

[c] Several studies, including de Almeida et al. 1988 and the Michigan Electric Options Study 1987, have concluded that up to half of all industrial motor input is for variable-load applications. This is consistent with Table 6-11, which shows pumps, blowers and fans, and compressors as dominating industrial motor use. Far more field data are needed, however, to verify these estimates. De Almeida et al. (1990) estimate that ASDs can save 18% of U.S. industrial electricity, or 133 TWh/yr.

[d] All thermal power stations have large pumps, blowers, and fans that consume most of the in-house electricity. Two ASD experts familiar with power-plant installations (Ralph Ferraro formerly of EPRI/PEAC and Jim Poole of Bechtel Corporation) support the ranges cited here. They maintain that ASDs are often useful even in baseload plants, and that they are applicable to many small motors in power plants in addition to the larger installations that are typically considered.

[e] While more than 60% of this sector's motor use (largely water and wastewater pumping) is variable load, parallel unequal systems may be more applicable than ASDs in some instances.

[f] Conservative extrapolation from 15 TWh EPRI estimate cited above for 1985.

process compressors operate virtually continuously in the U.S. (Lovins et al. 1989). These values yield an aggregate energy use of about 26 TWh/yr, or nearly 2% of all drivepower energy. Savings of 15–25% on this input would amount to 3.9–6.5 TWh/yr.

Conversations with users of such fast controls confirm the manufacturers' claims of typical paybacks of one to two years (based solely on energy savings and ignoring the equally important process optimization and service life benefits). Paybacks of two years or less correspond to a cost of saved energy of under $.03/kWh.

Other Controls

Sophisticated control systems on industrial refrigeration systems—including controls for suction and condensing pressure, defrost, and fan cycling—typically save about 10% of system energy use. Drivepower input in SIC 20 (food products) is about 44 TWh/yr (Figure 6-2). According to EPRI's industrial-sector cnd-use data (EPRI 1988a), half of the motor use in SIC 20, or 22 TWh/yr, is for refrigeration. Assuming that half of such systems can be retrofitted with better controls and that 10% savings are typical yields ~1 TWh/yr in savings.

Sequential (lead-lag) controls are applicable to perhaps 15–25% of compressed air systems and typically save 3–7%. Table 6-11 shows compressors consuming 116 TWh/yr. Subtracting the 26 TWh/yr in centrifugal and axial machines that could be served by fast controllers leaves 90 TWh/yr. If 15–25% of these systems could be fitted with lead-lag controls, savings would approximate 0.4–1.6 TWh/yr.

Perhaps 5% of industrial blowers and pumps are installed in parallel and can thus benefit from sequencing controls, with typical savings of 5–15%. Applying these values to the 360 TWh/yr shown in Figure 6-11 for these end uses suggests a savings potential of 0.5–1.5 TWh/yr.

Several control and design modifications can yield substantial savings in supermarket refrigeration systems, which now consume about 50 TWh/yr. Once load-reduction measures like glass doors and better insulation on display cases are dealt with, the use of multiplex parallel unequal compressors and floating condenser head pressure can save on the order of 3–6 TWh/yr nationally, at an average cost of $.03/kWh (Shepard et al. 1990). Note that these measures apply to new installations and renovations, not retrofits.

The savings potentials from all control technologies discussed above is summarized in Table 7-3 (next page). These estimates are

Table 7-3. U.S. savings potential and costs of savings from motor controls.

Control type	Energy savings (TWh/yr)	Cost of saved energy ($/kWh)
ASDs	76–316	$.01–0.05
PFCs	2	<$0.02
Fast compressor controls	3.9–6.5	<$0.03
Industrial refrigeration controls	1	?
Lead-lag control on air compressors	0.4–1.6	?
Sequencing controls on blowers and pumps	0.5–1.5	?
Supermarket refrigeration controls	3–6	$0.03
Total	87–335 (or 5.5–21% of motor input)	

conservative because many other process-specific control possibilities are not considered here.

Improvements to Synchronous and DC Motor Systems

As discussed in chapter 6, we estimate that synchronous and DC motors combined account for only about 4% of total motor input, or 63 TWh/yr. We allocate 55 TWh of this amount to the DC units, and 8 TWh to synchronous motors.

Synchronous motors are large, efficient (96–98%) units made to order. Because they are already so efficient, we assume no significant savings potential. Because they are carefully specified, we conservatively assume no penalty from oversizing. However, synchronous motors are subject to the same potential rewind damage as induction motors. We estimated that damage to be about 1% of input to induction motors and assume the same for synchronous equipment. Thus, perhaps 0.1 TWh/yr of excess losses could be corrected virtually for free by adopting better repair practices as the motor stock is replaced.

The savings potential of DC motors is considerably larger. Some DC motors are driven by motor-generator (M-G) sets, wherein a 90%-efficient AC motor turns an 85%-efficient generator to produce DC power for an 85%-efficient DC motor. These values are imprecise but representative, and they yield an overall efficiency of only 65%. Replacement of such systems wherever practical with AC motors controlled by ASDs can potentially improve efficiencies by tens of percentage points. And because rewinding costs more for DC motors than for AC units, the economics of replacing instead of repairing are even more attractive for DC systems.

M-G sets not only are inefficient but also create large idling losses in numerous applications. For instance, they are commonly found in traction drives for such uses as elevators, cranes, and hoists. When the drive is idle the DC motor shuts down, but in many cases the M-G set continues to run. In other cases, DC motor efficiency can be markedly improved by installing more sophisticated solid-state controls or using more efficient electronically commutated DC motors.

Losses from oversizing are probably not as big a problem with DC motors because they are more carefully sized due to their higher cost. Rewind damage in DC motors is probably comparable to AC models, where we estimate an overall 1% penalty.

Combining the essentially zero-cost 1% savings from correcting for past rewind damage in both DC and synchronous motors with the sizable savings available from replacing or better controlling DC motors, we assume that 5% of synchronous and DC drivesystem energy use can be saved for less than $.03/kWh. Again, the estimate is imprecise, but the term is so small that it matters little to the overall analysis.

Applying this 5% improvement to the estimated 63 TWh/yr used by synchronous and DC motors yields a savings potential of 3 TWh/yr.

Better Mechanical Drivetrain Equipment and Lubrication

A much overlooked area of inexpensive savings potential—improvements in drivetrain design and equipment lubrication and maintenance—can be significant. For example, as discussed in chapter 4, synchronous and cogged V-belts can often be substituted for conventional V-belts, with potential efficiency gains of 0.5 to 8 percentage points.

Proper selection of speed reducers (or increasers) is another critical area, since different types with the same capacity and design life and suited to the same task can vary in efficiency by up to 25 percentage points. Moreover, the most efficient option is not necessarily the most expensive. For example, a Dodge APG size 4 helical geartrain listing for $579 with 94% efficiency has comparable capacity, design life, and speed ratio to a Dodge Master WM40 worm gear drive costing about four times more and yielding 68% efficiency (Lovins et al. 1989).

Premium lubricants have yielded 3% to 20% energy savings in various devices, from wire-drawing machines (Ibáñez 1978) to automobiles (Gutman and Stotter 1984, Milton and Carter 1982) to gear reducers, compressors, and motors (Kent 1989). A U.S. Navy–sponsored case study of one specialty lubricant in a 350-hp compressor reduced electricity use by 2.6%, at a negative cost of saved energy, since the lubricant lasted four times longer, more than compensating for its higher price(Kent 1989).

Improved maintenance is also critical to reducing costly downtime and keeping efficiencies at optimal levels. An EPRI study stated that "The efficiencies of mechanical equipment in general can be increased typically 10 to 15 percent by proper maintenance" (Ibáñez 1978). A motor diagnostics program involving a vibration tester and a surge tester cut motor failures in one plant by about half (Kochensparger 1987).

In some cases less maintenance of the wrong kind is called for. Southwire Company's energy manager Jim Clarkson has observed a tendency of plant managers to have workers paint motors and other equipment to spruce up a facility when important visitors are expected. Since extra paint makes motors run hotter, thereby reducing their lifetime and efficiency, Clarkson only half in jest issued a policy requiring workers to first strip the old paint before painting any electrical equipment (Clarkson 1990).

Potential savings from improved drivetrain technologies and maintenance practices are difficult to estimate. Lovins and his colleagues, citing numerous case studies, suggest that optimal practice in these areas could save 3–10% of all drivepower input, with a cost of saved energy reduced to nearly zero by improvements in reliability and equipment life. Most of the cited studies of the belt, chain, and gearbox systems showed paybacks from energy savings alone of under two years. The EPRI study by Ibáñez estimates 10–15% savings from better maintenance alone.

It is very difficult, however, to extrapolate to full savings potential from case studies. First of all, the relative shares of various transmission types are poorly documented, as is typical lubrication and maintenance practice. Secondly, case studies, particularly from vendors, tend to be skewed toward the most attractive applications. Nevertheless, anecdotal evidence suggests that considerable room for improvement exists in this area. More objective and well-documented case studies are needed.

Given the lack of data, we attempt no independent estimate here. The Lovins et al. and EPRI/Ibáñez estimates are plausible but difficult to verify. To be conservative, we adopt a lower range of 3–7% as an almost certain opportunity for inexpensive savings.

Indirect Savings

Reduced Distribution Losses

An estimated 6% of energy input is lost in the distribution wires between the customer's meter and the motor terminals (see discussion of wire sizing in chapter 3). As net input at the meter decreases, distribution losses will fall in relation to the ratio of the square of the current.

HVAC Bonus

More efficient drivesystems produce less waste heat and thus reduce the need for cooling in buildings and factories. They also increase the heating load. Since space cooling uses three times the energy of space heating in the commercial sector and twice as much in the industrial sector, reducing internal waste heat released in those sectors results in a net HVAC bonus by allowing the cooling system to save more than the heating system must make up. Only in the residential sector, where (nationally) heating loads are nearly twice cooling loads, will reduced heat from motor systems add to space conditioning energy.

Some researchers have documented HVAC bonuses in the 40% range for commercial spaces in hot climates (California Energy Commission 1984, Linn 1987, Treadle 1987). A typical HVAC bonus in New England is 27% (Jackson 1987), the value we use as representative. The HVAC effects in other energy-use sectors are smaller and not as well documented. One estimate puts the average effect at about 4% in industry and at about minus 5% in residences (Lovins et al. 1988).

Applying these values to the sectoral shares of motor input energy shown in Table 6-11 (21% residential, 19% commercial, and 60% industrial plus other) yields an input-weighted HVAC bonus for waste drivepower heat of around 6.5%. To avoid double counting, this free savings in net space conditioning energy must be discounted by the proportion of savings achieved by prior measures, before being applied to the cumulative savings.

Adding Up the Savings

We now compile in Table 7-4 (next page) all of the savings estimates discussed above. Note that not all the savings are additive;

Table 7-4. Summary of savings potential.

1988 Input to U.S. Drivepower   1,574 TWh/yr
Allocated Thus:

Induction	1,511	
Synchronous	55	
DC	8	

	Savings (TWh/yr)	Remaining Input (TWh/yr)	Cost of Savings (cents/kWh)
Induction motors		1,511	
Replacement with high E motors	59		1.5–2.6
Elimination of past rewind damage	15		0
Correction of previous oversizing	8	1,429	0
Electrical tune-ups (1–5% savings)	14–72	1,357–1,415	?
Controls (5.5–21% savings)	75–298	1,059–1,340	1–5
DC and synchronous motors		63	
5% savings from all measures	3	60	<3
Remaining input to all motors		1,119–1,400	
Drivetrain, lubrication, and maintenance savings:			
3–7% from all measures on all motors	34–98	1,021–1,366	?
Indirect savings			
Reduced distribution loss[a]	24–55	966–1,342	0
Reduced HVAC effect[b]	13–24	942–1,329	0
Total savings	245–632		
Savings as % of original input	16–40%		

[a] 6% × initial input = initial distribution loss. Initial distribution loss ×
$$\left[1 - \frac{(\text{remaining input after efficiency measures})^2}{(\text{initial input})^2} \right] = \text{reduction in losses.}$$

[b] $6.5\% \times \dfrac{\text{remaining input}}{\text{initial input}} \times \text{cumulative savings} = \text{reduction in HVAC effect.}$

savings percentages are applied only to the input remaining to motors after savings from prior measures have been subtracted.

We thus estimate that 16–40% of all U.S. motor input energy (9–23% of all U.S. electricity) can be saved by full application of the measures described above, most of which cost less than three cents per saved kilowatt-hour. Most of this savings potential remains untapped.

Some savings require replacement of existing equipment, a process that will take years. Barriers from lack of information to lack of incentives must be overcome. Fortunately, some innovative programs have already begun with the aim of capturing drivepower savings. The lessons learned in these early efforts and ideas for further implementation and research strategies are discussed in the chapters that follow.

Chapter 8

The Motor Market

To design programs and policies that successfully promote motor system efficiency requires more than familiarity with the relevant technologies. It calls also for an understanding of the key players in the motor market, their motivations, and the challenges they face. Successful programs must appeal to the key decision-makers and fit into their current ways of doing business.

Given the many players involved, various approaches are needed, each one suited to particular market segments. For example, the decision-makers, decision processes, and economic criteria involved in purchasing motors for new applications differ from those involved in purchasing motors for replacement applications. Programs or policies can be better targeted if they reflect an understanding of the differences between these two markets.

How to tailor programs and policies to such differences is the subject of chapter 9. Here, we lay the foundation for that discussion by identifying the players, their interactions, and the factors that influence their decisions. Among the key players are

- End-users
- Motor manufacturers
- Motor distributors and repair and rewind shops
- Original equipment manufacturers
- Consulting engineers and design-build contractors
- Electronic motor control equipment manufacturers, distributors, and representatives
- Mechanical equipment manufacturers, distributors, and representatives
- Electric utilities, universities, government agencies, and trade associations.

183

The interactions among these players are illustrated schematically in Figure 8-1.

The material discussed in this chapter is based on a number of published reports (the most useful ones are listed in the references section at the end of this book), as well as on a series of discussions with knowledgeable people in the motor field, including manufacturers, distributors, end-users, and representatives of utilities and trade associations.

End-Users and Customers

Customer needs, attitudes, decision criteria, and decision-making processes vary widely. This section summarizes many of the important issues facing drivepower equipment users.

Decision-Makers

Maintenance staff often make purchase decisions for replacement equipment. Existing equipment typically is replaced or repaired without engineering analysis, and often is replaced with the same brand and model number. Only in the case of large motors (over approximately 250 hp) with high operating costs does engineering

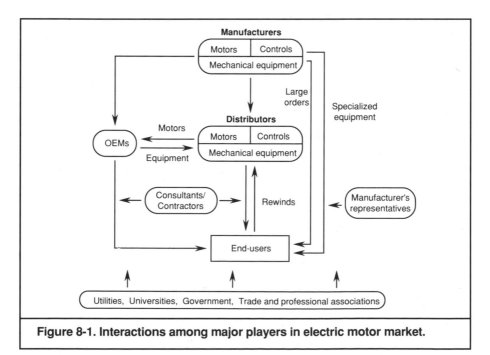

Figure 8-1. Interactions among major players in electric motor market.

or economic analysis usually precede decisions concerning replacement equipment.

By contrast, purchase decisions for new equipment and replacements for ASDs usually involve engineering staff or consultants. The level of engineering analysis varies widely but generally increases with the project's size. Sometimes engineers oversee bidding and installation. In other cases, engineers write specifications, and then the purchasing department takes over. Large companies are more likely to do a thorough analysis than small companies. For ASDs and major design decisions, the most prevalent criteria are reliability, features and performance, equipment and installation costs, and savings in operation and maintenance.

End-users typically obtain information on motors and related equipment from trade publications, manufacturer's representatives, distributors, contractors, and professional organizations.

Aversion to Downtime

In some manufacturing facilities, if a production line is shut down for several hours, the value of the lost production capacity can be greater than the value of many years of energy savings. End-users are thus very concerned about downtime and reliability, particularly of replacement equipment. For this reason, there is a need to document case studies on equipment reliability and to provide this information to end-users.

Concerns about downtime also cause customers to demand quick delivery of equipment and to avoid purchasing high-efficiency equipment if it will take longer to obtain. Downtime costs also make end-users reluctant to replace operating equipment, even if it is inefficient. Thus, the only time many end-users will consider high-efficiency equipment is during the brief period between the failure of old equipment and the hasty purchase of new components.

To minimize downtime, many companies stock spare motors and such mechanical drivetrain parts as gearboxes. To reduce the number of items that must be stocked, some companies standardize on a limited number of equipment sizes, such as 50- and 100-hp motors. For applications that require an intermediate size of, say, 75 hp, the next largest stocked motor is used. This practice results in systematic oversizing.

Customers commonly believe that motors under approximately 200 hp, and other drivepower components such as gearboxes, bearings, belts, chains, and lubricants are commodity items, meaning that models produced by different manufacturers are interchangeable. Given this outlook, purchase decisions are made based primarily on reliability, price, and availability, not on efficiency (A.D. Little 1980).

Replace or Rewind?

Because rewinding is generally more expensive than replacing a small motor and less expensive than replacing a large one, most end-users replace small motors and rewind large ones in the event of burnout. Different companies use different guidelines for deciding whether to replace or rewind. In most cases, motors of 5–10 hp or less are replaced (Seton, Johnson & Odell 1987b), while motors of 40 hp or more typically are rewound (Marbek 1987). Decisions on 15- to 30-hp motors vary extensively between firms. A few firms will routinely replace motors as large as 100 hp instead of rewinding them (Lovins et al. 1989). Replace-or-rewind decisions are generally made at the plant level, although a few large corporations have established guidelines for all their plants.

There are a few exceptions to the "replace small motors, rewind large motors" rule:

• Specialty motors and old, odd-dimension motors. Since new

Southwire Company's Rewind Policy

Southwire Company, a major wire and cable firm with annual sales in excess of $1 billion, has a general policy to replace all standard-efficiency motors of 125 hp or less instead of rewinding them. Southwire has calculated that replacement is cost-effective in nearly all applications. For larger motors the firm compares the costs and savings of rewound versus new motors and purchases a new motor if the net present value of savings over five years exceeds the financing costs. Furthermore, the company has a general policy to replace motors when their repair cost exceeds 40% of the purchase price of a new motor. Southwire has experienced reliability problems with rewound motors and is also concerned about efficiency losses from the rewind process. Hence, the firm prefers new motors unless the incremental cost of a new motor is excessive. Furthermore, the policy of buying large numbers of new high-efficiency motors has enabled the company to negotiate a price for high-efficiency motors that is only 5% above the cost of standard-efficiency motors. To achieve this price, Southwire generally buys all motors from one supplier, and expects that supplier to have high-efficiency motors in stock at all times (Clarkson 1990).

motors of these types are hard to find, these motors are generally rewound when they burn out.

- Situations where the old motor is damaged and cannot be rewound. In these cases the motor is replaced.

- Situations where considerable time can be saved by either rewinding or replacing an old motor. When order time for a new motor is long, the motor is usually rewound. When rewind shops are busy and cannot provide quick service, and replacement motors are readily available, a new motor may be purchased.

End-users select rewind shops primarily on the basis of quality and speed of service. To encourage competition, most end-users use more than one rewind shop (Seton, Johnson & Odell 1987b).

Adjustable Speed Drives

End-users often treat ASDs differently from other types of equipment because ASDs are a relatively new and unfamiliar technology. Problems in some early ASD systems have made some end-users distrustful of even the new, improved versions.

Due to their high price and high savings, ASDs usually receive engineering attention from either in-house or consulting engineers. However, it can be difficult and expensive to precisely determine how much energy and money a particular ASD installation will save. Many simplifying assumptions are often made, adding to the range of uncertainty for the end-user.

ASDs are not treated as a commodity product. There are important differences among models that warrant careful comparisons. Furthermore, many plants use only a particular brand of motor switchgear. In these cases, engineers will generally try to specify ASDs made by that manufacturer. Where the brand is not critical, particular features and capabilities can be specified and more than one product allowed to compete. This is particularly true in government facilities, where competitive bidding is generally required.

When an ASD unit fails, it can typically be repaired by replacing a circuit board rather than the whole unit. However, the rapid evolution of ASDs means that new features are regularly entering the market. Customers desiring new features sometimes will replace equipment for that reason. In the rare cases when replacement parts for early units are no longer manufactured, customers must buy new equipment when an existing component fails.

Maintenance Practices

Motor maintenance practices are generally limited to what is needed to keep equipment running, rather than to optimizing performance and saving energy. Most industrial plants and large commercial firms have a full-time maintenance staff who regularly lubricate (and often overlubricate) motors, listen for bearing noise (a sign of wear or misalignment), and check and tighten belts as needed. Few firms do any more sophisticated monitoring or maintenance work on motor systems.

Most large industrial and commercial firms have some type of tracking system for motor system maintenance involving log books, file cards on each motor (sometimes kept in a central file and sometimes attached to individual motors), or computer-based records.

Small firms without maintenance staff primarily rely on outside equipment contractors to maintain HVAC systems, conveyer belts, and other drivepower components. Maintenance frequency varies widely. Some firms schedule regular service calls. Others wait for problems before calling someone in. In many of these cases, the call is placed too late, as equipment has been damaged beyond repair and must be replaced.

Overall, motor maintenance practices in the U.S. are less than optimal. In Japan, on the other hand, the responsibility for maintaining individual motors is often assigned to specific mechanics, who receive extensive training. When a motor fails on the production line, the problem can be attributed to its mechanic. This provides a strong impetus to do preventative maintenance (Johnston 1990). Approaches for encouraging improved maintenance are discussed in chapters 9 and 10.

Several other factors, in addition to those related specifically to motor systems, influence most efficiency-related investment. Some of the more important ones are discussed below.

Limited Access to Capital

The typical end-user is more restrictive with capital than with operating funds (A.D. Little 1980, Comnes and Barnes 1987). Capital expenses are generally closely scrutinized and require approvals at multiple levels in a company. To minimize capital outlay, companies tend to choose the least expensive equipment that will do the job satisfactorily.

Operating funds, on the other hand, are relatively easy to obtain, since they are required for production. Operating budgets are typically based on expenses in previous years and are only seriously examined when out of line with expectations. Moreover, unlike capital costs, operating costs are paid with pretax dollars.

Payback Gap

It is a curious fact that most firms look for a simple-payback period of two to three years or less on energy projects and other operations and maintenance investments (Marbek 1987), even though longer paybacks are often considered when investing in new product lines. This difference, known as the payback gap, makes it difficult to implement all but rapid-payback energy-saving measures, although measures with longer paybacks will sometimes be considered as part of a major facility upgrade designed to improve the long-term competitiveness of the firm (Comnes and Barnes 1987). The payback gap is most pronounced when viewed from the societal perspective—individual firms pass up energy-saving investments with paybacks of three to four years, while utilities invest in power plants with economic returns equivalent to 10- to 20-year paybacks.

Low Priority Assigned to Energy Matters

For the average industrial firm, energy costs represent only a small percentage of total costs; labor and material costs are usually far greater. For example, A.D. Little (1980) estimated that motor system energy costs represent approximately 1% of sales for the average industrial firm. Energy costs usually represent a small proportion of average end-users' operating costs; consequently, motor and other energy-related operating costs are rarely examined in reviews of operating expenses. Breaking this logjam will require creative approaches, some of which are discussed in chapters 9 and 10.

Misplaced Program Emphasis

Since they have full-time maintenance staff or energy managers, large firms are more likely to be interested in energy efficiency. Even in firms with energy managers, however, motor systems historically have not received a lot of attention because of (often incorrect) perceptions that motor system improvements have high capital expense, low rates of return, and low percentage savings. Energy managers tend to focus on low capital cost measures with high savings (A. D. Little 1980). While this approach is reasonable during the start-up stages of an energy management effort, many firms have not moved beyond high-savings, low-cost measures. Moreover, many drivepower saving measures are relatively inexpensive.

Lack of Internal Incentives

In many companies, energy bills are paid for the company as a whole, not allocated to individual departments. This practice gives maintenance and engineering staff little incentive to pursue energy-saving investments, because the savings in energy bills show up in a corporate-level account, where they provide little or no benefit to maintenance and engineering decision-makers. As is discussed in chapter 10, mechanisms to improve internal incentives have been put into place in some facilities.

We have covered some of the main factors influencing the decisions of motor system users. Now we turn to another of the key players, the motor manufacturer.

Motor Manufacturers

In 1977, eight major manufacturers accounted for over 75% of the North American motor market (A.D. Little 1980). Since then, a few new manufacturers, including Toshiba, have entered the market and a few old ones have left (MagneTek bought both Century and Louis Allis, and Westinghouse contracted all but its large motor business to Reliance), but the total number of major manufacturers has changed little (see appendix B for a list of the major manufacturers and their product lines).

Some manufacturers such as General Electric make motors ranging from fractional horsepower up to custom motors of thousands of horsepower. Others are more specialized. Emerson and Dayton, for example, primarily make single-phase motors of less than a few horsepower, U.S. Motors makes motors ranging from 3/4 hp up to 350 hp, and Westinghouse's U.S. line starts at 250 hp. Types of motors sold by the different major manufacturers are listed in appendix B.

In general, motors below 250–350 hp are regularly produced and stocked by manufacturers, while larger motors are custom built to purchaser specifications, including desired efficiency levels. Within the 1- to 200-hp range, with one exception, all major manufacturers produce two lines of motors—a standard-efficiency line and a high-efficiency line. Some manufacturers produce high-efficiency motors as small as 1/2 hp, and as large as 350 hp.

High-efficiency three-phase motors are generally available in T-frames for 1,200-, 1,800-, and 3,600-rpm nominal speeds for both TEFC and ODP enclosure types. High-efficiency motors are produced in EXP (explosion-proof) enclosure types, C-face frames (used for pumps), and 900-rpm models; however, because consumer demand for these products is low, only a few manufacturers produce them. It is technically straightforward to produce high-efficiency versions of most types of motors; in fact, for a price premium, increases in efficiency can be ordered for nearly any motor. Thus, availability of high-efficiency motors primarily depends on manufacturer's perceptions of likely customer demand.

Efficiencies for both standard- and high-efficiency lines vary among manufacturers. High-efficiency models produced by one manufacturer are occasionally no more efficient than the standard-efficiency motors made by another manufacturer (see Table 2-3). Moreover, a given manufacturer might have the highest efficiency motor in one size class, but not in another. Thus, a user interested in purchasing the highest efficiency motors will need to purchase them from more than one manufacturer.

With standard-efficiency motors, manufacturers compete primarily on price (Seton, Johnson & Odell 1987a). Because price is so critical with standard motors, efficiencies vary by up to five points between manufacturers. In the high-efficiency lines, manufacturers compete on efficiency ratings, as well as price, so the spread in efficiency between different manufacturers is smaller, though still significant. This variation among manufacturers suggests that buyers should comparison shop even in the high-efficiency lines.

Manufacturers sell motors in two major ways: through local and regional distributors, and directly to large national companies and original equipment manufacturers (OEMs). OEMs are firms, such as air conditioning manufacturers, which incorporate motors into the equipment they make.

Most distributors are independent. That is, they have no formal link with the manufacturer(s) whose products they sell. A few manufacturers, including Reliance, have their own networks of regional distributors. Sales from distributors to customers are discussed later in this chapter.

Large national companies and OEMs have sufficient buying power to demand the lowest possible price from manufacturers. To achieve these low prices, manufacturers sell directly to large customers, avoiding any distributor markup. Depending on the size of the order and the customer, these prices can be 50–70% off of suggested list prices (Seton, Johnson & Odell 1987a).

Motor Distributors and Repair-Rewind Shops

Most motor distributors are small local operations with fewer than ten employees. Large regional distributors, though few, account for a disproportionate share of motor sales (specific figures are unavailable). Many distributors also repair and rewind motors. Conversely, most repair-rewind shops sell at least one line of new motor.

Motor distributors often sell related electrical equipment, like relays, circuit breakers, generators, and motor controls. A study of the Canadian motor market (Marbek 1987) found that, on average, regional distributors received 35% of their income from new motors, 25% from motor rewinds and repairs, and 40% from sales of other equipment. This study also found that approximately half of the distributors stock only one brand of motor.

Distributors generally stock common motor types and sizes, such as standard-efficiency ODP and TEFC motors from 1 hp to 100 hp or 200 hp. Less common sizes and such specialized lines as explosion-proof and hollow-shaft motors are usually handled as special orders. High-efficiency models are stocked by some distributors and are left to special order by others (data on how many distributors fall into each category are not available).

Because distributors generally stock only those motors for which there is significant customer demand, and because many customers will only buy a high-efficiency motor if it is in stock, market penetration of high-efficiency lines remains relatively low. Solutions to this problem are discussed in the next chapter.

Motor distributors primarily sell to small and medium customers including some small OEMs. The previously mentioned Canadian study (Marbek 1987) found that 64% of distributor sales are to end-users, 21% are to contractors and engineers, and 14% are to OEMs.

Manufacturers generally publish suggested list prices on motors, but only small orders to very small customers are actually sold at list price. Distributors generally sell motors at a discount that varies with the order size, with how valued the customer is, and with how many other distributors are competing for a particular order. Discounts are typically 30–50% (Seton, Johnson & Odell 1987a) although high-volume dealers can sometimes provide discounts as high as 60% (Stout 1990).

Dealers may compete intensely on orders that go out to bid. In these situations, unless high-efficiency motors are specified in the bid documents, pressure to come in with the low bid often means that the successful bid will be for standard-efficiency motors. Utility rebate programs (discussed in the next chapter) can help overcome this problem.

For the repair of existing motors, distributors compete primarily on speed and price. When a motor breaks down, a whole process line may be shut down, so customers generally want repairs done "the day before yesterday." As chapter 2 discussed, unless customers test each repaired motor, they cannot easily assess the quality of a repair job, so distributors do not usually compete on quality.

Due to the variety of products they carry, many distributors have neither the knowledge nor the time to provide detailed information on whether a high-efficiency motor is appropriate for a particular application. The Marbek study (1987) found that 25% of Canadian distributors surveyed had misconceptions about the technical reliability and applicability of high-efficiency motors.

Sales of high-efficiency products are also hampered by the way orders are handled. First, many orders are processed over the telephone, which provides little opportunity for a distributor to explain high-efficiency products. Second, orders are often placed by maintenance or purchasing staff who are concerned primarily about availability and price and have neither the technical background nor the interest to consider high-efficiency products. Third, most order-desk personnel at the distributors have only as much technical knowledge as is published in manufacturer's brochures and catalogs. Clearly, improved dealer education and support are needed so that dealers can better work with customers to improve motor system efficiency (specific ideas are discussed in the next chapter).

Original Equipment Manufacturers

OEMs incorporate motors into many types of equipment including pumps, compressors, fans and blowers, air handling and HVAC units, industrial machine tools, and conveyer systems.

A study of the U.S. motor market completed in 1980 estimates that OEMs account for about 80% of all motor purchases, including approximately 85% of motors of 1 hp or less, 80% of 1.5- to 5-hp motors, 75% of 7.5- to 20-hp motors, 65% of 25- to 50-hp motors, 45% of 60- to 125-hp motors, and 15% of motors over 125 hp (A.D. Little 1980). Viewed by sector, OEMs buy nearly all the motors used in residential equipment, most of the motors used in commercial buildings, and a sizable portion of industrial motors. They thus play a key role in the market penetration of high-efficiency drive-power equipment.

The reality is, however, that OEMs operate in a highly competitive market that encourages them to keep costs down. Because few customers purchasing OEM equipment are aware of or concerned about efficiency, OEMs generally use standard-efficiency components. Marbek's 1987 study of the Canadian motor market found that high-efficiency models constituted approximately 1.4%

of OEM motor purchases, compared with 3% of distributor sales and approximately 30% of sales directly from manufacturers to end-users. While sales of high-efficiency motors are generally higher

---

Compaq Computers Pushes OEMs to Improve Equipment Efficiency

The Compaq Computer Corp. is a large manufacturer of personal computers. It has over 20 facilities throughout the world and spends millions of dollars for energy each year. A large portion of energy use in Compaq facilities is for HVAC applications in offices and manufacturing clean-rooms. Compaq often purchases the most efficient HVAC equipment on the market because in the company's highly competitive market, minimizing long-term operating costs is an essential part of efforts to maintain or increase market share and profits.

However, for major equipment, Compaq is not always content to purchase even the most efficient equipment on the market. Its analyses showed that chillers and cooling towers with even higher efficiency could be produced, through changes such as improved motors and heat exchangers. Armed with this information, Compaq approached major chiller and cooling tower manufacturers about producing such units. Manufacturers were originally skeptical, but when Compaq went to Japan to purchase a cooling tower with a higher efficiency than any produced in the United States, domestic manufacturers began to produce the higher-efficiency units too. In fact, due to the competition, the price of this equipment to Compaq has declined. Efforts to obtain higher-efficiency chillers are still going on.

Compaq's former Energy Manager of Facility Resource Development, Ron Perkins, offers this advice for companies seeking to improve the efficiency of OEM equipment:

• Conduct some research (using technical bulletins) to estimate what efficiency improvements are possible.

• In approaching manufacturers, make the case that there will be a market for the new high-efficiency equipment, either due to the purchaser's own market size, or due to the fact that the purchaser is a leader in its industry or trade association.

• Publicize success stories and offer testimonials to responsive manufacturers. These efforts will make manufacturers more responsive in the future (Perkins 1989).

in the U.S. and have increased since 1987, the important point here is that OEMs are less likely to purchase high-efficiency motors than are distributors or end-users.

OEMs generally have engineering staffs who evaluate motors and other components and provide a list of acceptable products to the purchasing department. Besides basic equipment characteristics such as size and speed, the most important technical factor in motor evaluations is reliability; efficiency is usually not considered. For all but the largest motors, engineering staff are usually not involved in purchasing decisions after a list of acceptable products is developed. The purchasing department selects motors from the approved list based primarily on price and delivery terms (Marbek 1987, A.D. Little 1980).

In many cases, particularly with small equipment orders, OEMs are unable or unwilling to use high-efficiency motors, even when requested by the customer. The most common reasons are small differences in motor technical characteristics and difficulties in obtaining high-efficiency motors through supply channels geared to supplying large volumes of standard-efficiency motors.

These restraints do not mean that customers have no chance to influence OEM practice. Marbek also found that approximately half of the Canadian OEMs surveyed will use high-efficiency motors, in a small proportion of their products, either when specified by customers, or with some types of large, specialized industrial equipment.

While OEMs have been slow to use high-efficiency motors, some have adopted improved gears and belts. As was discussed in chapter 3, high-efficiency gears and belts are sometimes stronger than conventional equipment. Added strength can extend service life and reduce maintenance requirements—important marketing advantages in some product lines. Stronger materials also allow components to be downsized, thus reducing material costs at the same time that efficiency increases.

Consulting Engineers and Design-Build Contractors

Consulting engineers prepare designs and specifications and help oversee the bid and construction process, but they leave the construction work to in-house staff or outside contractors. Design-build contractors, on the other hand, handle both design and construction, thereby providing turnkey facilities. Design-build contractors are typically very large firms.

Consultants are generally hired by end-users to assist with large projects such as new plants, new production lines, and major new equipment such as large ASDs. These projects are infrequent and large enough to make hiring temporary consultants more cost-effective than hiring permanent staff.

Successful consulting practices are based on satisfied customers. For this reason, consultants tend to use similar decision criteria as end-users: they stress reliability and are notorious for oversizing motors so as to provide a wide safety margin (Van Son 1989). Consultants sometimes recommend high-efficiency equipment, but if the client resists the idea, the consultant will usually drop the suggestion. Even when high-efficiency equipment is specified, it is commonly one of the first items to be cut from a project if bids come in higher than expected (a common occurrence).

When equipment unfamiliar to installation contractors is specified—ASDs, for example—installation prices often include a "risk factor" to cover the cost of unanticipated problems. In essence, the customer is paying for the contractor to learn how to install and troubleshoot a new type of system.

Design-build contractors increasingly are being paid through fixed-cost contracts (Ontario Hydro 1988). A fixed-cost contract provides protection against cost overruns and thereby allows the end-user to correctly budget for a project; it also places great pressure on design-build contractors to keep initial costs down. Under these conditions, high-efficiency motors and other high-efficiency measures with increased costs are rarely specified.

OEM representatives (discussed in the next section) often provide design services similar to those provided by consultants. OEM representatives are not paid directly for this work: their fees are incorporated into the price of OEM equipment. The smaller the projects, the more likely OEM reps are to be the primary designers.

Control Equipment Manufacturers, Distributors, and Representatives

Manufacturers of ASDs and other electronic controls include some large firms (both motor manufacturers and independent control manufacturers) as well as many small specialty firms. Most manufacturers specialize in particular sizes or types of electronic controls.

Adjustable-speed drives and other electronic controls are generally sold through distributors or sales representatives, most of which handle more than one type of equipment. In addition to ASDs, for instance, they may also sell power factor correction or uninterruptible power supply equipment. Some handle more than one product line for a particular type of equipment. Distributors stock products while representatives generally place orders as needed with the manufacturers. Representatives and distributors will often do design work for end-users in an effort to sell equipment they represent.

Electronic controls are generally purchased in small quantities, so steep discounts are the exception rather than the rule. Because installation costs can be substantial, equipment and installation costs are often considered together in evaluating system economics.

Mechanical Equipment Representatives and Distributors

Manufacturers of gears, belts, chains, bearings, and lubricants also range widely in size. Large firms involved in mechanical equipment include motor manufacturers (Reliance is a major producer of gears), rubber companies (which manufacture belts), and oil companies (which manufacture lubricants). Many small, specialty firms also are involved in the manufacture of mechanical equipment. A growing number of foreign firms are exporting mechanical equipment, especially belts, to the North American market.

Mechanical equipment manufacturers primarily sell to smaller buyers through mechanical equipment distributors, and sell to OEMs and large end-users directly. Sometimes a motor distributor will also stock mechanical equipment. Smaller firms often use regional representatives to promote their products, with stocking and distribution handled by the central office.

Lubricants are generally sold by large oil companies through local petroleum product distributors. Petroleum product distributors do most of their business in the automotive field—commercial and industrial business represents only a small portion of their annual sales. Small lubricant companies, which manufacture premium-grade lubricants (high-efficiency and long-life), tend to sell through a network of product representatives.

Mechanical equipment is often priced in a similar manner as motors. Suggested list prices are published, but most purchasers receive a substantial discount that varies with size of the order, size of the customer, and other factors. With lubricants, published price schedules include volume discounts. These prices are generally adhered to except in bid or other special situations (Hudson 1989, Kent 1989).

Electric Utilities, Universities, Government Agencies, and Trade and Professional Associations

These groups have become increasingly active in the motor market. Many provide educational materials, seminars, and technical assistance on motor system issues. In addition, trade associations, such as NEMA, and occasionally government agencies and utilities develop standards for testing and labeling of motor system products.

Utilities have become increasingly involved in improving power factor, power quality, and efficiency, and in promoting use of electrotechnologies among customers. Utility efforts are primarily educational, although several utilities are providing rebates for purchase of high-efficiency motors and other energy saving measures.

Further information on the activities of these players in the motor market is provided in chapter 9.

Summary

The motor market is quite complex, in that it involves many different players and many different decision criteria. The market for replacement equipment is very different from the market for new applications. Key attributes of these two markets are summarized in Table 8-1. Planners and managers of programs to promote motor system improvements need to understand these different markets in order to shape programs that will successfully serve them.

The replacement equipment market involves frequent purchases of small quantities of equipment, primarily from local or regional distributors. Decisions are made quickly by maintenance or purchasing staff, primarily on the basis of availability, cost, and reliability. Engineering analyses of such replacement purchases are rare; old equipment, when it fails, is often replaced with identical components.

Programs designed for this market need either to reach the decision-maker at the time the purchase is made, or to promote standard policies that determine in advance which equipment to replace with high-efficiency models. The Southwire case study earlier in this chapter exemplifies how this can be done.

Purchases of equipment for new applications occur infrequently, but when they are made, they generally involve an engineering analysis. These analyses often balance reliability, initial cost, delivery terms, and performance and operating cost considerations. Due to

Table 8-1. Key attributes of the markets for replacement equipment and new applications.

	Replacement Equipment Market	New Application Market
Frequency of end-user purchase decisions:	Continual	Infrequent
Order size:	Generally small	Frequently medium or large
Decision-makers:	Maintenance and production staff	Engineering staff (sometimes at head office)
Time spent making decision:	Hours or days	Weeks or months
Key factors:	Availability, reliability, cost	Cost and reliability for commodity products; performance, price, and operating cost savings for noncommodity products and for design decisions
Engineering analysis:	Generally none, except sometimes for large motors and OEM equipment, or for setting general purchasing guidelines	Usually done by either in-house staff, outside consultants, or OEM reps
Purchased from:	Distributors or, for very large companies, manufacturers	Design-build contractors, OEMs, manufacturers directly, or, for small projects, distributors
External influencers:	Distributors, service shops	Engineering consultants, contractors, OEM reps

pressures to minimize initial costs, future savings are frequently heavily discounted. But, because new application decisions often involve large quantities of equipment, selection and design decisions are made over a period of months, and there is some time to reach the decision-maker.

Programs and policies to reach these different markets are discussed in the next chapter.

Chapter 9

Programs to Promote Motor System Efficiency Improvements

The opportunities for increasing motor system efficiency are very large but so are the barriers to such improvements. Programs designed to overcome these barriers and foster more efficient drivepower systems fall into five principal categories:

- Education
- Technical assistance
- Testing, labeling, and minimum efficiency standards
- Financial incentives
- Research, development, and demonstration projects

This chapter discusses some of the most successful, promising, and well documented of these programs.

Education Programs

Education efforts to date have emphasized high-efficiency motors, ASDs, and power-factor correction, although innovative work has taken place in other areas. Here we review some of the more noteworthy examples. Unfortunately, the effectiveness of these educational efforts has not been evaluated.

Seminars and Courses

The North Carolina Alternative Energy Corporation (NCAEC) and the North Carolina Industrial Extension Service (NCIES) both run a series of seminars for industrial firms on topics such as "Reducing Electrical Costs at Your Facility" and "Reducing Operating Costs of AC Motors with Adjustable-Speed Drives." The seminars, which last a full day each, cover fundamentals, case studies, and practice analyses. Seminars are promoted through professional organization newsletters and direct mail. Similar seminars have been offered by a number of electric utilities including the Bonneville Power Administration and Niagara Mohawk Power Corporation.

The NCAEC seminars originally targeted engineering and maintenance staff, but recent courses have also been aimed at managers, because management must be committed if efficiency improvements are to be implemented (Kellum 1989). The NCIES seminars are targeted at senior maintenance staff, and include a special inducement—when staff attend a seminar, they are eligible for an 80% subsidy on a technical assistance study. For example, if staff attend a seminar on compressors, then NCIES will hire a compressor expert to come out to the plant to work with the seminar attendees on assessing compressor improvement opportunitites in their plant (Johnston 1990).

The Association of Energy Engineers offers two seminars—"Motor Applications: Efficiency Improvement and Testing" and "Application of Variable-Speed Drives"—approximately once a year. The courses are directed at energy managers and emphasize practical applications issues (Oviatt 1989).

The Electrification Council, a nonprofit organization supported by electric utilities and electric equipment manufacturers, offers a course entitled "Motors and Motor Controls." This course is primarily used by utilities to train their own staffs and those of their industrial, commercial, and institutional customers. The course contains 12 two-hour units covering motor fundamentals, types of motors and motor controls, and preventative maintenance. Unfortunately, scant time is devoted to energy efficiency. The Electrification Council distributes the course in the form of an instructor's outline, text, and visual aids. The course sponsor supplies the instructor. The course has been available since 1959 and is updated approximately every five years (Bowles 1989).

Among other utilities, British Columbia (BC) Hydro and New England Electric System (NEES) offer training programs for motor dealers as part of motor rebate programs. The courses include technical information on efficiency improvement measures, program procedures, and techniques for marketing and providing technical assistance. The success of these training programs hinges on the rebates: the lure of financial rewards and increased sales is the reason for dealer interest and attendance.

Most current courses and seminars are offered sporadically, serve only a limited audience, and generally address limited topics. New efforts are needed that reach more people and cover additional topics, including motor maintenance practices and system optimization. Interest in courses and seminars is likely to increase significantly when the target audience has an incentive—like rebates—to attend.

Publications

Many organizations—particularly manufacturers, trade associations, and electric utilities—have issued publications relating to motor system efficiency. Perhaps the most comprehensive publications effort yet is that of BC Hydro. This utility has three types of publications on motor systems. Small brochures on topics such as adjustable speed drives and efficient compressed air systems include examples, case studies, and phone numbers to call for further information. "Guides to Energy Management," a series of technical briefs, are designed to be inserted in a notebook. Subjects include selection criteria and payback calculations for ASDs, ASDs for fans and pumps, and case studies of half a dozen specific ASD applications. Booklets, including "High-Efficiency Motors" and "Adjustable-Speed Drives," offer greater detail: the former explains why to change to high-efficiency motors, considerations when buying high-efficiency motors, and how to operate high-efficiency motors; the latter discusses drive types, features, advantages, selection considerations, economics, and harmonic distortion.

These publications are distributed primarily through personal contacts between the utility's industrial marketing engineers and large industrial customers. Customer reception has been good because the publications emphasize practical, applications-oriented information and examples from case studies. Other utilities are considering republishing some of BC Hydro's publications (Henriques 1989).

Motor distributors and utilities generally report that the most effective motor brochures combine catchy graphics with simple graphs or tables that help the reader determine the savings that high-efficiency motors can achieve in a particular application (see Figure 9-1, next page).

EPRI publishes an adjustable-speed drive directory, which is updated approximately every three years. The directory describes what ASDs are and what they can do for an end-user, and includes data useful for preparing ASD specifications. It also lists all ASD manufacturers according to the types and sizes of ASDs they produce. The directory is distributed to electric utilities and end-users. Other educational booklets on motor systems are put out by the National Electrical Manufacturer's Association (NEMA) and the National Electrical Contractor's Association (NECA).

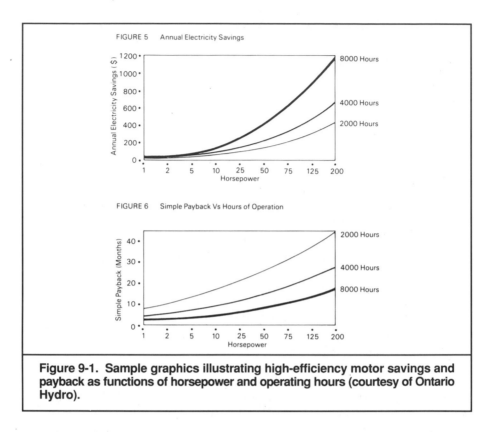

FIGURE 5   Annual Electricity Savings

FIGURE 6   Simple Payback Vs Hours of Operation

Figure 9-1. Sample graphics illustrating high-efficiency motor savings and payback as functions of horsepower and operating hours (courtesy of Ontario Hydro).

Periodicals covering motor efficiency issues and books on drivepower efficiency are listed in the annotated bibliography at the end of this book.

Databases

With personal computers widespread, computerized databases represent another type of publication. BC Hydro has prepared a database listing all three-phase induction motors sold in the province for sizes from 1/4 to 500 hp, speeds from 900 to 3,600 rpm, and enclosure types including ODP, TEFC, and EXP. The database is available as a diskette or printed copy to motor dealers and large customers (Henriques 1989). A sample printout from this database is shown in Figure 9-2. A similar database for motors sold in the United States has been prepared for the Bonneville Power Administration by the Washington State Energy Office (Litman 1990). Both databases include efficiency and power factor at 1/2, 3/4, and full load. The Washington database also includes list price.

HORSEPOWER: 50 SPEED: 1800 RPM ENCLOSURE: TEFC

| EFFICIENCY | | | MANUFACTURER | MODEL | FRAME SIZE | FULL LOAD POWER LOSS (Watts) | — POWER FACTOR — | | |
Full Load	3/4 Load	1/2 Load					Full Load	3/4 Load	1/2 Load
95.2	N/A	N/A	BALDOR	SUPER-E	326T	1,881	86.0	N/A	N/A
94.5	94.5	94.5	MARATHON ELECTRIC	HIGH EFFICIENCY	320T	2,171	85.0	83.0	76.0
94.5	94.5	94.5	US MOTOR	PREMIUM EFFICIENCY	326T	2,171	87.0	85.5	80.5
94.1	93.2	93.0	TECO	HIGH EFFICIENCY	326T	2,339	N/A	N/A	N/A
94.0	93.0	91.0	LEROY SOMER	ECO+	326T	2,339	86.0	85.0	83.0
94.0	94.0	93.0	US MOTOR	PREM EFF – U FRAME	365U	2,381	87.0	86.0	82.0
93.6	N/A	N/A	RELIANCE ELECTRIC	ENERGY EFFICIENT	326T	2,550	85.6	N/A	N/A
93.2	93.5	92.5	TOSHIBA	STD – U FRAME	326T	3,024	97.5	86.5	83.0
93.2	93.2	93.2	TOSHIBA	EQP-575V	326T	2,721	85.4	83.8	77.7
93.0	N/A	N/A	BALDOR	STD	326T	2,808	86.0	N/A	N/A
93.0	N/A	N/A	LEESON ELECTRIC MOTORS	STD	326T	2,808	87.2	N/A	N/A
92.8	93.0	91.7	CANADIAN GENERAL ELECTRIC	HIGH EFFICIENCY	326T	2,894	84.5	81.0	73.0
92.7	91.7	89.8	WEG	HIGH	326T	2,937	90.0	88.0	82.0
92.6	N/A	N/A	SIEMENS	N/A	N/A	2,981	N/A	N/A	N/A
92.5	92.6	92.1	WESTINGHOUSE	LIFE-LINE PLUS	326T	3,156	83.8	81.3	73.7
			POWER SMART RECOMMENDED EFF.						
91.7	92.4	92.4	TECO	STD	326T	3,376	91.0	90.2	86.5
91.8	92.2	91.8	DELCO FANUC	U FRAME	N/A	3,332	91.2	89.6	87.3
92.0	N/A	N/A	BALDOR	STD-U FRAME	365U	3,243	86.0	N/A	N/A
91.7	91.7	90.2	CANADIAN GENERAL ELECTRIC	STD	326T	3,376	84.0	81.0	71.5
91.5	91.4	89.4	WEG	STD	326T	3,465	88.0	86.0	80.0
91.3	91.5	91.0	US MOTOR	STD	326T	3,689	88.5	86.5	81.5
			BASE EFF. FOR COMPUTING REBATE						
91.1	91.3	90.5	ELEKTRIM	STD	326T	3,689	89.0	89.0	85.2
91.1	N/A	N/A	WESTINGHOUSE	STD	N/A	3,644	83.6	N/A	N/A
90.3	91.1	N/A	DELCO FANUC	STD	N/A	4,007	83.6	79.7	70.8
91.0	90.0	90.0	MARATHON ELECTRIC	STD-460/575V	326T/S	3,689	82.5	78.5	70.0
91.0	N/A	N/A	LINCOLN	STD	326T	3,689	N/A	N/A	N/A
91.0	N/A	N/A	RO-EL	STD	326T	3,689	91.0	81.0	72.0
90.5	90.5	90.0	GEC CANADA	STD	326T	3,915	84.0	84.0	78.0
90.5	90.0	89.0	HAWKER SIDDLEY/BROOK CROMPTON	STD	326T	3,915	86.0	85.7	80.0
90.2	90.0	90.0	TOSHIBA	STD	326T	4,053	87.2	85.0	77.0
90.0	90.0	88.0	LEROY SOMER	STD	326T	4,144	88.0	N/A	N/A
88.5	N/A	N/A	RELIANCE ELECTRIC	STD	326T	4,847	83.4	N/A	N/A

Figure 9-2. Sample page from BC Hydro's electric motor database (BC Hydro 1989b).

Technical Assistance Programs

Technical assistance programs primarily fall into two categories: calculational aids, and on-site assessments and assistance. As with education programs, there has been little or no follow-up evaluation of these programs' effectiveness.

Calculational Aids

One reason motor systems are not properly optimized is that decision-makers do not have the time or skills to do the necessary calculations. Calculational aids address this problem with the use of reference tables, slide rules, and, most frequently, computer programs. These aides generally estimate costs and savings for a given application, based on a limited amount of information gathered by the user, e.g., motor size, speed, operating hours, and efficiency.

Ontario Hydro distributes to motor distributors and customers a free spreadsheet program that calculates the energy savings, demand savings, return on investment, and simple payback of high-efficiency versus standard-efficiency motors. User inputs are motor horsepower, operating hours, efficiency, and a peak coincidence factor. A sample output from the program is shown in Figure 9-3 (Ontario Hydro 1989). A similar package is distributed by BC Hydro (Henriques 1989), and one is under development by the NCAEC (Flora 1990).

Computer programs are also available to assess the costs and savings of ASDs. For example, EPRI distributes two programs to assess the economic feasibility of retrofitting fan and pump motors with ASDs. One program, ASCON I, assesses individual pumps and fans from 7.5 hp to 2,000 hp over the range of temperatures and pressures normally encountered in industry. The program calculates the energy consumption for a pump or fan running at constant speed (using a throttling device such as a control valve or vane to control flow) and at variable speed (with the addition of an electronic ASD), over the specified range of operating conditions. The program determines the equipment, engineering, and installation costs of the ASD system and computes a simple payback period for the investment. Present worth and internal rate of return can also be calculated. ASCON II is designed to assess large fan and pump applications at power plants, incorporating detailed operating information (EPRI 1988b). BC Hydro has developed a somewhat simpler ASD program, which it distributes to dealers and large customers for free (BC Hydro 1989a).

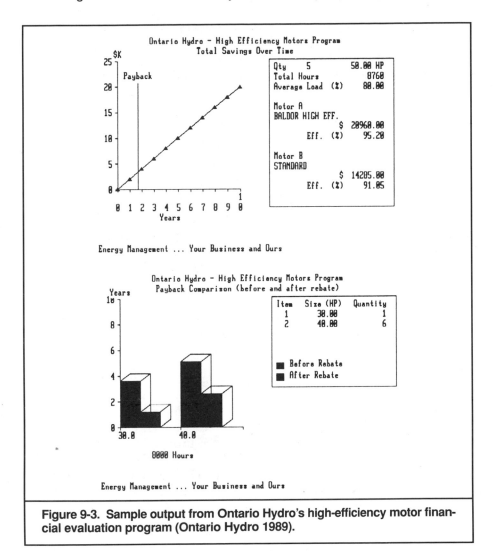

Figure 9-3. Sample output from Ontario Hydro's high-efficiency motor financial evaluation program (Ontario Hydro 1989).

Sizing programs for fans and pumps are produced by most major manufacturers. The user provides flow and either head or static pressure for the application, and the programs calculate optimal pump size, impeller diameter, and a pump curve for the application or, in the case of fans, optimal fan size, rpm, and a fan curve. Manufacturer representatives will run a sizing program for a prospective customer. Some manufacturers will provide copies of their programs to large engineering firms. While these programs generally do a good job of determining the optimal equipment for an application, there is still a tendency for some engineers to specify larger equipment in order to provide a safety factor. Also, since sizing programs are often provided by manufacturers, the program will recommend the best fan or pump produced by that manufacturer, but will not indicate whether a fan or pump produced by another manufacturer is a better choice.

Sizing is more complicated for fans and pumps driven by ASDs because the fan or pump must be optimized for multiple flow rates. Computer programs for optimizing such applications are now under development at EPRI and BC Hydro. Several fan and pump manufacturers are also reportedly working on such a program (Gudbjargsson 1990).

In addition to software designed to help with equipment purchase decisions, programs are available to track motor maintenance practices. Information on recurring maintenance tasks can be input into the computer, and work orders will be prepared by the computer according to a schedule set by the user. Systems of this sort range in cost from $400 to $250,000, depending on sophistication (Peele and Chapman 1986). Few data are available on how widely used this software is, but available evidence indicates that these systems are primarily used by large firms with at least a dozen maintenance workers on the payroll. For example, Ralston Purina reports a two-year payback on a sophisticated maintenance software system installed at their million-square-foot headquarters complex (Sperber 1989).

While a number of useful software tools are available, they have not reached many decision-makers. The best packages need to be better distributed, and more customers need to be trained to use them. Also, as is discussed in chapter 10, additional, user-friendly systems are needed in some areas. See appendix B for listings of some of the available software.

On-Site Assessments and Assistance

The simplest of these programs feature computerized audits furnished by utilities under the Commercial and Apartment Conservation Service Program (an audit program formerly mandated by the U.S. Federal government and currently required in several states). These audits, designed primarily for commercial facilities, often assess opportunities for using high-efficiency motors. Other motor system efficiency improvements are rarely considered.

The next step up in sophistication is technical assistance services that offer information and advice on efficiency improvements but do not provide a full engineering audit. For example, field engineers from the Snohomish Public Utility District in Everett, Washington, provide on-site technical analyses to industrial customers on specific facility and process conservation and load management improvements, including motor and motor control measures. In addition, specialized information is compiled and distributed to the most common industries in the service area. After one year, the program reached 35 out of approximately 400 eligible customers and resulted in energy savings of 757 MWh/yr (nearly 1% of the utility's industrial electricity sales). Data are not available on how much of these savings are due to motor system improvements (Pendleton 1989).

Detailed engineering analyses are provided by several industrial audit programs operated by utilities, state energy offices, universities, and private motor distributors. Carolina Power and Light provides free detailed audits to its large industrial customers (peak demand greater than 1 MW). From 1983 to 1989, approximately 200 customers received audits, resulting in demand reductions of about 75 MW. Audits cover all energy using systems in a plant, including motors, other electric equipment, and equipment that burns fossil fuels. Audits are provided by six to eight full-time industrial engineers who work out of the utility's central office. The utility has found that the best auditors are those who have worked as engineers within industry—they can best establish a rapport with the industrial customer. The program is marketed through personal contacts by utility field engineers and through word-of-mouth between industrial customers.

Only measures with a two-year payback or less are recommended because experience has shown that these are the only measures implemented by customers. Typically, if all recommended measures are implemented, customer demand is reduced by 10–15%. Follow-up surveys have found that approximately 50–60% of the audit recommendations are implemented, representing approximately 80% of the total MW savings included in audit recommendations (Castelow 1989). Recommendations for high-efficiency motors are implemented more often than other recommendations (there are no exact data on implementation rates for specific measures). The audit includes a motor survey for customers whose motors operate long periods of time. This survey includes spot metering of motor kW use and focuses on units that, upon failure, should be replaced with high-efficiency motors. Auditors recommend that customers mark candidates for replacement with yellow paint and instruct maintenance staff to purchase new motors when a "yellow dot" motor fails. Follow-up surveys indicate that this system works well in practice (Johnston 1990).

New York State's Energy Advisory Service to Industry program has since 1979 provided free energy audits to industries with up to 400 employees. Larger firms are not served because the state believes such firms have the resources to undertake audits on their own. Audits are provided by private contractors hired by the state. The $4–5 million/yr program is financed with oil overcharge funds and money from the federal State Energy Conservation and Energy Extension Service programs (Eggars 1989). Over the first ten years of the program, approximately 7,000 audits were conducted, with an estimated benefit-cost ratio of approximately 40 to 1 (Peat Marwick & Co. 1987).

NCIES also runs a unique and very successful on-site assessment service. As mentioned previously, NCIES runs seminars for senior maintenance staff on topics such as compressors, HVAC systems, and chillers and cooling towers. People who have attended a seminar are eligible for a technical assistance assessment during which a skilled engineer works in the plant with the seminar attendee to assess specific systems covered by the seminar. The engineer spends a total of one week on each assessment, including field work, write-up, and in-plant presentation. Reports emphasize measures with a simple payback of two years or less. Limited follow-up surveys indicate that about 60% of assessment recommendations are adopted by customers. NCIES pays 80% of the $2,000 technical assessment cost. In 1989, the program provided approximately 45 assessments, at a total cost of approximately $400,000 (including seminars, assessments, and other program activities), and resulted in an estimated $5.2 million in annual energy savings (Johnston 1990).

The Energy Analysis and Diagnostic Center (EADC) program is sponsored by the U.S. Department of Energy. EADCs at 13 universities around the country use faculty and students to provide about 300 free energy audits per year to small and medium-size manufacturers. Audits include examination of plant production, service, and HVAC functions. An evaluation of the program estimates that participants receive an average rate of return on their investments of approximately 200–300% (implying that they implement low-cost measures with rapid paybacks) and that the Federal government receives a return on its investment of approximately 40–70% (the government's return is in increased tax receipts, which result from reduced expenses and increased profitability of participating firms) (Kirsch 1989). Drivepower improvements account for less than 6% of the savings, which suggests that more could be done in the area of motor systems.

Walco Electric in Providence, Rhode Island, provides free motor analyses for customers and potential customers. Walco will prepare a motor inventory for a customer, noting motor power, speed, frame, enclosure, and efficiency, and will estimate the costs and savings of converting to new high-efficiency motors. The surveys identify applications where simple motor substitutions are not possible and where additional engineering or construction (like adjusting the speed of driven equipment or redrilling mounting bracket holes) will be required. Such add-ons are incorporated into the cost estimates. Walco is able to provide this service because rebates provided by the local utility make purchases of replacement motors attractive for many customers. The key to program success, according to Walco and the utility, is knowledgeable staff who can prepare motor inventories and analyses customers can depend on (Gilmore 1989). As a result of the motor audits and utility rebates, Walco Electric estimates that sales of motors have increased approximately 30% (Gordon 1990).

Equipment test services target common types of equipment prone to energy waste. Services are provided by trained technicians at costs considerably lower than a full engineering audit. For example, BC Hydro provides free airflow and leakage surveys of compressed air systems. The surveys assess leaks in the system, motor and compressor efficiency, system controls, and system pressure relative to compressed air needs. The test identifies the general location of leaks, estimates how much they are costing the customer, and suggests a leak reduction target. Follow-up leakage tests are provided three months after the initial assessment. If the target is achieved, the company gets an award, and the maintenance crew gets a free lunch and a door prize. In addition, to encourage regular tests of the compressor system, BC Hydro pays one-half the cost of a follow-up assessment one year after the original assessment. The hope is that by the third year, customers will undertake and finance annual assessments without utility involvement. The program is targeted at facilities with compressor systems of 100 hp or more. BC Hydro expects to reach most of the 300 targeted customers over a two-year period. The program also is providing the utility with extensive information on compressor systems, and the utility is using this information to design additional programs (Willis 1990).

Pump testing is another type of assessment being considered by several utilities. This service would identify oversized pumps and recommend corrective actions such as downsizing, installation of ASDs, or impeller trimming. The service would target such facilities as paper mills, chemical plants, and municipal water supply and treatment plants with major pump loads. Only pumps with motors of approximately 5 hp or more would be targeted (the savings available with smaller motors would probably not justify the expense). New England Electric System planned to offer this service (NEES and CLF 1989) but identified too few customers with large pump loads to justify the program (Stout 1990). BC Hydro is planning a pilot version of this program (Willis 1990).

New England Electric recently began offering two other programs—an industry expert service and an equipment loan service. The industry expert service hires engineers familiar with processes in specific industries to provide audits. Experts can identify conservation opportunities not apparent to "general practitioners." The equipment loan service provides equipment to monitor motor energy use, demand, power factor, and efficiency. Private contractors hired by the utility assist customers with the monitoring (Stout 1990). Other equipment, such as core-loss testers to determine if motor cores have been damaged during the rewind process, may be added in the future.

One final program worthy of mention is run by North Carolina Alternative Energy Corporation to encourage customers to hire energy managers. The NCAEC program encourages local governments to hire full- or part-time energy managers by guaranteeing that the energy manager will save his or her salary through energy savings, or NCAEC will pay the difference. NCAEC also provides training and technical assistance to the energy managers. In the pilot phase of the project, salary money paid by NCAEC amounted to only 3% of the total salaries paid to energy managers under the program. On the average, energy managers reduced energy bills by an amount equal to twice their salaries (Emmett and Gee 1986). Similar programs could be applied in industrial facilities.

These examples illustrate some innovative approaches to technical assistance. As the EADC program shows, many customers will implement technical assistance recommendations that provide rapid paybacks. Implementation rates are generally not as high for measures with moderate or high costs. For example, an audit contractor in one program estimated that a measure with a one-year simple payback is ten times more likely to be implemented than a measure with a two-year simple payback (Gustafson and Peters 1987). Similar results have been found for a variety of other audit programs (Nadel 1990a). Thus, technical assistance services, while important in themselves, must be complemented with other program approaches if a large proportion of the available savings is to be captured.

Labeling and Catalog Listings

In the late 1970s, the National Electrical Manufacturers Assocation established a voluntary motor labeling program for Design A and B polyphase induction motors of 1 to 125 horsepower. This labeling standard, revised in the early 1980s and made mandatory for NEMA members, is now widely observed by manufacturers in the United States and Canada, although labeling is spotty or nonexistent for motors less than 1 hp or greater than 125 hp, multi-speed motors, NEMA Design C and D motors, single-phase motors, and noninduction motors. Labels list the nominal and minimum efficiency (described in chapter 2). A sample label is contained in Figure 9-4.

Unlike appliances, where consumers compare models on a showroom floor, new motors are rarely seen by customers prior to purchase, so motor labels have limited value in the buying decision. However, labels may be useful in the used motor market by telling purchasers the motor's nameplate efficiency when the motor was new. In most cases, the effects of heat, dust, rewinds, and other factors will cause used motors to perform below nameplate efficiency, but the nameplate at least indicates the upper limit.

Unlike labeling, efficiency information in catalogs is critically important in the new motor market. Nevertheless, a review of 1989

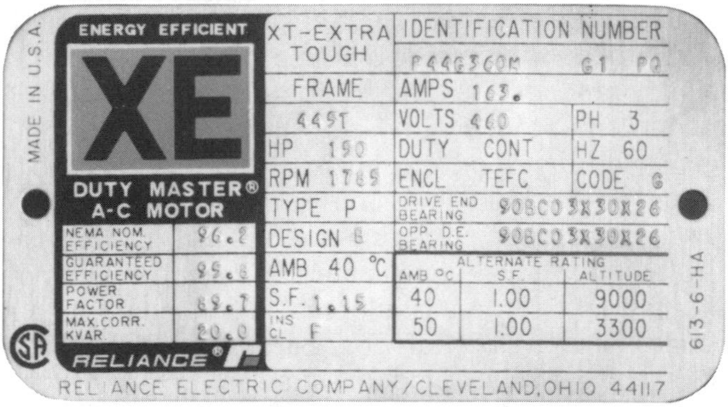

Figure 9-4. Sample motor nameplate showing nominal and minimum efficiency (courtesy of Reliance Electric).

catalogs for six major U.S. motor manufacturers shows that only one manufacturer provides efficiency ratings for both standard- and high-efficiency motors in the main listings in its catalog. Three manufacturers relegate efficiency ratings to an appendix or supplement, one lists the efficiency of only high-efficiency motors, and one does not provide any efficiency ratings.

NEMA should require complete efficiency data in all catalogs. Other desirable changes include extending the labeling program to all motors (not just Design A and B 1–125 hp) and narrowing the gap between adjacent nominal efficiency ratings. As discussed in chapter 2, the published nominal efficiency ratings represent the average measured efficiency of a group of identical motors rounded down to one of 51 preset values. Differences between preset values can be more than one efficiency point at the low end of the scale (see Table 2-7). The reason for the rounding is to allow for testing errors. In recent years, however, the accuracy of many types of test equipment has improved, which may allow for narrower gaps between adjacent preset values.

Efficiency Standards

A number of voluntary and mandatory efficiency standards for motors have been adopted and proposed over the years. These standards recommend or require motors to exceed specific efficiency levels, which vary with motor size and often with motor speed and enclosure type.

In 1989, NEMA published a standard that lists the nominal and minimum efficiency a motor must equal or exceed in order to be designated as energy efficient. The NEMA standard is met by nearly all motors produced by major U.S. motor manufacturers for their high-efficiency product lines and by some motors produced by several manufacturers for their standard-efficiency lines (see Figure 9-5). NEMA is working to have user organizations, electric utilities, government agencies, and state conservation programs adopt this definition for purposes of preparing motor specifications and determining eligibility for high-efficiency motor incentive programs (Raba 1990). The City of Seattle has adopted this standard as part of its building code (Seattle Dept. of Construction and Land Use 1991). New York State has also adopted this standard as part of its building code, effective 1991, with a more stringent standard scheduled to take effect three years later (New York State Energy Office 1991).

Several electric utilities have also established definitions for purposes of establishing eligibility for motor rebate programs. Utility

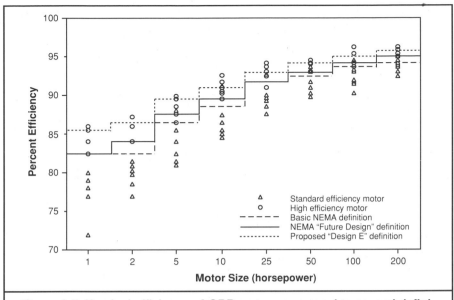

**Figure 9-5. Nominal efficiency of ODP motors compared to several defini-
tions of high efficiency established by NEMA.**

Note: This figure compares the efficiency levels of high-efficiency (○) and standard-efficiency (△) motors produced by eight major U.S. manufacturers to several
definitions of "high efficiency" established or being considered by the National
Electrical Manufacturers Association (NEMA). As can be seen, the basic NEMA
definition of "high efficiency" includes nearly all high-efficiency motors, as well as
several standard-efficiency motors. The NEMA "Future Design" definition includes
most but not all high-efficiency motors and practically no standard-efficiency
motors. The "Design E" proposed definition includes only the most efficient
motors now on the market.

definitions are usually more stringent than the NEMA definition.
A sample definition, developed by Ontario Hydro and BC Hydro, for
use by all Canadian utilities, is listed in Table 9-1 (page 217). Most
standard-efficiency motors, and even some high-efficiency motors
produced in 1989, did not meet this standard.

In 1990, NEMA recognized that many utilities and other parties
want a more stringent definition of "energy-efficient" than is contained in the official NEMA standard. To deal with this situation,
NEMA issued a "Suggested Standard for Future Design." This standard is listed in Table 9-2 (pages 218 and 219).

In addition, NEMA is also considering adoption of still higher
efficiency levels for "Design E" motors. These motors will have high
efficiency, but will also differ from the commonly used "Design B"

motors in some electrical and torque-speed characteristics. The draft "Design E" definition of high efficiency is summarized in Table 9-3.

ASHRAE and DOE have both adopted an Energy Conservation Standard for Commercial and Multifamily High-Residential Buildings. The DOE standard (DOE 1989) is mandatory for new federal buildings and voluntary for other buildings. The ASHRAE standard (ASHRAE 1990) is voluntary but typically gets incorporated into state building codes. Both the ASHRAE and DOE standards include minimum efficiency values for single-speed polyphase motors. The DOE standard, listed in Table 9-4 (page 221), is roughly similar in stringency to the NEMA standard. The ASHRAE standard matches the DOE standard beginning in 1992 and is slightly less stringent prior to 1992.

Both the DOE and ASHRAE standards apply the same efficiency requirement to a range of horsepower levels—the minimum efficiency for 50- to 90-hp, single-speed motors is set at 90.2%, for example. Motors at the low end of each horsepower range are less likely to meet the standard than motors at the high end of the range. For this reason, ASHRAE is now proposing to replace its version of Table 9-4 with a set of specific efficiency values for each size and type of motor. The proposed ASHRAE values are identical to those in Table 9-2 (ASHRAE 1991).

Mandatory efficiency standards for the United States and California were extensively studied in the late 1970s and early 1980s. In 1978, the U.S. Congress passed the National Energy Conservation Policy Act, one of whose provisions required DOE to evaluate the practicability and effect of minimum efficiency standards for motors and pumps. Under the act, DOE had authority to prescribe labeling standards for motors and pumps, but additional congressional authorization was needed to enact minimum efficiency standards. The analysis was prepared for DOE by Argonne National Laboratory and Arthur D. Little, Inc. The draft report, written in the closing days of the Carter administration, concluded that the benefits of motor efficiency standards were likely to be greater than the costs and that DOE should hence be authorized to enact efficiency standards for motors. Pump standards were not recommended because they did not appear to be significantly cost-effective (A.D. Little 1980).

The final DOE report (Argonne National Laboratory 1980), written early in the Reagan Administration, changed some of the draft report's assumptions and used higher costs for high-efficiency motors, higher discount rates (20% in the base case compared to 15% in the draft version), and higher assumed market penetration of high-efficiency motors in the absence of standards. Despite these changes, motor efficiency standards were found to have a benefit-cost ratio of 1.32 (benefits were estimated to be 1.32 times greater than costs). In addition to revising the original analysis, however, DOE employed a second model, which projected still higher market penetration of high-efficiency motors in the absence of standards. Under this second scenario, standards were not cost-effective. Faced with different projections of high-efficiency motor penetration in the absence of standards, DOE chose to use the second model until further data became available, and stated:

> DOE believes that the future market penetration of energy-efficient motors will be sufficiently near the economic market level to render the potential benefits of standards small in comparison to their costs. DOE intends to verify this assumption by measuring the actual market penetration.

Table 9-1. Nominal efficiency levels used for rebate program eligibility by Canadian utilities.

Motor Size (Hp)	2-Pole (~3600 rpm)	4-Pole (~1800 rpm)	6-Pole (~1200 rpm)	8-Pole (~900 rpm)
1.0	75.5%	82.5%	80.0%	74.0
1.5	82.5	84.0	85.5	77.0
2.0	84.0	84.0	86.5	82.5
3.0	85.5	87.5	87.5	84.0
5.0	87.5	87.5	87.5	85.5
7.5	88.5	89.5	89.5	85.5
10.0	89.5	89.5	89.5	88.5
15.0	90.2	91.0	90.2	88.5
20.0	90.2	91.0	90.2	89.5
25.0	90.5	91.7	91.3	89.6
30.0	90.8	91.9	91.4	90.7
40.0	91.4	92.5	92.3	90.6
50.0	91.9	92.7	92.3	91.3
60.0	92.4	93.2	92.9	91.6
75.0	92.5	93.5	93.1	92.8
100.0	93.0	93.7	93.5	92.7
125.0	93.6	93.9	93.6	93.4
150.0	93.8	94.3	94.2	93.4
200.0	94.3	94.5	94.6	93.9
250.0	95.0	95.0	95.0	95.0
300.0	95.0	95.0	95.0	95.0
350.0	95.0	95.0	95.0	95.0
400.0	95.0	95.0	95.0	95.0
450.0	95.0	95.0	95.0	95.0
500.0	95.0	95.0	95.0	95.0

Source: Ontario Hydro 1992.

DOE may, at such time as it determines its current conclusion to be wrong, revise its recommendation on the basis of the collected information.

Following completion of the DOE study, the California Energy Commission (CEC) examined motor efficiency standards. The CEC began by reviewing the final DOE report. This review (Burley and Leber 1981) concluded that DOE's own analysis found motor efficiency standards cost-effective at discount rates as high as 15% (an extremely high value, given that the agency should take a long-run, societal perspective). The review further noted that DOE monitoring of high-efficiency motor sales had just been canceled, and recommended that the CEC undertake these studies and consider enacting motor efficiency standards in California. However, the CEC analysis never took place because the commission was busy working on appliance efficiency standards and did not have the resources to examine motor standards as well (Messenger 1989).

Table 9-2. NEMA suggested standard for future designs of energy-efficient motors.

Open Motors at Full Load

Motor Size (Hp)	2-Pole (~3600 rpm) Nominal Efficiency	Minimum Efficiency	4-Pole (~1800 rpm) Nominal Efficiency	Minimum Efficiency	6-Pole (~1200 rpm) Nominal Efficiency	Minimum Efficiency	8-Pole (~900 rpm) Nominal Efficiency	Minimum Efficiency
1.0	—	—	82.5	81.5	80.0	78.5	74.5	72.0
1.5	82.5	81.5	84.0	82.5	84.0	82.5	75.5	74.0
2.0	84.0	82.5	84.0	82.5	85.5	84.0	85.5	84.0
3.0	84.0	82.5	86.5	85.5	86.5	85.5	86.5	85.5
5.0	85.5	84.0	87.5	86.5	87.5	86.5	87.5	86.5
7.5	87.5	86.5	88.5	87.5	88.5	87.5	88.5	87.5
10.0	88.5	87.5	89.5	88.5	90.2	89.5	89.5	88.5
15.0	89.5	88.5	91.0	90.2	90.2	89.5	89.5	88.5
20.0	90.2	89.5	91.0	90.2	91.0	90.2	90.2	89.5
25.0	91.0	90.2	91.7	91.0	91.7	91.0	90.2	89.5
30.0	91.0	90.2	92.4	91.7	92.4	91.7	91.0	90.2
40.0	91.7	91.0	93.0	92.4	93.0	92.4	91.0	90.2
50.0	92.4	91.7	93.0	92.4	93.0	92.4	91.7	91.0
60.0	93.0	92.4	93.6	93.0	93.6	93.0	92.4	91.7
75.0	93.0	92.4	94.1	93.6	93.6	93.0	93.6	93.0
100.0	93.0	92.4	94.1	93.6	94.1	93.6	93.6	93.0
125.0	93.6	93.0	94.5	94.1	94.1	93.6	93.6	93.0
150.0	93.6	93.0	95.0	94.5	94.5	94.1	93.6	93.0
200.0	94.5	94.1	95.0	94.5	94.5	94.1	93.6	93.0

Table 9-2 *(continued).*

Enclosed Motors at Full Load

Motor Size (Hp)	2-Pole (~3600 rpm)		4-Pole (~1800 rpm)		6-Pole (~1200 rpm)		8-Pole (~900 rpm)	
	Nominal Efficiency	Minimum Efficiency	Nominal Efficiency	Minimum Efficiency	Nominal Efficiency	Minimum Efficiency	Nominal Efficiency	Minimum Efficiency
1.0	75.5	74.0	82.5	81.5	80.0	78.5	74.0	72.0
1.5	82.5	81.5	84.0	82.5	85.5	84.0	77.0	75.5
2.0	84.0	82.5	84.0	82.5	86.5	85.5	82.5	81.5
3.0	85.5	84.0	87.5	86.5	87.5	86.5	84.0	82.5
5.0	87.5	86.5	87.5	86.5	87.5	86.5	85.5	84.0
7.5	88.5	87.5	89.5	88.5	89.5	88.5	85.5	84.0
10.0	89.5	88.5	89.5	88.5	89.5	88.5	88.5	87.5
15.0	90.2	89.5	91.0	90.2	90.2	89.5	88.5	87.5
20.0	90.2	89.5	91.0	90.2	90.2	89.5	89.5	88.5
25.0	91.0	90.2	92.4	91.7	91.7	91.0	89.5	88.5
30.0	91.0	90.2	92.4	91.7	91.7	91.0	91.0	90.2
40.0	91.7	91.0	93.0	92.4	93.0	92.4	91.0	90.2
50.0	92.4	91.7	93.0	92.4	93.0	92.4	91.7	91.0
60.0	93.0	92.4	93.6	93.0	93.6	93.0	91.7	91.0
75.0	93.0	92.4	94.1	93.6	93.6	93.0	93.0	92.4
100.0	93.6	93.0	94.5	94.1	94.1	93.6	93.0	92.4
125.0	94.5	94.1	94.5	94.1	94.1	93.6	93.6	93.0
150.0	94.5	94.1	95.0	94.5	95.0	94.5	93.6	93.0
200.0	95.0	94.5	95.0	94.5	95.0	94.5	94.1	93.6

Source: NEMA 1991a.

While neither DOE nor CEC monitored market penetration of high-efficiency motors in the absence of standards, a crude analysis can now be attempted by comparing penetration rates and average efficiency levels projected by DOE in 1980 for sales in 1988 with actual market data for 1988. This analysis is contained in Table 9-5 (page 222). The table shows that DOE overestimated by a factor of two to three the penetration of high-efficiency motors in the absence of standards. Likewise, actual average motor efficiencies in 1988 were lower than what DOE projected would occur with either standards or labeling, although they exceeded what DOE expected the market alone to produce. Thus, based on both market penetration and average efficiency indices, energy savings in the absence of standards were substantially less than DOE projected would occur with standards or labeling.

As a result of DOE's errors in analysis, and the fact that DOE did not monitor motor efficiency trends, as it had pledged in 1980, achievement of the substantial energy and dollar savings available from motor efficiency standards has been delayed by at least ten years. A rough estimate of the savings available from motor efficiency standards can be found in Table 9-6 (page 223). Over a 15-year period (a rough estimate of the typical life of a motor according to Gordon et al. 1988), national motor standards will save motor users an estimated $4.8 billion (1990 $) while reducing electric generating requirements by approximately 4,300 MW, capacity that would cost $5–10 billion to build. Given the large savings and favorable economics of high-efficiency motors (discussed in chapter 2), in 1991 elected officials in the U.S. and Canada became interested in minimum efficiency standards for motors and introduced enabling legislation.

In the U.S., this proposed legislation spurred negotiations between motor manufacturers and efficiency advocates, resulting in an agreement to jointly support modified legislation. This legislation would establish minimum efficiency standards for Design A and B, T-frame, foot-mounted, polyphase, squirrel cage induction motors of 1-200 hp. The minimum efficiency levels are identical to those in Table 9-2. The standards take effect five years after the legislation

Table 9-3. Full load efficiencies of Design E motors.

	Open Motors at Full Load							
Motor Size	2-Pole (~3600 rpm)		4-Pole (~1800 rpm)		6-Pole (~1200 rpm)		8-Pole (~900 rpm)	
(Hp)	Nominal Efficiency	Minimum Efficiency	Nominal Efficiency	Minimum Efficiency	Nominal Efficiency	Minimum Efficiency	Nominal Efficiency	Minimum Efficiency
1.0	—	—	85.5%	84.0%	82.5%	81.5%	77.0%	75.5%
1.5	85.5%	84.0%	86.5	85.5	86.5	85.5	78.5	77.0
2.0	86.5	85.5	86.5	85.5	87.5	86.5	87.5	86.5
3.0	86.5	85.5	88.5	87.5	88.5	87.5	88.5	87.5
5.0	87.5	86.5	89.5	88.5	89.5	88.5	89.5	88.5
7.5	89.5	88.5	90.2	89.5	90.2	89.5	90.2	89.5
10.0	90.2	89.5	91.0	90.2	91.7	91.0	91.0	90.2
15.0	91.0	90.2	92.4	91.7	91.7	91.0	91.0	90.2
20.0	91.7	91.0	92.4	91.7	92.4	91.7	91.7	91.0
25.0	92.4	91.7	93.0	92.4	93.0	92.4	91.7	91.0
30.0	92.4	91.7	93.6	93.0	93.6	93.0	92.4	91.7
40.0	93.0	92.4	94.1	93.6	94.1	93.6	92.4	91.7
50.0	93.6	93.0	94.1	93.6	94.1	93.6	93.0	92.4
60.0	94.1	93.6	94.5	94.1	94.5	94.1	94.1	93.6
75.0	94.1	93.6	95.0	94.5	94.5	94.1	94.5	94.1
100.0	94.1	93.6	95.0	94.5	95.0	94.5	94.5	94.1
125.0	94.5	94.1	95.4	95.0	95.0	94.5	94.5	94.1
150.0	94.5	94.1	95.8	95.4	95.4	95.0	94.5	94.1
200.0	95.4	95.0	95.8	95.4	95.4	95.0	94.5	94.1

Table 9-3 *(continued).*

Enclosed Motors at Full Load

Motor Size (Hp)	2-Pole (~3600 rpm) Nominal Efficiency	2-Pole (~3600 rpm) Minimum Efficiency	4-Pole (~1800 rpm) Nominal Efficiency	4-Pole (~1800 rpm) Minimum Efficiency	6-Pole (~1200 rpm) Nominal Efficiency	6-Pole (~1200 rpm) Minimum Efficiency	8-Pole (~900 rpm) Nominal Efficiency	8-Pole (~900 rpm) Minimum Efficiency
1.0	78.5%	77.0%	85.5%	84.0%	82.5%	81.5%	77.0%	75.5%
1.5	85.5	84.0	86.5	85.5	87.5	86.5	80.0	78.5
2.0	86.5	85.5	86.5	85.5	88.5	87.5	85.5	84.0
3.0	87.5	86.5	89.5	88.5	89.5	88.5	86.5	85.5
5.0	89.5	88.5	89.5	88.5	89.5	88.5	87.5	86.5
7.5	90.2	89.5	91.0	90.2	90.2	89.5	87.5	86.5
10.0	91.0	90.2	91.0	90.2	91.0	90.2	90.2	89.5
15.0	91.7	91.0	92.4	91.7	91.7	91.0	91.0	90.2
20.0	91.7	91.0	92.4	91.7	91.7	91.0	91.0	90.2
25.0	92.4	91.7	93.6	93.0	93.0	92.4	91.0	90.2
30.0	92.4	91.7	93.6	93.0	93.0	92.4	92.4	91.7
40.0	93.0	92.4	94.1	93.6	94.1	93.6	92.4	91.7
50.0	93.6	93.0	94.1	93.6	94.1	93.6	93.0	92.4
60.0	93.6	93.6	94.5	94.1	94.5	94.1	93.0	92.4
75.0	94.1	93.6	95.0	94.5	94.5	94.1	94.1	93.6
100.0	94.1	94.1	95.4	95.0	95.0	94.5	94.1	93.6
125.0	94.5	95.0	95.4	95.0	95.0	94.5	94.5	94.1
150.0	95.0	95.0	95.8	95.4	95.8	95.4	94.5	94.1
200.0	95.8	95.4	95.8	95.4	95.8	95.4	95.0	94.5

Source: NEMA 1991b.

Table 9-4. Minimum acceptable nominal full load motor efficiencies for single-speed polyphase motors in DOE voluntary energy standards for new buildings.

Horsepower	Efficiency
1–4	78.5%
5–9	84.0
10–19	85.5
20–49	88.5
50–99	90.2
100–124	91.7
125 and above	92.4

Source: DOE 1989b.

passes (seven years for motors or equipment that requires listing by United Laboratories [UL]). In addition the legislation calls for DOE to review the standards every five years, and to revise the standards to the highest levels that are "technologically feasible" and "economically justified." Furthermore, under the legislation, DOE will study minimum efficiency standards for fractional horsepower motors, and

Table 9-5. Projected and actual penetration rates and efficiencies for motors.

Source	Program	Year	% of Motor Sales That Are High Efficiency (by Hp)				Average Efficiency of All Motors by Hp Class			
			1–5	6–20	21–50	51–125	1–5	6–20	21–50	51–125
ANL model	Labeling & info	1985	2%	24%	45%	57%	85.9%	88.8%	91.4%	92.7%
	standards	1987	3%	100%	100%	100%	85.9%	90.0%	92.0%	93.0%
ISTUM model	Marketplace	1988	NA	62%	62%	62%	NA	86.5%	89.7%	92.2%
ACEEE	Actual experience	1988	8%	14%	19%	21%	80.4%	87.5%	91.3%	92.6%

Note: ANL model is the model used in both the draft and final DOE reports. ISTUM model was used in only the final report and presents a more optimistic view of high-efficiency motor penetration in the absence of standards.

Sources: ANL and ISTUM estimates from or derived from Argonne National Laboratory 1980. ACEEE estimates based on NEMA data on penetration of high-efficiency motors in 1988 and on the average efficiency of standard- and high-efficiency ODP and TEFC motors as listed in the 1988 technical literature from five major motor manufacturers.

Table 9-6. Estimated savings from motor efficiency standards in the United States.

Motor Horse-power	Annual Sales[1]	Sales for Which EEM Avail-able[2]	Avg. Motor Size[3]	Average Efficiency[4]		Avg. Annual Op Hours[5]	Average Annual kWh Savings Per Motor[6]	GWh Svgs[7]		MW Svgs[8]		Average Cost Premium for High Effic. Motor[9]	Net Dollar Savings —millions—[10]			
				Std.	High			1st Year	15th Year	1st Year	15th Year		1st Year	15th Year	15 yr Cumu-lative	15 yr Benefit Cost Ratio[10]
1–5	1,154,483	738,869	2.07	79.8	85.0	2,352	165	122	1,828	56	838	$43	($18)	$84	$496	2.3
6–20	470,211	300,935	11.9	86.3	90.7	2,928	868	261	3,917	96	1,443	$100	($8)	$211	$1,521	5.2
21–50	144,658	92,581	32.5	89.9	93.1	3,568	1,970	182	2,736	55	827	$223	($6)	$148	$1,066	5.3
51–125	70,298	44,991	86.7	92.0	94.7	4,163	4,977	224	3,359	58	870	$431	($2)	$186	$1,380	6.9
126–200	14,661	9,383	212	93.8	95.6	4,163	7,899	74	1,112	19	288	$1,534	($7)	$55	$361	3.1
Total	1,854,311	1,186,759						863	12,952	284	4,266	$98	($41)	$684	$4,824	4.5

Notes:

1. From Bureau of Census 1988.

2. Based on the following estimates: 87% of motors have a T-frame × 94% are TEFC or ODP × 98% are 2–6 pole × 80% are foot-mounted and of NEMA Type A or B. Figures from database compiled by Bill Gilmore, Walco Electric on over 1,000 motors in R.I. Foot-mount/Type A&B proportion estimated by Gail Katz, Momentum Engineering.

3. From Argonne National Laboratory 1980.

4. Average nominal efficiency for motor nearest in size to average motor size. Based on average nominal efficiency for 1,800-rpm ODP and TEFC motors as listed in Tables A-1 and A-2.

5. Average of values estimated by Arthur D. Little (1980) and by Xenergy (1989). Values from R.I. study of commercial and industrial motors (see note #2) are even higher.

6. Motor Hp × .746 kW/Hp × .75 avg. load × Operating hours × [1/(80% × std eff + 20% × high eff) − (1/high eff)]. 20% and 80% are estimates of present sales shares of high-eff. and std. motors in the U.S. (see chapter 6).

7. First-year savings are product of kWh savings/motor and annual sales to which stds. apply. 15th-year savings are 15 × 1st-year savings. For these simple calculations, growth in the motor stock over the next 15 years is ignored.

8. GWh savings / annual operating hours × 1.096 T&D loss factor × 1.23 reserve margin factor × 80% of motors assumed to be operating at time of maximum peak demand. T&D and reserve margin factors estimated from New England Power Pool data. Peak coincidence factor is an ACEEE estimate.

9. Average cost difference for motor nearest in size to average motor. Based on 1,800-rpm TEFC and ODP motors as summarized in Tables A-1 and A-2.

10. In 1990 $ based on $.06/kWh. Includes allowance for increased purchase price on 80% of motor sales that are presently std.-efficiency models. For this simple calculation, real fuel price inflation and discounting are ignored.

if feasible and justified, DOE will promulgate minimum efficiency standards for these motors (U.S. Senate 1992). As of summer 1992, the legislation had passed both the Senate and the House of Representatives, but differences between the two bills were still being worked out. Final passage is likely in the fall of 1992.

In Canada, in 1991, the Canadian Standards Association (CSA), working with the Canadian Electrical Association, developed recommended mimimum efficiency standards for Canada. The specific standards are summarized in Table 9-7. For these standards to become effective, individual provinces must adopt legislation referencing the CSA standard. British Columbia and Ontario were the first provinces to do this; their standards take effect September, 1992. The Canadian standards are not as strong as the U.S. standards, but the Canadian standards take effect much earlier than the U.S. standards. CSA hopes to revise their standard in several steps so that it matches the U.S. standard about the time the U.S. standard takes effect (Cohen 1992).

Table 9-7. Canadian Standards Association minimum quoted efficiency values.

Motor Size (Hp)	2-Pole (~3600 rpm)	4-Pole (~1800 rpm)	6-Pole (~1200 rpm)	8-Pole (~900 rpm)
1	74.0%	75.0%	75.0%	70.0%
1.5	76.5	77.0	77.0	72.5
2	78.6	79.0	79.0	74.5
3	80.5	81.0	80.9	76.5
5	82.0	82.5	82.5	78.4
7.5	83.3	84.6	84.0	80.1
10	84.5	86.0	85.2	81.6
15	85.5	87.4	86.3	83.0
20	86.5	88.3	87.2	84.0
25	87.4	89.0	88.0	85.1
30	88.0	89.5	88.6	86.0
40	88.5	90.0	89.2	87.0
50	89.0	90.5	89.6	87.8
60	89.4	90.9	90.1	88.5
75	89.7	91.2	90.5	89.2
100	90.0	91.6	90.9	90.0
125	90.4	91.8	91.3	90.5
150	90.6	91.9	91.6	91.0
200	90.8	92.0	92.0	91.4

These values have been adopted as minimum efficiency standards by the provinces of British Columbia and Ontario. Quoted efficiency in Canada is very similar to nominal efficiency in the U.S.

Source: CSA 1991.

Incentive Programs

Incentive programs are designed to overcome financial barriers to the purchase and use of efficient equipment. The most familiar kinds of incentives are gifts (typically limited to low-cost items like efficient light bulbs and showerheads), low- or zero-interest loans (common in early weatherization programs), targeted and generic rebates, and tax credits. Other approaches include equipment leasing, demand-side bidding, performance contracting, and linked design assistance and incentives. Most incentive programs are operated by utilities, although some states and the federal government have offered tax credits, rebates, performance contracts, and low-interest loans. Incentives targeted specifically at drivepower systems are discussed below, followed by general incentives.

Motor Rebate Programs

Approximately two dozen North American utilities offer rebates for high-efficiency motors. Rebates are typically about $10 per horsepower. Program eligibility levels and rebate formulas vary

Table 9-8. Sample motor rebate schedule.

Horsepower HP	Qualifying Efficiency*	Incentive $
1	84.0	$12
1.5	84.0	$18
2	85.4	$24
3	88.3	$36
5	88.5	$60
7.5	90.6	$90
10	90.8	$120
15	91.7	$180
20	92.4	$240
25	93.0	$300
30	93.4	$325
40	94.0	$375
50	94.1	$425
60	94.6	$475
75	95.0	$550
100	95.3	$675
125	95.3	$800
150	95.7	$925
200	95.9	$1175

*Nominal Efficiency

Source: Long Island Lighting Company.

Table 9-9. Typical utility motor rebate compared to customer costs in typical new purchase, rewind, and retrofit applications.

Motor Size	New Motor Cost		Rewind Cost	High-Efficiency Upgrade Cost			Typical Utility Rebate	Utility Rebate as a Percent of Upgrade Cost		
	Std. Eff.	High Eff.		New Motor	Instead of a Rewind	Replace Working Motor		New Motor	Instead of a Rewind	Replace Working Motor
1	$144	$183	$220	$39	−$37	$315	$10	26%	NA	3%
5	221	290	335	69	−45	422	50	72	NA	12
10	360	471	440	111	31	603	100	90	323%	17
25	744	968	755	224	213	1189	250	112	117	21
50	1405	1874	1175	469	699	2237	375	80	54	17
100	3398	4285	1610	887	2675	5141	625	70	23	12
200	6267	8294	2420	2027	5874	9150	1125	56	19	12

Notes:
Motor and rewind costs for TEFC motors (from Table A-1).
Upgrade cost for new motor is cost difference between standard and high-efficiency motor.
Upgrade cost instead of rewind is cost difference between rewind and new high-efficiency motor.
Upgrade cost to replace working motor is full cost of new high-efficiency motor plus labor cost (from Table A-1).
Utility rebate assumes $10/hp for first 25 hp and $5/hp thereafter.
NA = not applicable. Used where upgrade saves money, even before the rebate.

with motor horsepower and sometimes with motor speed and enclosure type. A typical rebate schedule is shown in Table 9-8.

Most programs target new applications or situations where a new motor will replace a failed motor. As Table 9-9 (page 226) shows, a $10/hp rebate covers much of the price difference between new standard- and high-efficiency motors and, in sizes up to about 30 hp, the difference beteen rewinding and a new high-efficiency motor. Such a rebate covers only a small portion of the cost increment for replacing instead of rewinding in large motors (greater than 30 hp), and an even smaller portion of the price premium involved in replacing a working motor.

Most motor rebate programs have received limited marketing and have had a modest impact on sales of high-efficiency motors. As Table 9-10 (pages 228 and 229) summarizing 16 rebate programs shows, most programs have cumulatively served less than 2% of eligible customers and have resulted in peak savings of less than 0.1% of their sponsoring company's total peak demand. Given that most programs target the new motor market, and that only a small portion of motors are replaced each year, low cumulative participation rates can be expected until a program has run for several years. However, even as a percentage of new motor sales, participation rates tend to be low. For example, Southern California Edison estimated that it provided rebates for 3% of motor sales during the one year it aggressively promoted its "A Rewarding Connection" program (Mayo 1989).

On the positive side, most programs have yielded inexpensive savings: $.001–.005/kWh and $79–822/kW saved. Given the fact that new baseload generating plants typically cost over $1,500/kW (New York State Energy Office et al. 1989) and that long-term marginal operations and maintenance costs for most utilities exceed $.04/kWh (Miller et al. 1989), motor rebate programs are likely to be cost-effective for most utilities.

While the average motor rebate program has had only limited participation, a few programs have done better. These programs, as well as several new programs that hold promise for higher participation rates in the future, are described below.

Niagara Mohawk ran a pilot program during 1986 in which 33% of targeted customers participated. The program targeted large customers with long operating hours, provided extensive personal attention, offered a free computer assessment of costs and savings, and provided high ($25/hp) rebates—sufficient to pay over half the cost of a new replacement motor in many applications. Most customers who did not participate were concerned about disruptions to production processes caused by the downtime required to change

Table 9-10. Summary of utility motor rebate program results.

Utility	State	Program	Time Period Start	End	Pilot or Full-Scale	Number Eligible	No. of Projects	Cum. Participation Rate	Est. Savings MW	GWh	1987 Peak Demand	Svgs as % of Pk	Expenses (1000s of $) Direct	Total	Based on Direct Util. Costs $/kW	or Total Costs	Util. Costs $/kWh
Bangr Hydro	ME	C/I Motor Efficiency	4/86	12/88	Pilot	~1750	97	5.5%	0.08	0.34	262	0.03%	$20	$23	$305	T	$0.007
BC Hydro	BC	High-Efficiency Motor Rebate	7/88	4/90	Full	142,779	309	0.2%	3.30	21.30	6,830	0.05%	$1,150	$1,400	$424	T	$0.006
CMP	ME	Motor Rebate	1986	12/89	Pilot	43,686	394	0.9%		2.04	1,455						
Jersey Cen.	NJ	Motor Rebate	6/87	12/88	Full	28,000					3,766		$43				
Met-Ed/GPU	PA	High-Efficiency Motor	1/87	12/87	Pilot	43,959			0.22	0.77	1,673	0.01%		$27	$122	T	$0.003
NEES	MA/RI	Lg. C&I Custom	1/88	6/89	Full	1890	23	1.2%	0.28		3,798	0.01%	$112		$401	D	
NEES	MA/RI	Energy Initiative	6/89	8/90	Full	~6000			1.51		3,798	0.04%	$1,028		$681	D	
Nevada Pwr	NV	En. Eff. Elec. Motor Rebate	4/89	2/90	Pilot	32,927	24	0.1%	0.03		1,740	0.00%	$4		$130	D	
NiMo	NY	Motor Rebate Pilot	5/86	12/86	Pilot	24	8	33.3%			5,403		$117	$144			
NSP	MN	C&I Motor Efficiency	1/87	12/89	Full	111,751	281	0.3%	0.91		5,543	0.02%		$350	$386	T	
Ont. Hydro	ON	High-Efficiency Motors	10/89	4/90	Full	1,176,000	250	0.0%	0.35		20,524	0.00%	$210		$600	D	
Palo Alto	CA	Partners Elec. Incentive	1985	3/90	Full	2,409	10	0.4%	0.16	0.77	182	0.09%	$29		$184	D	$0.005
PG&E	CA	Energy-Efficient Motor	1983	1983	Full	~25,000	431	1.7%			14,142		$1,273				
So. Cal. Ed	CA	Hardware Rebate	1/82	12/84	Full	393,754	177	0.3%	6.62	49.99	14,775	0.04%	$1,011		$153	D	$0.003
So. Cal. Ed	CA	A Rewarding Connection	11/86	9/87	Full	70,000			0.52	5.20	14,775	0.00%	$41		$79	D	$0.001
Wisc. Elec.	WI	Smart Money	6/87	3/90	Full	81,750	413	0.5%	2.40	17.00	3,810	0.06%	$777		$324	D	$0.006

(continued on next page)

Table 9-10 *(continued).*

Notes:
Time period is time period covered by data summarized in this table. The program may have operated for a longer period.
Number eligible is generally the total number of commerical and industrial customers.
Participation rate is number of projects divided by number eligible.
MW savings are absolute—they are not adjusted for coincidence with the system peak.
Direct costs are rebate costs; total costs include direct costs as well as indirect costs such as staff and marketing.
$/kW is program total costs divided by MW savings. Where only direct costs are available, this figure is used, but is noted with a "D" in the next column.
$/kWh is a levelized cost assuming a 15-year average motor life and a 6% real discount rate.
All data come from the individual utilities operating the programs listed.

motors (Niagara Mohawk 1987). While concerns about motor down-time are significant, continuous-run processes are routinely shut down for maintenance at some point during the year. These scheduled maintenance periods are ideal times to perform motor changeouts.

Pacific Gas and Electric (PG&E) achieved a 1.7% participation rate in less than one year with its Energy-Efficient Motor program targeted at medium and large C&I customers (annual electricity use greater than 100 MWh). Many additional customers were reached in subsequent years through a program that provided rebates for various measures including motors (motor participation figures are not available for this program). This program was promoted through mailings and extensive personal contacts with eligible customers and motor dealers. The program also included procedures to qualify applications over the phone, so that when a motor failed, the customer could quickly purchase a new high-efficiency motor and still qualify for a rebate (Calhoun 1984).

An alternative approach for dealing with failed motors has been used by Southern California Edison. That utility allows customers to receive rebates for a reasonable number of motors purchased for inventory stock. When an old motor fails, the new high-efficiency motor is already in the warehouse and can be quickly installed (Mayo 1989). Several other utilities have adopted this policy.

BC Hydro estimates that it influenced 15% of motor sales in its service territory after one year of operating its rebate program, and 30% of sales after two years (Habart 1990). This relatively high participation level was reached despite average industrial electricity prices of less than $.02/kWh. The utility's goal is to reach 50% of the motor market in the province.

The BC Hydro program combines education and incentives. As previously noted, the utility has developed an educational booklet for customers, computer software for dealers and large customers to use to estimate energy savings, a list of all dealers in the province supplying efficient motors, and a database of all motors sold in the province categorized by type and ranked from most to least efficient. In this program, rebates rise as motor efficiency increases above a base qualifying level. Rebates are available for motors as large as 5,000 hp. Rebates for motors just meeting the utility's efficiency standard range from $15 for a 1-hp motor to $3,000 for a 500-hp motor. For motors above 500 hp, rebates of $400/kW are paid for savings achieved relative to a motor with 95% efficiency. Program marketing emphasizes personal contacts between field representatives and large customers, consulting engineers, and motor suppliers. Seminars and trade shows have also been sponsored. Many motor suppliers are actively promoting the program (Henriques 1989).

Ontario Hydro's program resembles BC Hydro's. However, Ontario Hydro, in addition to paying rebates of $12/hp to customers, pays rebates of $3/hp (or $10/motor for motors of 3 hp or less) to dealers, to reimburse them for the cost of stocking high-efficiency motors and processing application forms. Initial response indicates that some dealers, including some of the region's largest, like the program, but that many dealers have yet to participate (Burrell 1990). A previous pilot program, which offered no dealer incentives, received very little interest from dealers. The pilot program did succeed, however in establishing an on-going relationship with the designers of a new Suzuki/General Motors plant, resulting in the installation of 9,000 hp of high-efficiency motors. This one plant accounted for 90% (horsepower-weighted) of the pilot program's rebates (Burrell 1990).

The current Ontario Hydro program includes an educational booklet for customers, a free computer program for estimating energy savings, and prepared marketing materials, including flip-charts, which are given by the utility to motor dealers for use with their customers (Burrell 1990).

New England Electric has run several multiple-end-use rebate programs over the past several years, targeted at medium and large C&I customers. Participation in the motor rebate component was just over 1% after 1.5 years, and most of the activity involved a few motor dealers who used the program as a cornerstone of their marketing efforts. Disappointed with the low participation, the utility in 1989 dramatically increased motor rebate levels (enough in many instances to pay the full cost of new high-efficiency motors), added generous rebates for adjustable-speed drives, simplified the rebate structure and application procedure, and held training seminars on the new program for motor dealers. With the new program, the utility is targeting retrofits of existing motors with high operating hours. A companion program pays lower rebates for motors installed in new construction projects. After one year, the new program has attracted approximately four times more participation than the old program (Stout 1990). The rebate schedule for the revised programs are listed in Table 9-11 (next page).

Nevada Power, Northern State Power, and Southern California Edison have run rebate programs for dealers as well as customers. These utilities' dealer rebates are generally low (respectively, $10/motor, $.50/horsepower, and credits that can be applied toward gifts) and have had little impact on program participation. Rebate

Table 9-11. New England Electric's motor rebates for retrofit and new construction applications.

HP	Minimum nominal efficiency	Motor incentives			
		Retrofits		New construction	
		Open drip-proof	Totally enclosed fan-cooled	Open drip-proof	Totally enclosed fan-cooled
1	82.0%	$80	$125	$25	$40
1.5	84.0%	$110	$150	$25	$50
2	84.0%	$120	$170	$30	$50
3	87.0%	$140	$200	$30	$50
5	88.0%	$180	$250	$35	$60
7.5	89.0%	$260	$320	$50	$85
10	90.0%	$310	$380	$65	$100
15	91.5%	$420	$520	$100	$130
20	92.5%	$520	$620	$100	$150
25	93.0%	$600	$700	$100	$175
30	93.0%	$650	$760	$125	$200
40	93.5%	$800	$1,170	$125	$250
50	94.0%	$950	$1,200	$150	$300
60	94.5%	$1,200	$1.400	$200	$425
75	95.0%	$1,470	$1,700	$200	$600
100	95.0%	$1,900	$2,000	$300	$750
125	95.0%	$2,000	$2,400	$500	$1,400
150	95.5%	$2,480	$2,650	$700	$1,500
200	95.5%	$2,900	$2,750	$1,000	$1,900
250	95.5%	$3,250	$3,400	$1,500	$2,000

program managers generally recommend higher rebates to dealers. The Southern California Edison program staff spent a year trying to develop relationships with dealers. A companion program for HVAC dealers received extensive interest, but motor dealers were not interested in participating (Tyre 1989, Gunn 1989, Mayo 1989).

Northern States Power (NSP) is trying not only dealer rebates but also the novel approach of providing higher rebates ($7/horsepower) for new motors that are replacing functioning motors and lower rebates ($2/horsepower) for new motors not replacing functioning motors. Participation in the new programs has been low, and the utility hopes improved marketing will stimulate further interest (Northern States Power 1988), although low participation could be due to the low rebate levels: the NSP rebate for replacement motors is less than what many utilities are paying for new, nonreplacement motors.

Boston Edison is planning a program based on the premise that if rebates for replacing existing motors approximate the cost difference between rewinding an old motor and buying a new high-efficiency motor (see Table 9-9), interest in the program might increase substantially (Boston Edison et al. 1990). This approach targets the motor rewind market and seeks to upgrade motor efficiency over the next 10–20 years, during which time most existing motors will face replacement or rewinding. To be most effective, such a program should involve motor rewind shops in program promotion, perhaps by giving them an incentive to sell efficient motors. For small and medium-size motors, the value of energy savings to the utility is likely to exceed the cost difference between the rewind and replacement options. The cost of new large motors (about150 hp and larger) is high relative to the small efficiency gains offered by high-efficiency units; consequently, to keep programs cost-effective, utilities might need to rebate only a portion of the cost difference between rewinding and replacement.

Pacific Gas & Electric (PG&E) has established a two-tier motor rebate program. In their program, motors meeting the efficiency values in Table 9-2 receive moderate rebates and motors meeting the efficiency values in Table 9-3 receive higher rebates. In this manner they hope to promote a shift towards the highest efficiencies now on the market. As of 1992, for most motor types and sizes, only three manufacturers produced motors that are eligible for the higher rebates (Fitzpatrick 1992).

Targeted Rebates for Other Motor Improvements

A few programs have promoted other kinds of motor efficiency improvements. The City of Palo Alto, California, offered incentives for motor downsizing, which few customers took advantage of (Davies 1989). Reasons for this lack of interest were never explored. Wisconsin Electric paid incentives of $100/hp for ASDs but became concerned that not all ASD applications were cost-effective to the utility and changed the program to pay incentives based on kW and kWh savings (the revised program is discussed later in this chapter) (Hawley 1990).

New England Electric may have found a way around the cost-effectiveness problem by limiting rebates to projects that pass a series of eligibility requirements. ASD rebates are paid per horsepower, with levels decreasing as system size increases (because larger ASD systems are less expensive per horsepower). The incentives are sufficient to pay the full cost of an adjustable-speed drive for many applications. Rebate levels and eligibility requirements are listed in Table 9-12 (next page) (Stout and Gilmore 1989). To qualify for a rebate, a motor application must have high operating hours under part-load conditions, or other features that make the

Table 9-12. New England Electric's ASD rebates and eligibility requirements.

Motor Size	Maximum Incentive
20 Hp or less	$300/Hp
25-100 Hp	$200/Hp
125 Hp or more	$125/Hp

Eligible Applications:
1. Variable air volume fans for commercial buildings or for commercial parts of industrial facilities (such as offices).
2. Chilled water pumps for HVAC systems that meet one of the following criteria:
   a. The pump runs in all seasons. Examples include:
      (1) Buildings with a water-side economizer
      (2) Pumps that supply dedicated cooling units for computer rooms
      (3) Other buildings where cooling is required all year
   b. Buildings with 24-hour occupancy. Examples include:
      (1) Hospitals
      (2) Computer centers
      (3) Prisons
3. Hot water pumps for HVAC systems that run in all seasons: for example, if reheat is required to meet ventilation or humidity control standards, such as in hospitals or some process areas.
4. Process pumps and fans that meet one of the following criteria:
   a. Average operation of at least 100 hrs/week operated at less than 80% of rated flow.
   b. Average operation of at least 70 hrs/week operated at less than 70% of rated flow.

efficiency improvement likely to be cost-effective for the utility and the customer. ASD applications that do not pass these screening tests are still eligible for custom rebates at levels based on projected energy savings (custom rebates are discussed below).

The NEES program also requires strict in-plant harmonic testing for ASD systems but, because of customer resistance, the utility plans to waive this requirement if the customer will isolate the ASD from the main electric system by means of a transformer. NEES is also considering testing individual ASDs and waiving in-plant harmonic tests for equipment with low harmonic levels (Stout 1990).

BC Hydro is in the process of developing incentives for compressor, pump, and fan system improvements. In each case, the utility starts with a pilot research program in which a limited number of systems are assessed, retrofit improvements are installed (at no or little cost to the customer), and savings are closely monitored. Based on the results of the pilot, options for full-scale programs are developed. For example, as a result of its compressor test program,

BC Hydro has discovered significant opportunities to improve compressor system performance through such measures as dryer surge controls, compressor sequencing controls, and booster compressors for high-pressure special applications. Incentive programs for some of these measures are being planned (Willis 1990).

Generic Rebates

Generic rebates reward customers for savings achieved through measures of the customers' choosing. They generally take one of four forms: payments per kW or kWh saved (either in the first year or over the measure life), payments as a percent of measure cost, or payments to bring the simple payback for a measure down to a specific level.

For example, Wisconsin Electric will buy down the cost of a measure to a 1.5-year payback. Incentives are capped at the utility's avoided cost or 50% of the measure cost, whichever is less. Customers submit project proposals that describe the equipment and include an analysis prepared by a licensed engineer of the energy savings that will result. After three years, over 100 custom industrial process improvements and more than 100 custom HVAC system improvements have gone through the program, including an estimated 25–50 ASDs. Wisconsin Electric has found that projects need careful screening, for while many projects are competently designed, some customers and vendors do not understand how to prepare the calculations, they overestimate energy savings, or they apply measures to inappropriate applications (Clippert 1989, Hawley 1990).

Bonneville Power Administration (BPA) and Central Maine Power (CMP) offer payments to industrial customers of, respectively, up to $.15/kWh saved and $.10/kWh saved in the first year. These utilities generally limit payments to levels necessary to bring the customer's simple payback to one to two years. Northeast Utilities (NU) had a similar incentive of $.10/kWh saved in the first year but now simply buys down the measure to a one-year simple payback. As in the Wisconsin Electric program, customers submit project proposals for approval by the utility. Motor system efficiency improvement measures have featured prominently. For example, of the first four projects in CMP's program, one involved replacement of 194 motors at a wood products manufacturing plant, one replaced 39 motors and pumps in a woolen mill, and one included installation of an ASD at a major paper manufacturing plant (CMP 1988). In the BPA program, out of 19 contracts awarded in the first year, six were for motor projects, including four ASD projects (BPA 1989). In the NU program, motor improvements account for the second largest share of energy savings (lighting improvements are first) (Morante 1990).

In some cases, utilities combine generic rebates with extensive technical assistance. For example, Puget Power and Light, Boston Edison, and Northeast Utilities offer programs that provide extensive assistance conducting audits, preparing bid specifications, and arranging for measure installation. Due to the extensive technical assistance involved, the programs serve only a limited number of customers each year (typically less than 100). These programs can be very popular with customers. For example, Puget has only advertised its program through word of mouth and at times has had a backlog of applications as long as two years (France 1989). The programs have promoted packages of conservation improvements, principally at commercial facilities. Energy savings have averaged 10–23%, at a cost to the utility of approximately $.02–.03/kWh saved (Nadel 1990a). The role of motor efficiency improvements in these programs has not been analyzed for any of the above programs.

To summarize, motor rebate programs have varied widely in effectiveness. The advantages and disadvantages of different approaches have been examined by Weedall and Gordon (1990) and by Nadel (1990a). The more successful efforts generally feature all or most of the elements and perspectives described below:

- An extensive education and technical assistance component.
- Marketing that emphasizes one-on-one personal contacts with dealers and large customers.
- Significant rebates for customers and dealers.
- Rebates for specific, widely applicable measures with clearly defined costs and savings are generally the easiest for customers to understand (for example, a $15 rebate for a 1-hp high-efficiency motor).
- Payments to bring the cost of a measure down to a specified payback level are more difficult for customers to understand and require considerable analysis to accurately estimate costs and savings. On the other hand, these incentives best advance a utility's long-term objective of encouraging medium- and long-payback measures, which customers would be unlikely to implement without utility assistance. This approach is perhaps best suited to large customers who can understand and are willing to work with complex programs.
- Payments per kW or kWh saved or as a percent of measure cost generally fall between the above two approaches in ease of customer comprehension. Such payments can be set to reflect the value of energy savings to the utility. On the other hand, they encourage rapid payback measures and discourage longer payback measures.

- Programs with the highest participation and savings generally combine substantial incentives with extensive assistance in assessing measures and facilitating their installation. Such programs tend to be more expensive than other approaches, although program costs are less than avoided costs for many utilities.

Loans

Loans can compensate for the limited access to capital that prevents many customers from investing in efficiency. Subsidized loans with zero or below-market interest rates are particularly attractive to customers. The New York State Energy Office complements its energy audit program with an Energy Investment Loan Program. The program buys down the interest rate (to 6–8.5%, depending on loan terms) of loans provided by local banks to businesses, farms, and owners of multifamily buildings. In two years, approximately 300 loans have been closed, totaling $23.5 million. Industrial projects account for approximately 20% of the loan dollars, primarily for HVAC and process improvements. Data on motor improvements are not available (Fenno 1989).

A number of utilities have offered loans to their customers. Wisconsin Electric and Puget Power and Light offer customers a choice between a zero-interest loan and rebates with the same cost to the utility. Over 90% of participating customers have selected rebates. Loans are thus less popular, but are useful for the minority of customers who lack investment capital. Furthermore, both utilities found the rebates easier to administer than the loans (Clippert 1989, France 1989).

Leasing

Leasing specific pieces of energy-saving equipment also obviates searching for investment capital. For example, the municipal utilities of Taunton, Massachusetts, Burlington, Vermont, and Gainesville, Florida, each will lease and install efficient lighting equipment in customers' facilities for a set payment per month. Lease payments, included in the monthly electric bill, are less than the value of monthly electricity savings. Thus the customer pays no money up front and saves from the first month forward. In some cases, the customer assumes ownership of the equipment when the utility has recovered its investment; in other cases, the customer remains a lessee, and the utility is responsible for replacing worn-out equipment (Desmond 1989, Buckley 1990, Donaldson 1989).

While leasing thus far has only been used for lighting, Alberta Power is now planning a motor leasing program. As the program is currently envisioned, the utility will lease high-efficiency motors and adjustable-speed drives to large customers. An unregulated subsidiary will provide leasing services, and the utility hopes to make a small profit on the arrangement (the utility reasons that risks are greater than with its core business of electric sales, and hence profits should be bigger as well). The subsidiary will purchase motors directly from manufacturers at prices typically paid by original equipment manufacturers, and it will offer a variety of lease arrangements (lease in perpetuity and lease with option to buy, for example) in order to respond to different customer needs. The program will especially target users of very large motors used for oil and gas drilling and pumping (Hotrum 1990).

New Construction Technical Assistance and Incentives

New construction embodies decisions that affect the energy use of a facility for many years to come. Incorporating efficiency measures when a facility is built is much less expensive than retrofitting them later because marginal capital, design, and installation costs are much lower for new construction. For these reasons, energy-saving opportunities in new facilities are often called "lost-opportunity resources": once the opportunity to acquire these resources inexpensively is lost, it may never come again.

Due to the unique energy-efficiency opportunities during new construction and to differences in the new construction and existing facility markets (discussed in chapter 8), several utilities and government agencies have targeted special programs toward the new construction market.

For example, BPA, the Tennessee Valley Authority, and Washington State Energy Office have all offered programs that provide extensive technical assistance to designers and developers of new commercial buildings. The programs include free consultations with experts in the energy-saving field and computer analysis of energy-saving options, including high-efficiency fans, pumps, motors, and adjustable-speed drives for ventilation and pumping applications. A review of these programs (Nadel 1990a) found that they were identifying many energy saving measures, some of which were being implemented. Program managers generally felt, however, that financial incentives would further encourage adoption of the recommended technologies.

Some programs do combine technical assistance with incentives for energy-efficient new construction. One example is BPA's Energy Edge program, which has produced buildings that, on average, use 21% less energy than those built with prevailing construction practices (this savings was calculated from computer simulations based on electric meter data—see Harris and Diamond 1989). Technical assistance includes brainstorming sessions on energy-saving options, computer modeling of each efficiency measure and of the interactive effects of multiple measures, consultation on particular technical issues, and assistance with building commissioning and on-going maintenance procedures. Financial incentives cover the difference in cost between standard- and high-efficiency construction. Building designers are reimbursed for the extra time spent on the project (Anderson and Benner 1988, Benner 1988). Similar programs are being run by Northeast Utilities (Benner et al. 1989) and New England Electric System (McAteer 1990).

While most new-construction programs have thus far concentrated on commercial buildings, a few are beginning to address industrial facilities. For example, BPA's Design Wise program provides engineering reviews of new facility and process line plans in order to make energy-saving recommendations. Reviews are provided by experts in a particular industry on retainer to the utility. The program also includes site visits to discuss new electrotechnologies (Little 1989).

Several utilities, including Wisconsin Electric and New England Electric System, allow industrial facilities to participate in their commercial new-construction incentive programs. Both programs pay incentives for measures exceeding standard construction practice in a specific industry. Customers propose projects and are reimbursed on the basis of kW and kWh savings. In the initial years of both programs, industrial process improvements were not actively promoted but both utilities plan to expand marketing and technical assistance to industrial customers in the future (Keneipp et al. 1990, New England Electric System and Conservation Law Foundation 1989).

Performance Contracting

Performance contracts typically involve private energy service companies (ESCOs), which contract with a utility or end-user to assess, finance, and install energy conservation measures. The ESCO takes as payment a share of the energy cost reduction, based either on engineering estimates or actual metered savings. In the latter case, the ESCO takes the risk that predicted savings will actually materialize.

A recent review of utility experience with performance contracting programs (Nadel 1990a) found that:

1. Most ESCOs concentrate on the largest customers and the most lucrative energy-saving measures (particularly lighting and

cogeneration). ESCOs sometimes install high-efficiency motors, but other motor system improvements are rarely pursued.

2. Programs with high participation rates and high savings per participating customer, such as those offered by Boston Edison and Commonwealth Electric, generally pay ESCOs amounts close to the utility's avoided costs.

3. The few side-by-side comparisons available indicate that other program approaches will achieve greater participation than ESCO-based programs. For example, New England Electric System has operated both a performance contracting program and a customer rebate program for several years. Participation rates were 40% greater and kW savings 165% greater for the rebate program, despite the fact that the performance contracting program paid an average of 45% higher incentives; operated six months longer; served larger customers (it targeted customers with billing demand greater than 500 kW, whereas the rebate program had a 100-kW threshold); and served about twice the total customer kW demand (Hicks 1989).

4. Performance contracting programs generally cost the utility more than utility-operated programs promoting the same measures because the utility must directly or indirectly cover ESCO overhead and profit.

5. Many utilities that offer or have offered performance contracting programs have either phased out these programs or chosen to complement them with other types of programs. Nevertheless, performance contracting programs can be useful for customers who do not have the time, money, or expertise to implement energy-saving measures on their own, and for utilities who prefer that outside contractors deliver energy services, thereby saving utilities from having to administer the programs themselves (although managing ESCOs can present its own problems).

Performance contracting is particularly attractive to government agencies, which often lack both the capital to finance and the expertise to assess and implement cost-effective conservation measures on their own. Performance contracting results in the public sector have been mixed. Success is most likely when an experienced and reputable ESCO is involved and when customers are sophisticated negotiators and knowledgeable about performance contracting. ESCOs are not interested in working with all customers, however, nor are all customers interested in working with ESCOs; these situations call for other approaches (DeWitt and Wolcott 1986).

In contrast to the U.S. experience, performance contracting has succeeded in achieving large energy savings, at least in multi-family housing, in France, where energy service companies rose to prominence after World War II in response to the need for skilled technicians to manage heating systems in multifamily housing. French ESCOs are primarily energy management companies, not financial companies, and staff are primarily engineers and techni-cians, not financial analysts. French ESCOs almost always are responsible for ongoing operations and maintenance at the facilities they serve; they work in a facility over the long term. Often they are paid a fixed fee, so each franc saved adds directly to their bottom lines. Due to their long-term perspective and the dominance of engi-neers, French companies are often interested in special projects that pose engineering challenges and have long payback periods (de la Moriniere 1989). This is the type of perspective needed to optimize motor system efficiency. It may be worth exploring whether the French experience in multifamily buildings can be adapted to North American commercial and industrial facilities.

Bidding

In the past few years bidding programs under which utilities request proposals from outside parties to supply demand-side and supply-side resources have attracted much interest. Bids are selected on the bases of price and other factors. One purpose of bidding programs is to allow the marketplace to determine the price of savings, although the tendency is for bids to approach the utility's avoided costs, which is where bid prices are typically capped (Estey 1989).

Several issues must be addressed in developing bid programs: the resources to be solicited and sectors to be targeted, bid evaluation criteria, bid ceiling prices, and mechanisms for measuring the sav-ings. Goldman and Hirst (1989) address these issues at length.

Demand-side bidding programs, though still few, are prolifer-ating rapidly. Most bids have been submitted by energy service companies for large C&I projects, although some large C&I cus-tomers have also submitted bids (Estey 1989). These programs appear able to achieve significant energy savings. By the end of 1989, Central Maine Power had signed contracts totaling 1.5% of its peak demand through its Power Partners and Efficiency Buyback programs. Some of these projects involve high-efficiency motors and adjustable-speed drives (Linn 1990).

While definitive conclusions await further experience, Goldman and Hirst (1989) conclude that demand-side bidding programs, though valuable, have a limited role to play in a utility's overall demand-side management strategy because (1) the U.S. energy services industry remains undeveloped; (2) transaction costs in the residential and small commercial sectors are high; and (3) bidding is inappropriate for some types of programs, such as design assistance for new construction and other informational programs. Bidding is likely to contribute most significantly to savings in the large commercial and industrial sectors, where motors are heavily used; for smaller facilities, other strategies are likely to prevail.

Tax Credits

Tax credits and accelerated depreciation allowances have been used in many countries to promote energy-saving investments. For example, from 1978 to 1982, U.S. businesses could take a tax credit equal to 10% of the capital cost of energy-saving investments. Similarly, Japan allows a 7% tax credit on energy-saving investments plus a special depreciation allowance equivalent to 18% of the acquisition price of energy-saving measures (Furugaki 1988).

An analysis of the U.S. program (Alliance to Save Energy 1983) found that the credit had little impact on energy investments made by U.S. industry—most firms who took the credit would have made the same investments if no credit had been available. This study also examined the hypothetical effects of higher tax credits, including a 40% tax credit and a 70% repayable credit, which would be repaid over a series of years. The analysis concluded that these latter two credits would have only a limited impact on project economics— that is, the internal rate of return for a project with a 2.5-year simple payback rose from 33% in the no-credit case to approximately 45% in the high-credit cases. While the impact of high credits cannot be determined with any certainty, the authors of the study conclude that "energy tax credits are relatively ineffective in inducing industrial firms to undertake additional conservation investment."

Research, Development, and Demonstration Programs

Most research, development, and demonstration (RD&D) work on new motor system technologies is done by private firms, ranging from industry giants to small start-ups. This work is largely carried out by companies in the United States, Europe, and Japan, and covers diverse fields such as materials, electronics, motors, and ASDs. Most RD&D work is done as a strategic effort performed by private companies to ensure the competitiveness of their products.

In addition, government agencies and electric utilities, through EPRI and the associated Power Electronics Application Center (PEAC), conduct drivepower RD&D, often in conjunction with industry. For example, the U.S. Department of Energy has worked with private manufacturers on the design of permanent-magnet motors, reluctance motors, and amorphous metal motors (these technologies are briefly discussed in chapter 2) (Comnes and Barnes 1987). Similarly, EPRI and PEAC are conducting research on a number of motor-related technologies including permanent-magnet motors, high-efficiency motor rewinds, manufacturability of more efficient motors, square-wave motors (motors optimized to run on the square-wave power produced by some ASDs), improved efficiency ASDs, and clean-power ASDs (those with less harmonic distortion) (Smith 1990).

Much of the recent applications-related RD&D activity has involved ASDs and, to a lesser extent, improved production processes and motor efficiency in rewind and part-load applications. For example, EPRI has conducted two ASD application and demonstration projects, one for very large motors (2,000 hp and up) used in power plants (Oliver and Samotyj 1989), and one for industrial applications of fans and pumps ranging from 5 to 1,250 hp (Poole et al. 1989). Some of these projects are discussed in chapters 4 and 5. The North Carolina Alternative Energy Corporation has conducted drivepower RD&D in wood-dust collection systems, industrial heat pumps, compressed air systems, variable-airflow fans for lumber dry kilns, and rewind and part-load applications (NCAEC 1989). Consolidated Edison has worked on ASDs for HVAC systems and other commercial applications (Consolidated Edison 1989). And BC Hydro's efforts have focused on fan, pump, and compressor systems (Schwartz 1990).

Many of these projects are used to develop full-scale programs. For example, BC Hydro has an active education program on ASDs that draws heavily from case study projects it has sponsored (Schwartz 1990). Based on pilot and demonstration projects, incentives for compressor efficiency improvements are now being developed (Willis 1990). BC Hydro hopes to use its fan and pump demonstrations to develop full-scale program ideas as well (Donnelly and Gudbjargsson 1990).

Utility- and government-funded RD&D has also stimulated commercialization of advanced energy-saving projects. For example, EPRI recently teamed up with the Carrier Corporation to produce a high-efficiency variable-speed heat pump that incorporates a permanent-magnet motor with an integral ASD—a first in the HVAC industry (Moore 1988). Likewise, EPRI's ASD demonstrations have probably speeded up market acceptance of these products, although definitive data on the effects of the demonstration projects are not available.

In addition to this extensive RD&D work, additional work is needed in order to advance the state of the art and to increase market share for commercial high-efficiency products. Specific recommendations are presented in chapter 10.

Conclusions

- Most programs address high-efficiency motors, and a handful include ASDs. Few programs address other technologies or application-related savings opportunities.

- Education and technical assistance programs appear to have had little impact in improving motor system efficiency. Such programs are never evaluated, however, so their effectiveness cannot be fully assessed.

- Research, development, and demonstration programs can build valuable experience from which full-scale programs can be developed.

- Rebates for high-efficiency motors are the most popular kind of incentive for drivepower efficiency improvement. Most motor rebate programs can benefit from increased packaging, marketing, and incentive levels. Incentives for ASDs and other motor system efficiency improvements are still in their infancy—some promising efforts are underway, but much work remains to be done. While most activity in recent years has emphasized rebate programs, a number of other program approaches such as technical assistance/incentive programs, new construction programs, and leasing merit serious consideration.

- Efficiency standards for motors were previously rejected because it was believed that the market would achieve the same savings. Available data indicates that this is not true. In the absence of standards, market barriers will severely restrict the penetration of high efficiency motors and hence standards merit the renewed consideration they are now receiving.

No single program or policy is likely to overcome all the barriers to motor efficiency improvements. To achieve even half the available savings will likely require a combination of complementary approaches, including variations on those discussed in this chapter as well as new, creative ones. Some specific recommendations are discussed in the next chapter.

Chapter 10

Recommendations

Full cost-effective application of the technologies and practices discussed in this book can reduce national drivepower energy use by 14–38%. While these potential savings are large, many barriers to their capture exist. This chapter recommends programs and policies to remove barriers and advance the implementation of energy-saving measures and practices. It also discusses research and development needs, both to help implement existing technologies and practices, and to advance the state of the art in efficient motor systems.

Program and Policy Options

Education, Technical Assistance, and Demonstration Programs

More education is needed on many fronts. Motor system efficiency should be incorporated into engineering curricula, junior engineers need one-on-one field training with experienced engineers, and practicing professionals should have ready access to continuing education programs. Most training programs to date have concentrated on high-efficiency motors. While these efforts are useful, expanded efforts are needed in numerous other areas. We cite these educational needs throughout our discussions below of other programs and policies.

Adjustable-Speed Drives and Other Controls

Although ASDs and other controls are the largest potential source of motor system energy savings, most end-users and many consulting engineers lack adequate information on ASD performance, economics, and applications. This information gap should be filled with handbooks and training courses that emphasize practical applications of ASDs and other controls and by one-on-one technical assistance. Existing programs are limited in number and scope. Many are one-shot efforts without follow-up.

In particular, customers need assistance monitoring load profiles in fan and pump operations. Such monitoring is important for accurately predicting ASD savings. To facilitate monitoring work, lending of test equipment and training in its use should be established by utilities, government agencies, and other interested parties. Such a program is now being set up by New England Electric System for its commercial and industrial customers (NEES and CLF 1989).

ASD demonstration projects employing appropriate monitoring instruments are also needed to collect detailed data on costs, savings, and reliability, and to overcome the concerns of skeptical end-users. Reliability is a major concern of end users but is seldom documented in demonstration projects. Such efforts should focus in particular on industrial process applications, because demonstration projects to date have emphasized utility power plant and commercial HVAC applications. Demonstration projects on small motors at power plants also are needed, since prior utility demonstrations have emphasized motors of 1,000 hp or more (Poole 1989). Hourly electric loads should be recorded, as utilities need load shape information to estimate the economic benefits of ASD incentive programs.

Intensive publicity campaigns should distribute results of demonstration programs to as large an audience as possible. Such publicity can take many forms: campaigns backed by industry associations, brief nontechnical summaries with attractive graphics, more extensive technical summaries, videotapes, and seminars built around case studies of demonstration programs.

Motor Repair and Replacement Practices

Many motors that could be cost-effectively replaced are rewound and motor efficiency can suffer from poor rewind practices. Motor users need education programs and instruction materials that focus on the economics of rewind versus replacement decisions, on the advantages of core-loss testing of motors before and after baking, on monitoring procedures that ensure burned-out motors are replaced when justified, on how to select a rewind shop that uses nondamaging rewind procedures, and on procedures for purchasing and inventorying motors so as to reduce the cost of new units and minimize downtime during replacements.

Some utilities are already promoting sound procedures for rewinding and replacing motors. The Carolina Power and Light program (discussed in chapter 9), in which motors meriting replacement are marked with a yellow dot, is exemplary. Southwire's policy (see chapter 8) of buying only high-efficiency motors and rewinding nothing under 125 hp is another good example of ways to keep costs down and procedures simple. Adding high-efficiency motors to inventory (either on site or at the local distributor) allows new high-efficiency motors to be quickly installed when an old motor fails.

Rewind shop operators as well as customers should be educated on procedures for minimizing core damage, including lower bake-oven temperatures and the use of such low-temperature processes as the Thumm-Dreisilker method discussed in chapter 2. Rewind shops also should be encouraged to test all motors for core damage before and after rewinding. Utilities could provide incentives for the purchase and use of the necessary test equipment. A certification program for rewind shops is another option worth exploring: to be certified, shops would have to use approved procedures; compliance could be determined by randomly submitting pretested motors to each shop for rewinding. If rewind damage were discovered, follow-up tests on additional motors and corrective measures, if required, would be pursued.

Optimal System Design

Efficient, cost-effective performance requires careful integration of motors, controls, electrical cables, drivetrains, and driven equipment. Such optimization is essentially a lost-opportunity resource—it needs to occur when the system is first designed or is being modified for other purposes. Optimization cannot be achieved piecemeal.

All too often, optimization is neglected because of inadequate time or expertise. Optimizing a system requires a knowledge of electrical and mechanical engineering, computer optimization techniques, and practical experience with the particular systems and processes affected. Few individuals have the requisite skills. To address these problems, utilities, government agencies, and other appropriate parties should provide free or low-cost, one-on-one technical assistance employing interdisciplinary teams of skilled motor system designers. To access the needed skills, utilities and government agencies will generally need to hire experienced private consultants. Also, EPRI, NEMA, DOE, and other national organizations should develop training-for-trainers programs.

Development of so-called expert systems (sophisticated computer programs that mimic an expert by going through complex decision procedures) could help to improve motor system designs and evaluate existing installations. Short of full-scale expert systems, increased efforts are needed to develop and distribute user-friendly software that allows engineers to better size fans, pumps, and other equipment, particularly for variable-speed applications, and to more easily estimate the energy savings and payback of high-efficiency motors, ASDs, and other motor system improvements. While some software has been produced, most of it is not widely distributed, and much of it is difficult to use.

Another aspect of system design meriting increased attention is the systematic addition of safety margins at each stage in the design process, resulting in dramatically oversized equipment. While safety margins have their place, engineers need improved training on equipment selection and control procedures that provide a safety margin without affecting efficiency.

Improved O&M Procedures

Training courses and manuals for maintenance staff—and for their bosses, who decide how much staff time should be allocated to maintenance—can go a long way toward improving maintenance practices in the field. If maintenance staff and their supervisors understand how costly it is to lose a percentage point or two in efficiency from inadequate O&M and appreciate how much down-time from bearing failures or other problems can be reduced by better O&M, these personnel are more likely to upgrade their maintenance practices.

One focus of this training should be encouraging users to measure motor efficiency when they first get a motor and then regularly thereafter, especially before and after repairs. This practice benefits users by helping them to spot problems and helps build a database that program and policy planners can access.

For example, Owens-Corning Fiberglas Corporation regularly tests motors to identify units that are likely to fail and need repair. Motors also are tested before and after repair to evaluate the repair job. The program has cut the motor failure rate significantly, improved motor repair quality, and reduced repair costs (Kochensparger 1987).

Additional mechanisms for improving O&M include lending libraries of test equipment and training in the use of (and perhaps incentives to purchase) maintenance software. As was noted in chapter 9, good maintenance software is available primarily for users of many motors; more work on the development and distribution of inexpensive, user-friendly software for smaller users is needed.

Testing, Labeling, and Standards

Motor testing and labeling can help consumers make informed

choices about motor efficiency. In the United States, 1- to 125-hp, three-phase, squirrel cage induction motors of NEMA Designs A and B are tested and labeled. This practice should be extended to other types of motors, including single-phase motors, motors of less than 1 hp or more than 125 hp, noninduction motors, and multispeed motors. Motor catalogs and promotional literature should publish the efficiency ratings (many catalogs presently omit this data, particularly for standard-efficiency motors). Manufacturers also should publish efficiency ratings for all motors at 0%, 25%, 50%, 75%, and 100% of load. They currently list values at only 50%, 75%, and 100% load, making the performance of a motor that will run below 50% loading difficult to evaluate.

Changes to the U.S. efficiency testing procedure may also be called for. As chapter 2 explained, published efficiency ratings in the United States are rounded down to one of 43 preset figures. For example, 90.2% and 89.5% are two adjoining preset efficiency values. The spread between these two values supposedly reflects typical testing tolerances. If several motors of the same model test at an average efficiency of 90.0%, the published nominal efficiency will be rounded down to the nearest preset value of 89.5%. This practice can mask considerable efficiency differences between motor models. For example, a 100-hp motor running 6,000 hours per year at 75% load uses $125 more worth of electricity each year (at $.06/kWh) if its efficiency is 89.5% instead of 90.0%. We recommend that the use of narrower efficiency increments be investigated, based on results of new, more accurate test instruments and modified test procedures. For example, some have argued that the Canadian test procedure, developed in 1985, contains less room for interpretation error than the older NEMA procedure (BC Hydro 1988).

Furthermore, the guaranteed efficiency of a motor is by definition two preset efficiency increments less than its nominal efficiency, regardless of the manufacturing tolerances employed by a particular manufacturer. Procedures for calculating minimum efficiency should be revised to reflect the manufacturing tolerances of the specific make and model number being rated. For example, if a particular manufacturer employing above-average quality control procedures can produce 95% of its motors within one preset efficiency increment, then the minimum efficiency should be only one preset increment below the nominal efficiency. As we go to press, NEMA is considering a proposal to make the minimum efficiency one preset increment below the nominal efficiency, effective at a future date yet to be determined (Raba 1990).

While the need for revised testing and labeling is clear, the question of who should make these changes remains. The industry may want to do so voluntarily, under the auspices of NEMA or another national organization. If the industry does not voluntarily move in this direction, the U.S. Department of Energy should step in.

Improvements in the design and enforcement of standards are also needed. For example, the Electrical Apparatus Service Association's (EASA) voluntary standards for motor rewinds should be made mandatory for EASA members, subject to periodic spot-checking, and revised to include before-and-after core-loss testing. Moreover, the issue of proper bake-oven temperature should be revisited by EASA, EPRI, or some other party.

The Institute of Electrical and Electronics Engineers (IEEE) has standards on harmonic distortion generated by electronic equipment, including ASDs. These standards are intended to minimize adverse impacts of harmonics on other equipment. They specify maximum harmonic distortion limits for end-user equipment, the levels varying with the characteristics of the end-user's electric service. The standards provide only limited guidance as to whether a particular piece of equipment will cause harmonic distortion problems in a particular application.

We recommend that IEEE, NEMA, or some other suitable organization establish maximum harmonic levels for specific pieces of equipment. At present utilities, end-users, and manufacturers are uncertain whether installing a specific piece of equipment will cause harmonics problems in a particular application. Equipment-specific maximum harmonic levels could be based on the probability of such problems. For example, the basic standard could be set at a level that would avoid problem-causing harmonics in 95% of facilities. More and less stringent standards could also be set for facilities particularly sensitive or insensitive to harmonics. A ratings scale could be established—"I" for equipment with a 1% probability of causing harmonics related problems, "II" for equipment with a 5% probability, and "III" for equipment with a 10% or 20% probability. The New England Electric System is now planning research in this area (Stout 1990). Such ratings for specific equipment will help users select equipment by clarifying the risks involved and will provide specific performance targets for manufacturers.

Motors are a textbook case of high-efficiency equipment sales significantly lagging their cost-effective potential. High-efficiency motors in 1990 have a market share of only about 20% (see chapter 6), despite the fact that they are generally cost-effective on a life-cycle basis at duty factors as low as 250 hours per year (see chapter 2). A very small portion of drivepower input goes to motors that operate this few hours per year (see Tables 6-7, 6-8, and 6-9).

Based on these considerations we recommend that federal governments adopt minimum motor-efficiency standards. DOE began but then abandoned this process ten years ago and the time to revisit the issue is long overdue. Congress should enact pending legislation. As testing and labeling expands to fractional horsepower motors, efficiency standards for these products should be considered as well.

Incentive Programs

Most current programs have had little success reaching customers and promoting savings, although these programs have yielded some important lessons. Some promising new programs have recently begun. Based on these experiences, a number of recommendations for future efforts can be made:

Motor Rebates

Motor rebate programs should be offered by additional utilities because these programs arc likely to be cost-effective in most applications. Rebates are particularly critical in the absence of standards. Available evidence indicates that programs with the following attributes do best:

- They are easy for dealers and end-users to understand and participate in.
- They are marketed through regular personal contacts with decision-makers. Often it takes several visits before a facility or dealer agrees to participate.
- They are coordinated closely with dealers, including inducements for dealers to stock and market high-efficiency motors.
- They provide high incentive levels to the extent justified by a utility's avoided costs and its conservation and load-management objectives. If a utility does not need additional capacity for many years, the utility may focus only on new installations of motors and controls and on repair and rewind decisions. Incentives need only be high enough to pay the incremental cost between standard- and high-efficiency motors (for new motors) or between rewinding and replacing an old motor (for failed motors). If a utility needs capacity in the short term and wants to encourage customers to replace existing inefficient motors, it will probably need to offer higher incentives.
- They include an aggressive education component for both dealers and end-users (as is done in the BC Hydro and Ontario Hydro programs discussed in chapter 9).

Generic Rebates and Other Incentives

Incentives other than motor rebates are also needed. Many of these were described in chapter 9 but have yet to be widely adopted. Options deserving greater attention include:

- Generic rebates per kWh or kW saved or to reduce the simple payback for a measure to a specified time period.
- Rebates per horsepower for ASD applications meeting specific criteria.
- New construction incentives tied to the incremental cost of improved designs and equipment.
- Leasing of efficient equipment, whereby lease payments are less than monthly energy savings and the lease payments are included on the customer's electric bill.
- Internal shared savings programs in which departments within a company or agency can keep a portion of the money they save and use it to fund further improvements and staff bonuses. For example, the State of Washington has a program under which employees involved in energy conservation programs share 25% of the savings resulting from the measures they implement, and the other 75% goes to the state (Lannoye 1988).
- Shared savings financing of drivepower installations, in which utilities or their contractors pay for new equipment and recover their costs through a share of the savings.
- Nonfinancial incentives, including marketing assistance, publicity, and prizes such as free trips offered to motor equipment dealers for reaching sales targets for efficient equipment.
- Technical assistance and perhaps incentives for correct motor sizing.
- Inducements to test for core losses, and replace motors with significant core damage.
- Free technical assistance or computer programs to track and schedule maintenance activities.
- A utility-subsidized motor maintenance service in which outside contractors would perform ongoing preventive maintenance and efficiency tune-ups in customer facilities.

Program and Policy Evaluation

Programs and policies too often fail to include plans for assessing how well they work. Evaluations that are conducted are rarely documented and disseminated so others can learn from the experience.

New programs and policies should from the beginning contain an evaluation component that compares the actions of program participants with those of a control group of nonparticipants so as to determine participation rates and savings attributable to the program.

Research, Development, and Demonstration

Our recommendations are grouped in two areas: equipment and data.

Equipment

High-efficiency motors are generally available for foot-mounted, three-phase, integral-horsepower, Design B induction motors with ODP and TEFC enclosures. Improving the efficiency of other induction motors, including single-phase units and additional polyphase models, should now have priority because these motors represent more than 25% of U.S. motor electricity use (Nadel 1990b).

Development should continue on magnetic core materials, permanent-magnet and reluctance motors, and other innovative designs. Each of these technologies has the potential to improve motor efficiency by 1–3% compared to currently available high-efficiency induction motors. Projected paybacks for these improvements are less than two years for a normal duty cycle (Comnes and Barnes 1987). Also, work is needed to optimize motors for variable-frequency power, as produced by ASDs.

RD&D on improved rewind practices is needed as well. The EASA study on the effects of conventional rewinds on motor core loss did not yield firm conclusions. New work, preferably funded by EPRI or DOE, should be conducted to definitively determine the relationship between burnout oven temperature and subsequent motor performance. Side-by-side comparisons of the costs and performances of alternative rewind techniques, including chemical stripping and the Thumm-Dreisilker method, should be conducted, and new low-temperature rewind methods should be explored.

Work is also needed to develop new ASD designs featuring improved torque and speed characteristics, higher efficiency, improved power factor, increased applicability, and reduced harmonics and interference problems. Projected developments in microelectronics and power electronics, including smart power integrated circuits and low-loss power switches, have the potential to reduce the size of ASDs so they may be easily packaged with the motor. Such miniaturization and packaging will reduce equipment and installation costs. For example, in late 1990 an American manufacturer introduced a washing machine featuring a small ASD incorporated into the motor. Adding integrated controls to ASDs will make it easier to build ASDs into control loops, a practice that can further cut equipment and installation costs. In addition to improving the technology, research should focus on ways to reduce manufacturing and installation costs. All steps in the manufacture and installation of ASDs should be examined for ways to cut costs without compromising reliability. For more on ASD research needs, see Greenberg et al. 1988 and de Almeida et al. 1990.

Better and less costly diagnostic equipment and techniques are needed. Many existing approaches are time consuming and expensive, and require special equipment. As diagnostics become simpler and cheaper, maintenance staff will be more likely to use them.

Data

Data on motor operating hours, load profiles, and actual efficiencies are spotty, sometimes dated, limited to specific regions, gathered under different methodologies, and often based as much on guesswork as on field analysis. A comprehensive field study on a representative sample of facilities is needed to estimate the potential savings across a wide range of measures—from high-efficiency motors and ASDs to better equipment sizing and maintenance—in a variety of end uses.

Such information, preferably stored in a database, will allow engineers, utilities, and other motor efficiency practitioners to better focus their efforts and develop rules of thumb and expert systems. Follow-up technical assistance (and perhaps incentives for measure implementation) might encourage participation in the data-gathering effort. By offering different post-survey follow-up services to different samples of customers, researchers could compare the effectiveness of various program strategies. Such a study would be a large undertaking, probably requiring involvement by the U.S. Department of Energy and the electric utility industry. Equipment manufacturers and user groups should also be involved. Such a study could start with one or two industries, then expand, until all major industries are addressed.

Disaggregated sales data for high-efficiency motors and ASDs are not now readily available. Information on the total number of these devices sold by size class would be very useful to program designers and policymakers. The U.S. Census Bureau and NEMA presently collect basic data on motor sales but do not separate high-efficiency motors from standard-efficiency motors. ASDs are combined with other types of electronic equipment, not tracked separately. NEMA and the Census Bureau should collaborate to gather and disseminate this information.

Conclusions

Tapping the riches of the drivepower gold mine will not be easy. While new and better hardware is welcome, we have only begun to take advantage of the advanced motors; controls; drivetrains; and monitoring, maintenance, and repair systems already available. We need far more and far better data on the existing motor inventory to identify and quantify the highest-potential savings opportunities. Field data on duty factors, load profiles, and suitability for retrofitting are musts. The lack of good information calls as well for interdisciplinary education and training, involving those who make, sell, specify, buy, and use drivepower technologies. To insure that no one buys uneconomic and wasteful equipment, we need better efficiency-testing of motors, inclusion of the results in motor catalogs, and minimum-efficiency standards for the most common types of motors. Also, in-plant testing is needed to make sure motor systems maintain their efficiencies over time. Utility or government rebates and technical assistance are vital, as are internal incentives that reward employees for saving energy. Demonstration programs are needed to prove the performance of new technologies. Finally, research is needed on new hardware and on program and policy options.

With roughly 60% of U.S. electricity use at stake, and a similar share in most other industrialized countries, motor systems can be fairly characterized as the mother lode of energy savings. While the challenges are large, the potential rewards are vast, in terms of energy and financial savings, economic competiveness, and environmental protection. We hope this book helps motor users everywhere to reap these important benefits.

Table 10-1. Summary of recommendations.

	WHO IMPLEMENTS
TRAINING, EDUCATION, AND TECHNICAL ASSISTANCE Provide practical training courses and handbooks on ASDs and other controls. Undertake education for users on improved monitoring, maintenance, and repair practices. Encourage before/after testing of all rewound motors. Provide technical assistance on system optimization. Provide extensive publicity on demonstration projects. Better incorporate motor system efficiency material into engineering curricula. Expand one-on-one field training for junior engineers.	Universities, engineering firms, trade associations, utilities, and government agencies.
Develop expert systems for system optimization. Develop and distribute low-cost, user-friendly maintenance software.	Engineering firms, software companies, perhaps with money from utilities and government.
CERTIFICATION AND REGULATION Make EASA's voluntary guidelines for motor rewinds mandatory for EASA members.	EASA.
Consider a rewind shop certification program.	EASA, utilities, government.
Expand NEMA testing and labeling requirements to motor types not presently covered.	NEMA.
Publish efficiency ratings in all catalogs.	Motor manufacturers, NEMA.
Develop narrower ranges for calculating nominal and minimum efficiency for motors.	NEMA.
Develop harmonics ratings for individual makes and model numbers.	Motor manufacturers, NEMA, IEEE.
Enact minimum efficiency standards for motors.	Federal and state governments.
Disaggregate high-efficiency motors and ASDs in statistical reports on equipment sales.	NEMA and Census Bureau.
PROGRAMS Expand utility incentive programs: More utilities. Improved education and marketing. Higher incentives for customers and dealers. Greater focus on measures other than high-efficiency motors. Include inducements for proper sizing, rewinding, system optimization, core-loss testing, monitoring, and maintenance. Greater attention to program evaluation.	Utilities and government agencies.

(continued on next page)

Table 10-1 (*continued*).

	WHO IMPLEMENTS
PROGRAMS (*continued*)	
Start-up test equipment lending sevices.	Utilities and government agencies.
Establish internal incentive programs so individuals can keep a share of the savings they achieve.	End-users w/ utility, government technical assistance.
RESEARCH, DEVELOPMENT, AND DEMONSTRATION	
Conduct a major field survey to collect data on motor operating hours, load profiles, and actual efficiencies, and estimate available savings across a wide range of measures.	EPRI, DOE, manufacturers, trade associations, other government agencies.
Research proper bake-oven temperatures, alternative rewinding techniques.	EASA, EPRI, DOE.
Commercialize high efficiency motors for additional motor classes.	Motor manufacturers.
Continue development of improved core materials, permanent magnet and reluctance motors, and other new designs.	Motor manufacturers, EPRI, DOE.
Develop better and less costly diagnostic equipment and techniques.	Test equipment manufacturers, EPRI, DOE.
Undertake ASD demonstration projects for industrial applications.	EPRI, utilities, government agencies.
Continue to improve ASD capabilities and efficiency while reducing costs, harmonics, and interference. Further integrate ASDs into OEM equipment.	ASD manufacturers, EPRI, and DOE.

Economics

Motor users and electric utilities each have reasons for installing efficient motors and drives: the motor users will save money on operating costs, and the utilities can purchase a demand-side resource that can be less expensive than new generating facilities. This appendix presents some general economic concepts used by each party, followed by technology-specific discussions and worksheets for evaluating the economics of drive-power investments.

Economic Concepts

Motor users and utilities use very different criteria to make decisions about investments. The perspective of the motor user is discussed first.

Motor User Perspective

Firms have choices about where to spend money to produce a return on their investment. Each option, including the purchase of energy-efficient equipment, must compete for scarce capital with other potential investments. Therefore, the economic analysis of efficiency investments should be formatted in the same way as the analysis of other capital investments, so that all options can be compared on an equal basis.

The return on investment needed to get a company to purchase energy-efficient equipment varies among firms. Many companies select products or make capital investments solely on the basis of least first cost. However, the most common method used by equipment buyers to evaluate conservation investments is the simple payback, or the time that it will take for the savings to pay back the cost of the investment. Simple payback is calculated by dividing the incremental cost of the efficient equipment by the value of the expected annual energy savings. For example, if an efficient motor costs $500 more than a standard motor and is expected to save $400 per year, the simple payback will be 1.25 years.

The use of the simple payback introduces some errors into the calculation by assuming that inflation is zero and utility rates are constant. It also ignores the life of the measure. A device with a six-month payback may seem like a good investment, but is not if it lasts only eight months. Because of the short payback requirements of most motor users, however, and the relatively low cost of installing efficient motors and drives, the errors in simple payback analysis are generally minor.

A related analysis, called return on investment (ROI), is used by some motor users. This method looks at the percent of the investment returned annually. For example, if an efficient motor costs $200 more than a standard motor and is expected to save $100 per year, it returns 50% of the investment annually. In general, this method produces results equivalent to the simple payback analysis and, additionally, is capable of evaluating a payment stream.

Within the format of a simple-payback or ROI analysis, the cut-off value for investments varies from company to company. However, survey data and discussions with motor users suggest that very few companies will invest in conservation improvements with paybacks exceeding three years, and many companies need to see a payback of two years or less (Marbek 1987).

Another method, called life-cycle cost analysis, calculates the total present cost of owning and operating the equipment over the life of the equipment, assuming that there is a time value of money. That is, future costs and savings are discounted back to the present, so that the cost of different options over the life of the equipment can be compared on an equal basis. This method presents the most accurate picture of investment options over the long term. In practice, however, very few companies use this method to make decisions about efficient equipment since most companies are typically looking only at short-term gains.

It is interesting to compare the results of different techniques for evaluating the economics of conservation measures. Assume that an energy-efficient motor costs $1,200 more than a standard motor and is expected to save the owner approximately $600 per year. The simple payback for this investment is $1,200/$600, or two years. The return on investment is $600/$1,200, or 50% per year. The motor is expected to be in service for 15 years. Over that period, it will save the owner $9,000. The present value of the energy savings depends on the discount rate used for the analysis. Using a real discount rate of 10%, the present value of these savings is $4,563, or almost four times the incremental cost of the efficient motor. If a real discount rate of 6% is used, the present value is $5,827, or almost five times the incremental cost of the equipment. Discount rates are user-specific; the higher the discount rate, the more the user values money in hand today over a stream of future savings.

Since simple payback is the most prevalent method used by companies, it will be used here in the tables and examples on the economics of efficient equipment.

Utility Rates

The economics of efficiency investments of course hinges on utility rates. Most utilities charge commercial and industrial customers for energy use (kWh) and peak demand put on the system (kW). The ratio of these two charges is highly utility specific and can produce bills where the demand component accounts for up to 45% of the total.

Most utilities also levy an additional charge if the power factor falls below a certain level (typically 0.85 to 0.95 depending on the utility). Again, there is a range of charges for power factor, although this charge rarely exceeds 5% of the total bill.

Utility Perspective

Utilities regularly evaluate investments based on life-cycle costs. They are used to purchasing power plants and transmission equipment that may take 15 years or more to pay back, and they apply the same kind of long-term analysis to conservation investments. This policy stands in strong contrast to the average energy user's requirement that a conservation investment pay back in under three years. This difference in perspective is referred to as the payback gap, a concept discussed in chapter 8.

In evaluating conservation investments, a utility compares the costs it avoids by making the investment (fuel, operation and maintenance, and the avoided cost of new generating facilities) to the amount it has to pay directly for the energy savings plus program implementation and administrative costs (typically 20–30% of direct costs). If the present value of the expenses avoided over the life of the conservation investment exceed the cost of the investment, the measure is attractive for the utility.

Avoided cost figures will often vary with off-peak and on-peak periods. Consider a utility that has an avoided cost of $.08/kWh for peak periods and $.03/kWh for all other times. It is considering giving a rebate for an efficient 10-hp, 1,800-rpm TEFC motor in a new installation. The specifics of the project include:

Rebate level:	$100
Program cost:	$ 30
Total utility cost:	$130
Total motor operation:	4,000 hrs/yr
On-peak motor operation:	1,000 hrs/yr
Off-peak motor operation:	3,000 hrs/yr
Motor load:	75% of rated load
Engineering motor life:	15 yrs
Expected operational life:	10 yrs
Utility real discount rate:	7%
Power savings (col. D, Table A-1):	0.32 kW
Energy savings on peak:	320 kWh/yr
Energy savings off peak:	960 kWh/yr
Annual dollar savings:	$54.40

The net present value of the energy savings over the ten-year period, using a real discount rate of 7%, is $382. Since this exceeds the amount that the utility will pay to capture the savings ($130), the rebate is cost-effective for the utility.

On a more complex level, utilities running conservation programs assume that some of the conservation would have taken place without an incentive. These customers, known as "free riders," receive rebates that cannot be credited toward conservation produced by the utility program.

Using the above example, if 20% of the efficient 10-hp motors that received a rebate would have been purchased in the absence of that rebate, the effective cost to the utility would be the same but the energy savings would be 20% less. The new avoided cost for the investment would be $306, which is still greater than the $130 spent by the utility and therefore still cost-effective.

Another popular analytic approach is to compare the cost to the utility per kWh saved (often called the cost of saved energy or CSE) to the utility's avoided cost per kWh. The CSE can be calculated using the following formula:

$$CSE = \frac{\text{Present value of utility costs to achieve savings}}{\text{Annual kWh savings} \times \text{present value factor}}$$

where the present value factor is the present value of annual payments of $1, made for the life of the measure, assuming a specific discount rate. Present value factor tables can be found in many economics and business textbooks, or one can use the "present value" function in most computer spreadsheet programs. In the example above, the CSE is as follows:

$$\text{CSE} = \frac{\$130 \text{ utility cost}}{1,280 \text{ kWh saved/yr} \times 7.024} = \$0.013/\text{kWh}$$

The cost of saved energy ($.013/kWh) is well below the utility's avoided costs of $.03/kWh (off-peak) and $.08 kWh (on-peak), so the investment is cost-effective for the utility.

Having covered some general economic principles and analytic methods, we present in the following sections worksheets and tables to illustrate how to evaluate specific installations.

Energy-Efficient Motors

Energy-efficient motors can be installed when a new motor is purchased, in lieu of rewinding an existing motor that has burned out, or as a retrofit replacing an operating standard-efficiency unit.

The relevant cost for financial analysis depends on the type of installation. When a new motor is purchased, the incremental cost of an energy-efficient model over a standard unit is the value to be used in calculations. When an efficient motor is installed instead of rewinding a burned out motor, the actual cost for the efficiency improvement is the cost difference between rewinding the old motor and purchasing a new efficient motor. When an EEM is installed as a retrofit, the costs of the efficiency gains include the full purchase price of the new efficient motor plus the labor to remove the old motor and install the new one. We ignore the salvage value of the retiring motor because it should be scrapped, not re-used, and the scrap value of the metal is small.

Motor Costs

Tables A-1 and A-2 (next two pages) present the costs for new efficient motors, new standard motors, and motor rewinds. These tables are based on some generic assumptions about the type of motor, the location, and the duty factor, and can be used in the worksheets on motor economics. When available, however, actual motor costs for specific applications should be used in calculations to more accurately reflect the economics of a specific project.

The critical factors influencing the purchase price of motors are, in order of importance, size, discount structure, speed, enclosure type, and efficiency. For most electrical equipment, manufacturers

Table A-1: Costs and performance of TEFC EEMs, standard motors, and rewinds.

A	B	C	D	E	F	G	H	I	J	K	L	M
hp	avg. effic. new TEFC std.	avg. effic. new TEFC EEM	new-motor TEFC kW savings (75% load)	average cost TEFC std.	average cost TEFC EEM	avg. effic. stock & rewound TEFC	TEFC retrofit & rewind kw sav. (75% load)	rewind cost TEFC	labor cost retrofit	baseplate adapter cost, U to T frame	baseplate adapter cost, pre-stnd. to T frame	adjustable sheave cost
1	76.8	84.0	0.063	144	183	74.8	0.082	220	132	12	37	10
2	81.1	85.3	0.068	169	216	79.1	0.10	245	132	12	37	10
3	81.4	88.8	0.17	178	251	79.4	0.22	290	132	18	61	25
5	83.9	89.0	0.19	221	290	81.9	0.27	335	132	18	56	25
7.5	84.7	90.8	0.33	301	389	82.7	0.45	375	132	21	70	45
10	86.4	91.0	0.32	360	471	84.4	0.47	440	132	21	69	45
15	87.1	92.0	0.51	498	646	85.1	0.74	500	221	28	100	65
20	88.3	92.7	0.60	615	787	86.3	0.89	610	221	28	127	65
25	89.7	92.9	0.53	744	968	87.7	0.89	755	221	39	127	76
30	90.4	93.2	0.55	888	1147	88.4	1.0	860	221	39	127	76
40	90.6	93.6	0.80	1143	1499	88.6	1.4	1020	363	50	164	98
50	91.5	93.8	0.76	1405	1874	89.5	1.4	1175	363	71	164	98
60	91.8	94.4	1.0	2173	2663	89.8	1.8	1295	363	92	177	115
75	92.1	94.6	1.2	2725	3270	90.1	2.2	1500	363	92	273	115
100	91.9	95.0	2.0	3398	4285	89.9	3.3	1610	856	102	custom	154
125	92.1	95.1	2.4	4464	5957	90.1	4.0	1820	856	custom	custom	custom
150	93.1	95.4	2.2	5338	6937	91.1	4.2	2125	856	custom	custom	custom
200	94.0	95.6	2.0	6267	8294	92.0	4.6	2420	856	custom	custom	custom
250	94.3	95.6	2.1	8239	10398	92.3	5.3	2915	856	custom	custom	custom

Note: Columns B and C are the averages of the full load nominal efficiencies of the motors supplied by the same eight manufacturers as in Table 2-3 (Litman 1990). Column D is calculated from B and C, assuming 75% load. Columns E and F are trade prices. Column G is two percentage points lower than column B (one point for rewind damage and one point because standard motors were less efficient in the past than they are today). Column H is calculated from C and G. Column J is from NEES 1990. Columns K, L, and M are from Gilmore 1990. Column L pertains to motors built before industry-wide U-frame sizes were developed in 1952.

Table A-2. Costs and performance of ODP EEMs, standard motors, and rewinds.

A	B	C	D	E	F	G	H	I	J	K	L	M
	avg. effic. new ODP std.	avg. effic. new ODP EEM	new-motor ODP kW savings (75% load)	average cost ODP std.	average cost ODP EEM	avg. effic. stock & rewound ODP	ODP retrofit & rewind kw sav. (75% load)	rewind cost ODP	labor cost retrofit	baseplate adapter cost, U to T frame	baseplate adapter cost, pre-stnd. to T frame	adjustable sheave cost
hp												
1	76.3	83.6	0.064	132	164	74.3	0.084	209	132	12	37	10
2	78.5	84.6	0.10	159	198	76.5	0.14	233	132	12	37	10
3	80.6	87.8	0.17	144	184	78.6	0.22	276	132	18	61	25
5	83.2	88.2	0.19	179	231	81.2	0.27	318	132	18	56	25
7.5	85.3	89.9	0.26	240	327	83.3	0.38	356	132	21	70	45
10	86.3	90.4	0.30	300	388	84.3	0.45	418	132	21	69	45
15	87.2	91.7	0.47	397	527	85.2	0.70	475	221	28	100	65
20	88.1	92.4	0.59	497	647	86.1	0.89	580	221	28	127	65
25	88.9	93.0	0.69	590	762	86.9	1.0	717	221	39	127	76
30	89.4	93.0	0.74	688	874	87.4	1.2	817	221	39	127	76
40	90.2	93.8	0.96	869	1111	88.2	1.5	969	363	50	164	98
50	90.9	93.9	1.0	1021	1278	88.9	1.7	1116	363	71	164	98
60	91.4	94.6	1.2	1282	1583	89.4	2.1	1230	363	92	177	115
75	91.9	94.8	1.4	1608	1925	89.9	2.4	1425	363	92	273	115
100	91.9	94.8	1.9	2090	2494	89.9	3.3	1530	856	102	custom	154
125	92.3	94.8	2.0	2465	2991	90.3	3.6	1729	856	custom	custom	custom
150	92.7	95.1	2.3	3209	3933	90.7	4.3	2019	856	custom	custom	custom
200	93.5	95.6	2.6	3909	4949	91.5	5.2	2299	856	custom	custom	custom
250	94.6	95.7	1.8	4674	6986	92.6	4.9	2769	856	custom	custom	custom

Note: Columns B and C are the averages of the full load nominal efficiencies of the motors supplied by the same eight manufacturers as in Table 2-3 (Litman 1990). Column D is calculated from B and C, assuming 75% load. Columns E and F are trade prices. Column G is two percentage points lower than column B (one point for rewind damage and one point because standard motors were less efficient in the past than they are today). Column H is calculated from C and G. Column J is from NEES 1990. Columns K, L, and M are from Gilmore 1990. Column L pertains to motors built before industry-wide U-frame sizes were developed in 1952.

have a published "list" price that is typically 30% to 50% above the trade price actually paid by industry. High-volume customers can often negotiate discounts of up to 50% off the trade price. In this appendix, we have made the simplifying assumption that all motors are purchased at the industry trade price. The trade price is approximately 10–30% higher for high-efficiency motors than for standard motors.

The slower a motor, the more material it requires and the higher its capital cost. Because more than 50% of the motors sold are rated at 1,800 rpm, the tables in this section are based on this operating speed.

Enclosure type also influences the cost of a motor. Totally enclosed fan-cooled motors require more material and must operate in more severe environments and therefore cost more than open drip-proof models.

Other Costs for Motor Replacement

U-frame motors can usually be replaced with T-frame motors of the same rating. However, the mounting holes do not line up for the two frame types, so a conversion baseplate must be used. Depending on the motor size, the adapter plate can cost between $10 and $100, with $25 as the most typical value. Using an adapter also adds an hour of labor to the installation.

As stated in chapter 2, many motors now in service are over-sized for the application. One way to save money on an efficient replacement is to install a new motor that is smaller than the original unit. In this case, however, both the mounting system and the shaft may be a different size than the original equipment. This change in the physical size of the equipment generally will force the user to install an adapter plate for mounting the motor and to change pulleys or put in a sleeve for attaching the existing pulleys to the new shaft.

Motor starters typically include protective devices designed to disconnect the motor if it is overloaded. Most starters use a tem-perature-activated switch that is warmed by a set of protective devices known as thermal overload elements or heaters. The heaters are sized according to the motor's full load input current rating, and will typically need to be changed if the motor is downsized.

Rewind Costs

The cost of rewinding depends on the extent of the repair and on local labor rates. The simplest version of a rewind replaces the motor windings only. Most rewinds also install new bearings, so the tables in this appendix include both winding and bearing costs, based on stripping in a burnout oven.

Energy Savings

The energy an EEM will save depends on the efficiency of the EEM compared to that of the standard motor the EEM serves in lieu of (including any degradation to the standard motor from aging and past repairs), on the magnitude of the load, and on the number of hours of operation. The efficiency values listed in Tables A-1 and A-2 are for 1,800-rpm TEFC and ODP standard- and high-efficiency motors, averaging across the offerings of the major manufacturers serving the U.S. market. The difference in energy use for these two sets of products was used to calculate the savings for new or retrofit applications. The rewind savings values in the tables assume that the motor being replaced lost one percentage point in efficiency from previous rewinds and an additional percentage point due to age (the average efficiency of a standard new motor has increased approximately 1.8% over the past ten years [Lovins et al. 1989]). Of course, if more specific values are available, they should be used in the calculations. Instead of relying on the average efficiencies listed in Tables A-1 and A-2, for example, one can use manufacturer-specific data from Table 2-3 for certain 1,800-rpm models or consult a motor catalog or the manufacturer for the efficiency of a specific motor. Note also that 3,600-rpm motors tend to be one or two points more efficient, and 1,200-rpm motors one or two points less efficient, than 1,800-rpm models (see Figure 2-9 [top]).

The savings values in Tables A-1 and A-2 assume that the motors run at 75% of the rated load. In most cases, the motor user is unlikely to know the actual load on the motor. However, if this load is known, the worksheet allows the user to adjust the calculation accordingly.

Estimating Costs and Benefits

Calculating Savings from EEMs

The following completed worksheets illustrate how to estimate the costs and benefits of installing an EEM (1) in a new application, (2) as an alternative to rewinding an old U-frame motor (an installation that requires an adapter baseplate), (3) as a retrofit for an in-service motor, and (4) in a downsizing application. Immediately following the examples are two blank worksheets: one (Worksheet A-1) to be used for the first three cases and the other (Worksheet A-2) to be used for downsizing applications. You may wish to photocopy these blank forms so that you will have an ample supply of worksheets for your own purposes.

Example #1 involves buying a new 20-hp TEFC motor that will operate at 85% load and 4,000 hrs/yr with utility rates of $.063/kWh and $88/kW-yr. Because the motor is a new installation, the relevant cost for analysis is the difference in price between a standard- and a high-efficiency motor.

Example #2 analyzes the choice between rewinding a 50-hp, U-frame, ODP motor and buying a new high-efficiency motor. The motor will operate 4,000 hrs/yr at utility rates of $.06/kWh and $70/kW-yr. Because the operating load is not known, the calculation assumes 75% loading. The conversion from a U-frame to a T-frame model requires an adapter baseplate and, in this case, new heaters.

Example #3 calculates the economics of replacing an operating standard-efficiency 30-hp TEFC motor with a high-efficiency unit. The motor runs 4,000 hrs/yr at unknown loading, with utility rates of $.07/kWh and $50/kW-yr. No baseplate adapter is required, but the cost of labor to remove the old motor and install the new one must be counted. We assume no heater replacement is needed, but a new pulley must be installed.

Calculating Savings from Downsizing

As a general rule, downsizing a motor can be cost-effective if the existing motor is operating at less than 40% of rated load. When a motor is running this lightly loaded, the combination of the energy savings that accrue from eliminating the reduced efficiency at low loads with the potential capital cost savings due to using a smaller motor (in new installations and, in some cases, instead of rewinds) will provide a good return on investment for downsizing. (Columns F and I in Tables A-1 and A-2 show that it is cheaper to buy motors below about 15 hp than it is to rewind them. Rewinding is less costly for motors larger than about 15 hp.) However, downsizing a motor that operates above 40% load is often not cost-effective, because the cost of installing the new motor, including adapter plates, new pulleys, and heaters, plus the efficiency loss from operating a smaller, less efficient motor, outweigh the savings from running the new, smaller motor at higher loading. Because large motors maintain their efficiency better at low loads than do small motors, the 40% cutoff is only a general guideline. Downsizing decisions should be evaluated for each specific application.

Example #4 (page 272) evaluates the economics of replacing a standard-efficiency 40-hp TEFC motor with a 15-hp high-efficiency unit when the 40-hp motor runs at only 11-hp load.

Example #1: Installing an EEM in a New Application

Application: New motor __X__ Rewind _____ Retrofit _____
1. Motor size (hp) __20__
2.ᵃ Operating hours per year _4,000_
3.ᵇ Operating load (if known) _85%_
4.ᵃ Utility rate (energy) (a) $ _0.063_ per kWh
               (demand) (b) $ _88_ per kW-yr
5. Motor enclosure ____ ODP (Table A-2) _X_ TEFC (Table A-1)
6. Enter kW savings at 75% load from Table A-1 or A-2 __.60__
    (column D for new installations, column H for retrofit or rewind)
7. Adjust kW savings to actual operating load, if known
    (_.60_) × (_85_%)/(75%) = _0.68_ kWᶜ
    line 6   line 3
8. Calculate kWh savings
    (_0.68_) × (_4,000_) = _2,720_ kWh
    line 7   line 2
9. Annual cost savings
    (_2,720_) × (_0.063_) + (_0.68_) × (_88_) = $ _231_
    line 8   line 4(a)   line 7   line 4(b)
10. Costs (from Table A-1 or A-2)
    (a) $ _787_ New efficient motor (column F)
    (b) $ _0_ Motor baseplate (U-frame, column K; pre-NEMA,
            column L)
    (c) $ _0_ Labor (retrofit only, column J)
    (d) $ _0_ Heaters ($20 if needed)
    (e) $ _0_ Pulley (if needed, column M)
subtotal _787_
    (f) $ _615_ Cost of alternative (standard motor, column E;
            rewind, column I)
11. Calculate net cost:
    Add lines 10(a) through 10(e), then subtract line 10(f) = $ _172_
12. Simple payback (_172_)/(_231_) = _0.74_ years
             line 11   line 9

ᵃ If possible, break out operating hours, load profile, and rates by on- and
  off-peak times, using a separate worksheet for each segment.
ᵇ If load is under 40%, consider downsizing and refer to Worksheet A-2.
ᶜ If actual load not known, use line 6 value.

Example #2: Installing an EEM as an Alternative to Rewinding

Application: New motor _____ Rewind __X__ Retrofit _____
1. Motor size (hp) __50__
2.[a] Operating hours per year _4,000_
3.[b] Operating load (if known) _N/A_
4.[a] Utility rate (energy) (a) $ _0.060_ per kWh
       (demand) (b) $ _70_ per kW-yr
5. Motor enclosure _X_ ODP (Table A-2) ____ TEFC (Table A-1)
6. Enter kW savings at 75% load from Table A-1 or A-2 __1.7__
   (column D for new installations, column H for retrofit or rewind)
7. Adjust kW savings to actual operating load, if known
   (_____) × (_____%)/(75%) = __1.7__ kW[c]
     line 6   line 3
8. Calculate kWh savings
   (_1.7_) × (_4,000_) = _6,800_ kWh
     line 7   line 2
9. Annual cost savings
   (_6,800_) × (_0.060_) + (_1.7_) × (_70_) = $ _527_
     line 8   line 4(a)   line 7   line 4(b)
10. Costs (from Table A-1 or A-2)
    (a) $ _1,278_ New efficient motor (column F)
    (b) $ ___71___ Motor baseplate (U-frame, column K; pre-NEMA,
             column L)
    (c) $ ___0___ Labor (retrofit only, column J)
    (d) $ __20__ Heaters ($20 if needed)
    (e) $ ___0___ Pulley (if needed, column M)
subtotal _1,369_
    (f) $ _1,116_ Cost of alternative (standard motor, column E;
             rewind, column I)
11. Calculate net cost:
    Add lines 10(a) through 10(e), then subtract line 10(f) = $ _253_
12. Simple payback (_253_)/(_527_) = _0.48_ years
                     line 11   line 9

[a] If possible, break out operating hours, load profile, and rates by on- and
off-peak times, using a separate worksheet for each segment.
[b] If load is under 40%, consider downsizing and refer to Worksheet A-2.
[c] If actual load not known, use line 6 value.

Example #3: Installing an EEM as a Retrofit for an In-service Motor

Application: New motor _____ Rewind _____ Retrofit __X__
 1. Motor size (hp) __30__
 2.[a] Operating hours per year _4,000_
 3.[b] Operating load (if known) _N/A_
 4.[a] Utility rate (energy) (a) $ _0.07_ per kWh
                    (demand) (b) $ _50_ per kW-yr
 5. Motor enclosure _____ ODP (Table A-2) _X_ TEFC (Table A-1)
 6. Enter kW savings at 75% load from Table A-1 or A-2 _1.0_
    (column D for new installations, column H for retrofit or rewind)
 7. Adjust kW savings to actual operating load, if known
    (_____) × (_____%)/(75%) = _1.0_ kW[c]
     line 6   line 3
 8. Calculate kWh savings
    (_1.0_) × (_4,000_) = _4,000_ kWh
     line 7   line 2
 9. Annual cost savings
    (_4,000_) × (_0.07_) + (_1.0_) × (_50_) = $ _330_
      line 8   line 4(a)   line 7   line 4(b)
10. Costs (from Table A-1 or A-2)
    (a) $ _1,147_ New efficient motor (column F)
    (b) $ ___0___ Motor baseplate (U-frame, column K; pre-NEMA,
                  column L)
    (c) $ _221_ Labor (retrofit only, column J)
    (d) $ ___0___ Heaters ($20 if needed)
    (e) $ _76_ Pulley (if needed, column M)
subtotal _1,444_
    (f) $ ___0___ Cost of alternative (standard motor, column E;
                  rewind, column I)
11. Calculate net cost:
    Add lines 10(a) through 10(e), then subtract line 10(f) = $ _1,444_
12. Simple payback (_1,444_)/(_330_) = _4.4_ years
                     line 11   line 9

[a] If possible, break out operating hours, load profile, and rates by on- and
  off-peak times, using a separate worksheet for each segment.
[b] If load is under 40%, consider downsizing and refer to Worksheet A-2.
[c] If actual load not known, use line 6 value.

Example #4: Downsizing as a Retrofit for an In-Service Motor

1. Current motor size __40__ hp
2. Current motor load (a) __11__ hp; (b) __28__ %[a]
3. Proposed motor size __15__ hp
4. Estimated efficiency of current motor running at current percent of rated load, using Figure 3-5 and, if speed is not 1,800 rpm, Figure 2-9 (top) __83__ %
5. Motor enclosure _____ ODP (Table A-2) __X__ TEFC (Table A-1)
6. Efficiency of the proposed smaller motor operating at calculated percent of rated load, from column C of Table A-1 or A-2 or from Table 2-3 for specific 1,800-rpm models __92.0__ %
7. Motor operation __4,000__ hrs/yr[a]
8. Utility rate[a] (energy) (a) $ __.06__ per kWh
   (demand) (b) $ __70__ per kW-yr
9. Calculate change in power requirements

$$[0.746 \text{ kW/hp} \times (\underbrace{\_\_11\_\_}_{\text{line 2a}})] \times [\frac{1}{(\underbrace{\_.83\_}_{\text{line 4}})} - \frac{1}{(\underbrace{\_.92\_}_{\text{line 6}})}] = \underline{\ 0.97\ } \text{ kW}$$

10. Calculate energy savings
$$(\underbrace{\_0.97\_}_{\text{line 9}}) \times (\underbrace{\_4,000\_}_{\text{line 7}}) = \underline{3,880} \text{ kWh/year}$$

11. Cost savings
$$[(\underbrace{\_3,880\_}_{\text{line 10}}) \times (\underbrace{\_.06\_}_{\text{line 8a}})] + [(\underbrace{\_0.97\_}_{\text{line 9}}) \times (\underbrace{\_70\_}_{\text{line 8b}})] = \$ \underline{\ 300\ }$$

12. Costs (from table A-1 or A-2)
   (a) $ __646__ New efficient motor (column F)
   (b) $ __120__ Motor baseplate[b]
   (c) $ __221__ Labor (retrofit only, column J)
   (d) $ __20__ Heater ($20 if needed)
   (e) $ __65__ Pulley replacement (column M)
   Total cost $ __1,072__
13. Simple payback ($\underbrace{\_1,072\_}_{\text{line 12}}$)/($\underbrace{\_300\_}_{\text{line 11}}$) = __3.6__ years

[a] If possible, break out operating hours, load profile, and rates by on- and off-peak times, using a separate worksheet for each segment.
[b] Baseplate costs are application-specific in downsizing installations.

Worksheet A-1

Application: New motor _____ Rewind _____ Retrofit _____
 1. Motor size (hp) _____
 2.ª Operating hours per year _____
 3.ᵇ Operating load (if known) _____
 4.ª Utility rate (energy) (a) $ _____ per kWh
                    (demand) (b) $ _____ per kW-yr
 5. Motor enclosure ____ ODP (Table A-2) ____ TEFC (Table A-1)
 6. Enter kW savings at 75% load from Table A-1 or A-2 _____
    (column D for new installations, column H for retrofit or rewind)
 7. Adjust kW savings to actual operating load, if known
    (_____) × (_____%)/(75%) = _____ kWᶜ
      line 6   line 3
 8. Calculate kWh savings
    (_____) × (_____) = _____ kWh
      line 7   line 2
 9. Annual cost savings
    (_____) × (_____) + (_____) × (_____) = $ _____
      line 8   line 4(a)   line 7   line 4(b)
10. Costs (from Table A-1 or A-2)
    (a) $ _____ New efficient motor (column F)
    (b) $ _____ Motor baseplate (U-frame, column K; pre-NEMA,
                  column L)
    (c) $ _____ Labor (retrofit only, column J)
    (d) $ _____ Heaters ($20 if needed)
    (e) $ _____ Pulley (if needed, column M)
subtotal _____
    (f) $ _____ Cost of alternative (standard motor, column E;
                  rewind, column I)
11. Calculate net cost:
    Add lines 10(a) through 10(e), then subtract line 10(f) = $ _____
12. Simple payback (_____)/(_____) = _____ years
                      line 11   line 9

ª If possible, break out operating hours, load profile, and rates by on- and
  off-peak times, using a separate worksheet for each segment.
ᵇ If load is under 40%, consider downsizing and refer to Worksheet A-2.
ᶜ If actual load not known, use line 6 value.

Worksheet A-2

1. Current motor size _____ hp
2. Current motor load (a) _____ hp; (b) _____ %[a]
3. Proposed motor size _____ hp
4. Estimated efficiency of current motor running at current percent of rated load, using Figure 3-5 and, if speed is not 1,800 rpm, Figure 2-9 (top) _____%
5. Motor enclosure _____ ODP (Table A-2) _____ TEFC (Table A-1)
6. Efficiency of the proposed smaller motor operating at calculated percent of rated load, from column C of Table A-1 or A-2 or from Table 2-3 for specific 1,800-rpm models _____%
7. Motor operation _____ hrs/yr[a]
8. Utility rate[a] (energy) (a) $ _____ per kWh
   (demand) (b) $ _____ per kW-yr
9. Calculate change in power requirements

$$[0.746 \text{ kW/hp} \times (\underset{\text{line 2a}}{\underline{\hspace{1cm}}})] \times [\frac{1}{(\underset{\text{line 4}}{\underline{\hspace{1cm}}})} - \frac{1}{(\underset{\text{line 6}}{\underline{\hspace{1cm}}})}] = \underline{\hspace{1cm}} \text{ kW}$$

10. Calculate energy savings

$$(\underset{\text{line 9}}{\underline{\hspace{1cm}}}) \times (\underset{\text{line 7}}{\underline{\hspace{1cm}}}) = \underline{\hspace{1cm}} \text{ kWh/year}$$

11. Cost savings

$$[(\underset{\text{line 10}}{\underline{\hspace{1cm}}}) \times (\underset{\text{line 8a}}{\underline{\hspace{1cm}}})] + [(\underset{\text{line 9}}{\underline{\hspace{1cm}}}) \times (\underset{\text{line 8b}}{\underline{\hspace{1cm}}})] = \$ \underline{\hspace{1cm}}$$

12. Costs (from table A-1 or A-2)
    (a) $ _____ New efficient motor (column F)
    (b) $ _____ Motor baseplate[b]
    (c) $ _____ Labor (retrofit only, column J)
    (d) $ _____ Heater ($20 if needed)
    (e) $ _____ Pulley replacement (column M)
    Total cost $ _____
13. Simple payback $(\underset{\text{line 12}}{\underline{\hspace{1cm}}})/(\underset{\text{line 11}}{\underline{\hspace{1cm}}}) = $ _____ years

[a] If possible, break out operating hours, load profile, and rates by on- and off-peak times, using a separate worksheet for each segment.
[b] Baseplate costs are application-specific in downsizing installations.

Calculating the Economics of
Efficient Motors: Utility Perspective

Many motor purchase decisions are made in a hurry, when a motor fails. When this occurs, it is difficult to evaluate the exact economics of improving motor efficiency since the efficiency of the original, failed motor can no longer be measured. As a result, utility programs often provide fixed rebates for motors based solely on the size of the motor in order to expedite the purchase of an efficient motor when an existing unit needs to be replaced or rewound. To ensure the cost-effectiveness of its investment, a utility will sometimes set minimum bounds on the operation of the motor (such as a minimum number of operating hours per year).

Utilities use avoided cost data along with estimates of energy savings to set rebates and define applications where incentives can be offered. In general, the steps include the following:

1. Determine the avoided cost (sometimes differentiated for different seasons and/or time periods such as on- and off-peak).

2. Determine the average operating life of the measure, allowing for equipment that is removed before the end of its rated life because of production line remodeling or other reasons.

3. Look at the cost for efficient motors for the three cases: new motor purchases, installing efficient motors instead of rewinds, and retrofitting with efficient motors.

4. Estimate the utility cost savings for each application for different operating hours and loads. This calculation involves multiplying the kWh and kW savings by the utility's avoided cost per kWh and/or kW. Alternatively, the cost of saved energy (in $/kWh) can be calculated for each application.

5. Compare the avoided cost to the utility with the incremental cost or cost of saved energy for the application. When costs for the applications are expressed in terms of $/kWh, avoided costs per kWh need to be adjusted to incorporate avoided capacity costs. This is generally done by taking avoided capacity costs (expressed in $/kW-yr) and dividing by the average annual operating hours of the capacity in question (approximately 8,000 hours for baseload capacity). The resulting value is added to the avoided costs per kWh, to yield an estimate of capacity-adjusted costs per kWh.

An example of these calculations is provided on pages 262 and 263. Alternatively, in Figures A-1 through A-6, the cost of saved energy is calculated for high-efficiency motors, for different applications (new motors, rewinds, and retrofits) and operating hours. In this analysis, we assume that the rebate equals 100% of the measure cost, the values of which are listed in Tables A-1 and A-2. We also assume that the utility pays marketing and administrative costs equal to 30% of the rebate costs and that the average motor remains in use for 15 years. This simple analysis does not differentiate between peak and off-peak periods. In this analysis, whenever the cost of saved energy is less than a utility's avoided cost, it will be cost-effective for the utility to offer a 100% rebate for a particular measure and application. As can be seen in the figures, if a utility has a long-run avoided cost of $.05/kWh (a typical value), it will generally be cost-effective to the utility to pay 100% of the measure costs in new motor or rewind situations; full payment in retrofits is cost-effective for ODP motors of 5–200 hp, and for TEFC motors of 7.5–40 hp, when motors operate at least 4,000 hours per year.

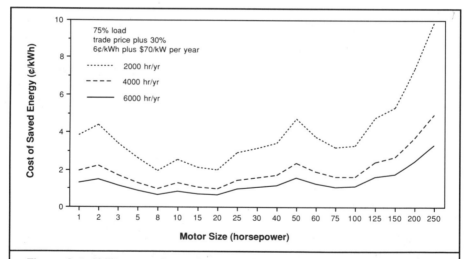

Figure A-1. Utility cost of saved energy for new 1,800-rpm TEFC motors as a function of motor size.

Note: The utility perspective is represented by increasing the incremental difference in trade prices by 30% to reflect program costs. Calculations assume a 7% real discount rate and a 15-year motor life. A 75% coincidence factor is applied to the demand charge.

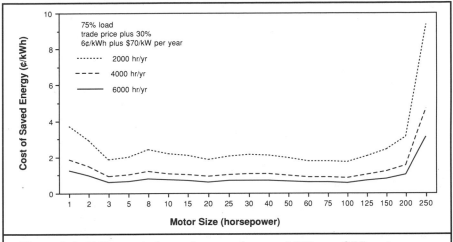

Figure A-2. Utility cost of saved energy for new 1,800-rpm ODP motors as a function of motor size.

Note: The utility perspective is represented by increasing the incremental difference in trade prices by 30% to reflect program costs. Calculations assume a 7% real discount rate and a 15-year motor life. A 75% coincidence factor is applied to the demand charge.

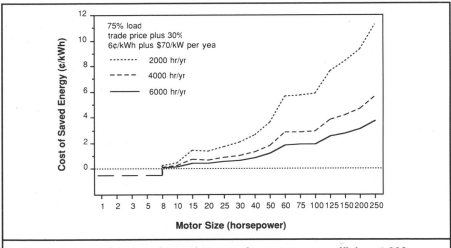

Figure A-3. Utility cost of saved energy for new energy-efficient 1,800-rpm TEFC motors used instead of rewinding existing failed motors.

Note: The utility perspective is represented by increasing the incremental difference in trade prices by 30% to reflect program costs. Calculations assume a 7% real discount rate and a 15-year motor life. A 75% coincidence factor is applied to the demand charge. At 7.5 hp and below, the incremental cost of the EEM is negative.

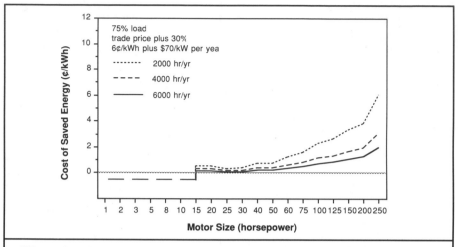

Figure A-4. Utility cost of saved energy for new energy-efficient 1,800-rpm ODP motors used instead of rewinding existing failed motors.

Note: The utility perspective is represented by increasing the incremental difference in trade prices by 30% to reflect program costs. Calculations assume a 7% real discount rate and a 15-year motor life. A 75% coincidence factor is applied to the demand charge. Below 15 hp, the incremental cost of the EEM is negative.

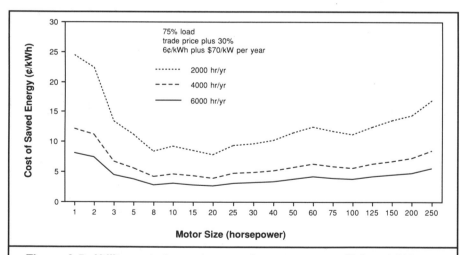

Figure A-5. Utility cost of saved energy for new energy-efficient 1,800-rpm TEFC motors used to retrofit in-service motors.

Note: The utility perspective is represented by increasing the incremental difference in trade prices by 30% to reflect program costs. Calculations assume a 7% real discount rate and a 15-year motor life. A 75% coincidence factor is applied to the demand charge.

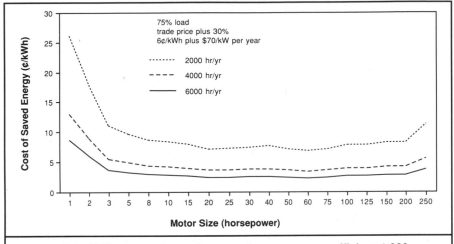

Figure A-6. Utility cost of saved energy for new energy-efficient 1,800-rpm ODP motors used to retrofit in-service motors.

Note: The utility perspective is represented by increasing the incremental difference in trade prices by 30% to reflect program costs. Calculations assume a 7% real discount rate and a 15-year motor life. A 75% coincidence factor is applied to the demand charge.

Adjustable-Speed Drives

As discussed in chapter 4, ASDs for variable-torque loads—like centrifugal pumps and fans—cost 10–20% less than ASDs that drive constant-torque loads—such as conveyors—because the latter require heavier duty electronics that can withstand the full motor inrush current.

ASDs are similar to motors in that there is typically a list price that has little meaning in the marketplace and a trade price that is the actual purchase price for a low-volume user. ASD trade prices appear in Table 4-3 (page 121).

Most of the potential for energy savings from ASDs is in applications of centrifugal fans and pumps. Therefore, most of the costs used in this section are based on variable-torque controllers suitable for centrifugal equipment. One example is included of converting an older DC drive system to an AC adjustable-frequency drive that requires constant-torque equipment.

Two major factors besides the ASD affect its installed cost: the other equipment required to make the ASD a usable part of the system, and the options ordered with the unit.

Consider the example of a pump being installed to control the pH of the fluid in a basin by adding caustic. The base installation (without the ASD) includes a pH sensor in the basin, a feedback controller that takes the reading from the sensor, compares it to the desired pH, and sends out a signal to a control valve, and the control valve itself, which changes the flow to the basin. All of the components required to control the flow of caustic by varying the speed of the pump are already planned for this system, so there is no additional cost for using the ASD except for the ASD itself.

A second application involves a factory that has three well pumps. When the plant is operating, each pump runs continuously, producing water pressure that varies between 50 psi and 100 psi, depending on the water use at any given time. The minimum pressure required to keep the equipment at the factory supplied with water is 50 psi.

In this application, there is currently no sensor that sees the complete picture of the water flow needed by the plant. Each pump rides up and down its own pump curve so that the pressure in the system increases as the flow decreases. As a result, the costs for installing an ASD must include the costs of installing a pressure sensor in the plant at a central or critical location, of providing a feedback controller (which takes the reading from the sensor, compares it to the desired pressure, and sends out a signal to the ASD), and of the ASD itself on one of the pumps.

The costs for adding a feedback control loop are specific to the project, since they depend on the type of sensor and the type of controller that are required. A pressure sensor for an industrial process application will cost between $300 and $1,200, depending on the location, the environment, the range, and the brand. A pressure sensor used in an HVAC duct might only cost $50. Dual sensors (a primary unit and a back-up unit) are sometimes used for reliability in critical industrial applications.

An inexpensive stand-alone controller for a process application will cost between $250 and $400, while a stand-alone, self-tuning process controller can cost up to $1,600. An application in a plant where there is a central programmable logic controller (PLC) can have no incremental cost if there are extra channels available to run the control circuit without adding any input/output hardware. Again, components that control HVAC equipment are far cheaper than industrial process control systems.

Most ASDs include options for control panels with different enclosures and different features such as switches, safeties, overloads, and pilot lights. In addition, most manufacturers can either include the control panel in the same physical location as the controller or wire the ASD so that it can be controlled from a remote station.

Historically, most manufacturers recommended that ASD users purchase an isolation transformer to keep harmonics generated by the ASD from entering the electric distribution system. Modern ASD technology has reduced the harmonic components emitted by the ASDs, particularly for smaller (under 150 hp) units to the point where an isolation transformer is not necessarily recommended. However, these transformers are still frequently installed if there is any question of power quality, either because of the harmonics emitted by the ASD or because of the possibility of damage to the ASD by transients in the electric distribution system. An isolation transformer adds approximately 10% to 20% to the cost of an ASD.

Other less frequently seen options are as follows:

- ASDs sometimes emit radio-frequency (RF) noise. RF suppression is sometimes an option.
- Automatic restart after a power failure or a motor trip is also offered by some manufacturers as an option.
- Some ASDs have circuits that substitute for a process controller.
- Some ASDs have signal outputs so that more than one ASD can be set to track together for applications such as conveyors.

The capital costs for ASDs used in this section include a basic unit operating at 480 volts and an attached control panel. These data are shown in Table 4-3 (page 121). The total costs used in the following examples include materials and labor and whatever else is needed to integrate the ASD into the system. Such costs are too application-specific to be listed in a "cookbook" table.

Calculating ASD Energy Savings

The amount of savings from the use of adjustable speed drives on centrifugal machines such as pumps and fans is dictated by both the variation in flow for the system and the way the system is currently controlled. As a result, the actual savings are site-specific. Nevertheless, there are some general conditions that offer some clues as to whether a specific application is likely to be cost-effective. Several case studies have been used to help the reader screen for applications where some potential exists. The case studies include:

1. A pump that provides variable flow into a long pipeline so that most of the energy of the liquid as it exits the pump goes into overcoming the frictional losses in the pipeline.
2. A pump that provides intermittent flow into a long pipeline so

that most of the energy of the liquid as it exits the pump goes into overcoming the frictional losses in the pipeline.

3. A pump that feeds into a header that supplies a number of faucets, where the pressure in the pipe is allowed to vary with the demand.

4. A fan that supplies air to a variable-air-volume (VAV) system in a commercial building.

5. The replacement of an older DC drive system, used for speed control, by an AC system with ASDs.

Example #1: Pump with variable flow into a long pipeline.

A pump is used to transfer wastewater from an intermediate catch basin in a factory to a sewage treatment plant. Under current operation, dirty water enters the basin at a varying rate depending on how the plant is operating. There is a throttle valve on the outlet from the pump that controls the flow to maintain a constant level in the catch basin. The system operates continuously at varying flow rates. The system has the following characteristics:

Pump: Cornell 6NHP
Rated Pump Flow: 1,200 gpm @ 55 feet of pressure, known as the total dynamic head (TDH)
Actual Pump Flow:

% of rated flow	% of time at each flow
51% to 60%:	20%
61% to 70%:	20%
71% to 80%:	20%
81% to 90%:	20%
91% to 100%:	20%

To calculate the energy savings from adding an ASD, complete the following steps:

1. Calculate the system curve that shows the pressure drop in the piping system. This calculation can be done by using a table or a formula to establish the pressure drop through the piping at different flow rates. In this example, the pipe is 6,800 feet long and 10 inches in diameter. The losses in the pipe are as follows:

Flow rate (gpm)	Losses per 100 ft	Total losses
200	.028 ft	1.9 ft
400	.099 ft	6.7 ft
600	.213 ft	14 ft
800	.370 ft	25 ft
1,000	.569 ft	39 ft
1,200	.811 ft	55 ft

2. Draw the system curve on the pump curve by plotting the pressure drop at each given flow rate calculated in step 1 and connecting the points (Figure A-7, next page).

3. Estimate the power needed for each flow band from the pump curve for the throttled system. In this case, the power needed for the given flow rate is equal to the power required by the base pump curve (or the pump curve when the pump is operating at full speed). For example, looking at the pump curve, when the actual flow is 85% of the design flow (1,200 gpm x 0.85 = 1,020 gpm), the power required at the pump shaft is 21 hp (since this flow rate falls about one-fifth of the way from the 20-hp and 25-hp lines). The electrical power required to drive the shaft is 21 hp times the conversion factor from hp to kW (.746) divided by the motor efficiency at the estimated motor load. For a standard 25-hp motor, this would be approximately (21 hp) × (0.746 kW/hp)/0.897 (from Table A-1) = 17 kW. Note that the pressure produced by the pump exceeds the pressure required by the system at most flow rates. This extra pressure represents wasted energy that is released across the valve to control the flow to the required level.

This method can be used to estimate the power needed at each flow rate as follows:

Flow rate	Gpm at midpoint	HP throttled	Throttled power (kW)
51% to 60%:	660	18	14.9
61% to 70%:	780	19	15.7
71% to 80%:	900	20	16.6
81% to 90%:	1,020	21	17.4
91% to 100%:	1,140	22	18.2

4. Estimate the power needed for the system using an ASD by estimating the energy needed for each flow range to follow the system curve and adding the losses for the ASD. For example, if the pump was only turning fast enough to overcome the frictional losses in the piping at 85% of rated flow, the pump shaft would be rotating at about 1,020 rpm (based on the point where the system curve crosses a line of equal motor speed at 1,020 gpm). Looking at the pump curve, the power needed at the pump shaft at this speed and flow would be about 15 hp. The electrical power required to drive the shaft is 15 hp times the conversion factor from hp to kW (.746) divided by the efficiency of the motor and the ASD at the estimated motor load and ASD speed. (See Figure 4-9 [page 116] for typical ASD efficiencies as a

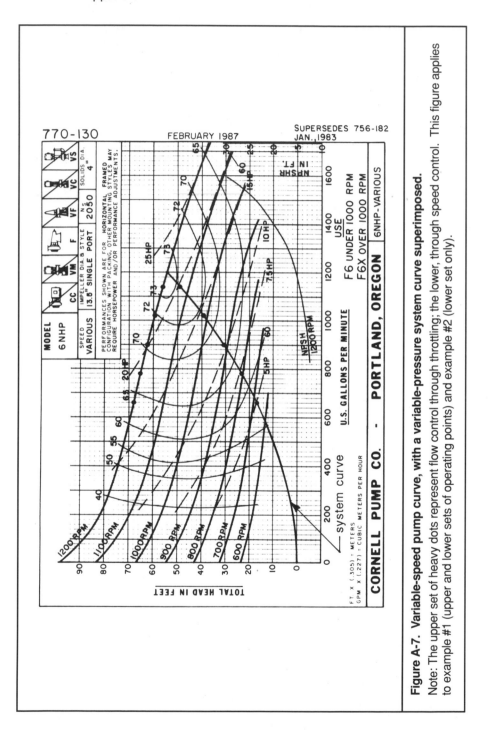

Figure A-7. Variable-speed pump curve, with a variable-pressure system curve superimposed.

Note: The upper set of heavy dots represent flow control through throttling; the lower, through speed control. This figure applies to example #1 (upper and lower sets of operating points) and example #2 (lower set only).

function of speed; the present example is a variable-torque load). For a 25-hp motor, the required electrical power would be approximately

$$(15 \text{ hp}) \times (0.746 \text{ kW/hp})/0.897/0.95 = 13 \text{ kW}$$

The above method can be used to estimate the power needed at each flow rate as follows:

Flow rate	Gpm at midpoint	Hp with speed control	ASD effic.	Input Power w/ASD(kW)
51% to 60%:	660	5	0.89	4.7
61% to 70%:	780	7	0.92	6.3
71% to 80%:	900	10	0.94	8.8
81% to 90%:	1,020	15	0.95	13.1
91% to 100%:	1,140	20	0.96	17.2

5. Estimate the total energy savings due to the ASD by calculating the power savings for each flow range (the value derived in step 3 minus the value derived in step 4) and multiplying by the number of hours in that flow range (20% of full time operation or 1753.2 hours per year at each flow range):

Flow range	Power reduction	Energy savings/yr
51% to 60%:	10.2 kW	17,883 kWh
61% to 70%:	9.4 kW	16,480 kWh
71% to 80%:	7.8 kW	13,675 kWh
81% to 90%:	4.3 kW	7,539 kWh
91% to 100%:	1.0 kW	1,753 kWh
Total:		57,330 kWh

Note that the energy used at full flow increases slightly (compared to the original system) due to the inefficiencies of the ASD: from 18.6 kW with wide-open throttle to 19.4 kW with ASD at 100% speed. At 95% flow (used in this example for the range of 91% to 100% flow), the ASD still saves energy. If a system demands full flow for a significant fraction of its operating time, the ASD losses can be eliminated by operating the ASD in bypass mode (which requires the installation of a bypass switch).

6. The dollar savings for the above at $.07/kWh equals $4,013.

7. The cost is $6,900 for a retrofit, $6,200 for new construction.

8. The simple paybacks are 1.7 years for retrofit and 1.5 years for new construction.

In this example, there is no change in the flow rate when the control valve is changed to an ASD. The energy savings comes from the reduction in pressure from the pump for all of the flow ranges. This pressure reduction is substantial since the pump output can follow the system curve, which has a low pressure requirement at low flows.

Because the base system is already designed with a level sensor, a feedback controller, and a control valve, the cost for the installation includes the cost of the ASD plus some labor time for wiring the unit.

The project is cost-effective because of the relatively low cost and because there is a sufficient period when the system operates at low flows to produce significant savings and justify the investment.

Example #2: Pump with intermittent flow into a long pipeline.

A pump is used to transfer wastewater from an intermediate catch basin in a factory to a sewage treatment plant. Under current operation, dirty water enters the basin at a varying rate depending on how the plant is operating. The basin has a control that turns the pump on and off to maintain the level between a high and low set-point. When all of the lines in the plant are operating, the pump runs 95% of the time. When only one line is running, the pump runs 50% of the time. The system operates continuously with the pump cycling to meet the varying flow rates. The system has the following characteristics:

Pump: Cornell 6NHP
Rated Pump Flow: 1,200 gpm @ 55 feet TDH
Actual Pump Flow:

% of rated flow	% of time at each flow
51% to 60%:	20%
61% to 70%:	20%
71% to 80%:	20%
81% to 90%:	20%
91% to 100%:	20%

To calculate the energy savings, complete the following steps:

1. Calculate the system curve that shows the pressure drop in the piping system using the same calculation as in the first example.

2. Draw the system curve on the pump curve (Figure A-7, page 284).

3. In the base case (existing equipment) the pump either operates at the rated flow or is shut off. As a result, the power needed for each flow band can be estimated by taking the energy use at full flow and

multiplying it by the percent of time the pump is operating to meet the flow band. For example, the power at full flow is:

$$(22.5 \text{ hp}) \times \frac{(0.746 \text{ kW/hp})}{0.897} = 18.7 \text{ kW}$$

The average power needed at 85% of full flow is:

$$(18.7 \text{ kW}) \times (0.85) = 15.9 \text{ kW}$$

This method can be used to estimate the power needed by the existing system at each flow rate as follows:

Flow rate	Power
51% to 60%:	10.3 kW
61% to 70%:	12.2 kW
71% to 80%:	14.0 kW
81% to 90%:	15.9 kW
91% to 100%:	17.8 kW

4. Estimate the power needed for the system with an ASD by estimating the energy needed for each flow range to follow the system curve and adding the losses for the ASD.

The same calculation was done for example #1 and yielded the following estimates of the power needed at each flow rate:

Flow rate	Power
51% to 60%:	4.7 kW
61% to 70%:	6.3 kW
71% to 80%:	8.8 kW
81% to 90%:	13.1 kW
91% to 100%:	17.2 kW

5. Estimate the total energy savings from the ASD by calculating the power savings for each flow range (line 3 minus line 4) and multiplying by the number of hours in that flow range. Again, the continuously operating system runs 20% of the time (1,753.2 hours per year) at each flow range:

Flow range	Power reduction	Energy savings/yr
51% to 60%:	5.6 kW	9,818 kWh
61% to 70%:	5.9 kW	10,343 kWh
71% to 80%:	5.2 kW	9,117 kWh
81% to 90%:	2.8 kW	4,909 kWh
91% to 100%:	0.6 kW	1,052 kWh
Total:		35,239 kWh

6. The dollar savings for the above at $.07/kWh equals $2,467.

7. The cost is $8,600 for a retrofit, $7,400 for new construction. The costs are higher for this application than for the system in example #1 since they include a flow sensor and a feedback controller that is not needed for the base system.

8. The simple paybacks are 3.5 years for retrofit and 3.0 years for new construction.

In this example, there is a trade-off between running the pump continuously at reduced pressure and flow with the ASD and running the pump for limited periods at full flow. In the base case, the pump cycles and operates at a relatively high efficiency when it runs. There is some savings from allowing the water to flow at lower rates, producing a lower pressure drop, but these savings are not as large as in the first example since there is no deliberately wasted energy for a control valve in this system. Note that running the pump continuously at reduced speed instead of cycling will reduce wear on the pump, motor, and associated electrical equipment.

The paybacks may not be attractive for many motor users but will be for some users and for most utilities.

Example #3: Pump with variable flow into a header.

A pump is used to supply water to a factory with multiple water uses. Under current operation, the water is stored in a tank and pumped into the plant by a single pump, which pressurizes the system. Since the water demand varies, the pump will ride up and down the pump curve so that the pressure in the system varies from 55 feet of head to 70 feet of head, and thus requires a relatively high minimum pressure. The system operates continuously at varying flow rates and has the following characteristics:

Pump: Cornell 6NHP
Rated Pump Flow: 1,200 gpm @ 55 feet TDH
Actual Pump Flow:

% of rated flow	% of time at each flow
51% to 60%:	20%
61% to 70%:	20%
71% to 80%:	20%
81% to 90%:	20%
91% to 100%:	20%

To calculate the energy savings, complete the following steps:

1. This system has very large pipes feeding the plant. As a result,

the pressure drop is minimal at all flow rates, and the pressure required by the system is constant regardless of the flow. This pressure is 55 feet of head.

2. Draw the pressure requirements on the pump curve (Figure A-8, next page). Note that this system curve is a straight line at 55 feet of head instead of the more typical system curve (as in Figure A-7).

3. From the pump curve, estimate the power needed for each flow band for the existing, throttled system. In this case, the power needed for a given flow rate is equal to the power required by the base pump curve (that is, the pump curve when the pump is operating at full speed). Note that this is the same calculation for the base system in example #1:

Flow rate	Power
51% to 60%:	14.9 kW
61% to 70%:	15.7 kW
71% to 80%:	16.6 kW
81% to 90%:	17.4 kW
91% to 100%:	18.2 kW

4. Estimate the power needed for the same system using an ASD by estimating the power needed for each flow range to maintain a constant pressure of 55 feet and adjusting for the losses in the motor and ASD:

Flow rate	Gpm at midpoint	Hp with ASD	ASD effic.	Power with ASD(kW)
51% to 60%:	660	14.0	0.95	12.2
61% to 70%:	780	16.0	0.95	13.9
71% to 80%:	900	17.5	0.95	15.3
81% to 90%:	1,020	20.0	0.96	17.2
91% to 100%:	1,140	21.5	0.96	18.5

5. Estimate the total energy savings due to the ASD by calculating the power savings for each flow range and multiplying by the number of hours in that flow range:

Flow range	Power reduction	Energy savings/yr
51% to 60%:	2.7 kW	4,734 kWh
61% to 70%:	1.8 kW	3,156 kWh
71% to 80%:	1.3 kW	2,279 kWh
81% to 90%:	0.2 kW	350 kWh
91% to 100%:	−.3 kW	−526 kWh
Total:		9,993 kWh

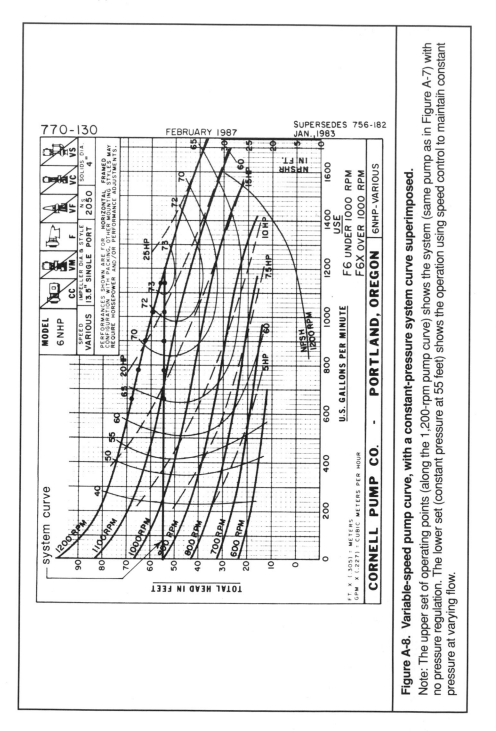

Figure A-8. Variable-speed pump curve, with a constant-pressure system curve superimposed.

Note: The upper set of operating points (along the 1,200-rpm pump curve) shows the system (same pump as in Figure A-7) with no pressure regulation. The lower set (constant pressure at 55 feet) shows the operation using speed control to maintain constant pressure at varying flow.

6. The annual dollar savings for the above at $.07 equals $700.

7. The cost is $8,600 for a retrofit, $7,400 for new construction. The costs are higher for this application than for the system in example #1 since they include a flow sensor and a feedback controller that is not needed for the base system.

8. The simple paybacks are 12.3 years for retrofits and 10.6 for new construction.

In this example, the energy savings are produced by controlling the pressure to the minimum acceptable level for the application instead of letting it increase at low flow rates. In other words, the wasted energy is the extra pressure between the pump curve and the system requirement of 55 feet of head generated by the pump at low flow rates. The savings for this application are lower than in the two previous examples because there is no reduction in pressure requirements at low flow rates, and the pressure generated by the pump only narrowly exceeds the pressure needed by the system. In general, it will not be cost-effective to install an ASD on systems that require a high minimum pressure. As noted in example #1, if the system is expected to operate much of the time at close to full flow, the ASD can at those times be operated in bypass mode (at additional installed cost and control complexity).

Example #4: ASD on a variable-air-volume fan in a commercial building.

Many commercial buildings have air-handling systems in which the air volume is varied to meet the cooling demand in the building. These systems, known as variable-air-volume (VAV) systems, have boxes that serve each thermal zone in the building, with the airflow to that zone adjusted by a local thermostat. Inlet vanes have historically been used on the supply fan to match the air flow to the output of the boxes. While the power requirement of a fan with inlet vanes decreases as the flow decreases, the power does not fall off as fast as the flow because the inlet vanes reduce the efficiency of the fan. For example, the power required by a typical inlet vane system at 50% flow is about 75% of full power.

There are also many commercial buildings with constant-volume systems that may be good candidates for conversion to VAV using an ASD. Such systems include terminal-reheat and dual-duct configurations. Note that converting such simultaneous heating and cooling systems to VAV saves significant heating and cooling energy as well as ventilating energy.

In general, it is difficult to do a hand calculation on the impact of installing an ASD on a constant-volume system or instead of an inlet vane or discharge damper on a VAV system, because the impact depends on the building's thermal characteristics, which change with time. For example, in the morning, waste heat from lights may be absorbed by the cool mass inside the building and help to make the space more comfortable. Later in the day, as the building's mass charges up, that same waste heat may serve to overwarm the space. Even with a constant outside temperature of, say, 50°F, a building might thus need heating in the morning and cooling in the afternoon.

Because of such issues, the easiest way to evaluate the savings from installing an ASD on an air handler will be to run a thermal model (such as DOE2, Trak-Load, or ASEAM; see appendix B) of the building and look at the relative energy use for ventilation using different assumptions for controlling the air flow. Such simulations will also estimate the heating and cooling savings made possible by converting constant-volume systems to VAV.

Assume that running this type of a model produces the following for a 100,000-square-foot office building:

Base fan energy use:	160,000 kWh/yr
Fan energy use with an ASD:	96,000 kWh/yr
Energy savings:	64,000 kWh/yr
Dollar savings @ $.07/kWh:	$ 4,480

The economics of the project to the motor user will depend on the building. If the building is a three-story suburban office building with two large (50 hp) supply fans, the installed cost will be about $17,000, and the simple payback will be 3.8 years.

If the building is a 12-story office building in a downtown core area with a floor-by-floor air handler system and twelve 7.5-hp motors, the installed cost will be $35,000, and the simple payback will be 7.8 years.

Note that the economics of the project to the utility may be very different from that of a project with the same customer economics in the industrial sector because there are differences in the value of energy savings during different time periods. Specifically, the projects in industry in the first three examples can be assumed to be driven by process requirements, which are typically consistent during the day and during the year. However, the use of an ASD on a VAV system will save energy when the building cooling load is low, typically in the winter and in the early morning hours in the summer. If a utility has a summer peak, the use of an ASD on a VAV system may not yield any energy savings during the system peak unless the system is oversized (which is common).

**Example #5: DC drive system to be replaced
with AC variable-speed drives.**

An older plant that manufactures metal widgets uses a 5-hp DC motor to drive the shaft on each of 20 milling machines with 5-hp DC motors. The plant currently has a motor-generator (MG) set where an AC motor drives a DC generator to supply DC power to the machines. The motor on the MG set runs for 4,000 hours per year and draws 110 kW. The MG set needs to see a constant load in order to regulate the voltage, so it is designed so that the DC generator sees an artificial resistive load if all of the DC motors are not in operation. The plant uses all 20 machines on the day shift (2,000 hours per year) but only five machines on the swing (evening) shift.

Use the following steps to calculate the energy savings:

1. The current energy use is (110 kW) $\times$ (4,000 hrs) = 440,000 kWh. Note that there are no savings during the swing shift when fewer machines are running, since the MG set must see a constant load to properly regulate the DC voltage.

2. Calculate the expected energy use based on 15, 5-hp motors running 2,000 hours per year and 5, 5-hp motors running 4,000 hours per year. Assuming the motors operate at 50% load, and the overall efficiency of the motor and ASD together is 70%, the total energy use is:

   (0.5)(15)(5 hp)(.746 kW/hp)(1/0.70)(2,000 hours) +
   (0.5)(5)(5 hp)(.746 kW/hp)(1/0.70)(4,000 hours) = 130,000 kWh/yr.

   The total use for the proposed system will be 130,000 kWh per year.

3. The energy savings is 310,000 kWh per year.

4. The dollar savings at $.07/kWh is $21,700.

5. The cost for the retrofit (assuming that constant-torque ASDs are needed for the application) is about $70,000, which includes the cost for new motors.

6. The simple payback is 3.2 years.

   In general, AC drives provide speed control at a lower energy premium than DC drives, particularly when the DC drives are the older style systems that use motor-generator sets. In addition, AC motors have much lower ongoing maintenance costs for rewinding and repair.

Oversized Wiring as a Conservation Measure

As discussed in chapter 3, it is often cost-effective (in new installations and remodels but not retrofits) to install cable larger than required by code because larger wire has lower losses. In general, it is cost-effective to substitute larger wire for motors that operate for long periods of time at close to full load, particularly if larger wire can be installed in the same size conduit. Two case studies will be used to illustrate where using oversized wiring is cost-effective.

Example #1:

A 100-hp, three-phase, 480-volt motor is being installed at a distance of approximately 500 feet from the motor control center (MCC). The motor is expected to run for two shifts, five days per week (approximately 4,000 hours per year). Using the tables in the NEC:

Full load current:	124 amps
Base wire size:	1/0 XHHW copper[a]
Base conduit:	1.5"
Estimated loss @ 100% load:[b]	2.8 kW, 11,000 kWh
Estimated loss @ 75% load:	1.6 kW, 6,400 kWh
Base cost, wire:	$1,980
Base cost, conduit:	$2,185
Base cost, total:	$4,165

It is proposed that the wire size be increased to 2/0 XHHW, which would require 2" conduit:

Estimated loss @ 100% load:	2.3 kW, 9,200 kWh
Estimated loss @ 75% load:	1.3 kW, 5,200 kWh
Proposed cost, wire:	$2,200
Proposed cost, conduit:	$2,525
Proposed cost, total:	$4,725
Incremental cost:	$560
Simple payback @ $.07/kWh, 100% load:	4.4 years
Simple payback @ $.07/kWh, 75% load:	6.7 years

---

[a] American Wire Gauge copper wire sizes #8, #6, #4, #3, 1/0, 2/0, 3/0, and 4/0 have respective diameters of 0.129, 0.162, 0.204, 0.229, 0.325, 0.365, 0.410, and 0.460 inches or 3.264, 4.115, 5.189, 5.827, 8.252, 9.266, 10.40, and 11.68 mm. These diameters are for solid conductors; stranded wire has larger overall diameters to yield the same net cross-sectional area of copper.

[b] Based on loss = $3I^2R$ (because each of three phases has same loss). Resistance (R) in ohms/1,000 feet of wire at 75°C (167°F) for the various wire diameters is as follows: 0.12 for 1/0 copper; 0.10 for 2/0 copper; 0.077 for 3/0 copper; and 0.062 for 4/0 copper. Because these values are for 1,000 feet of wire, R equals ½ these values for the 500-foot runs used in this example. To adjust these resistance values for other temperatures, use the formula
$$R_2 = R_1 (1 + 0.00323 [T_2 - 75]), \text{ where}$$
R_2 is the new resistance, R_1 is the given resistance, and T_2 is the new temperature in degrees Celsius.

Because using larger wire in this case requires larger conduit, the paybacks are longer than many consumers are willing to accept, although such paybacks might be acceptable to a utility.

Example #2:

A 125-hp, three-phase, 480-volt motor is being installed at a distance of approximately 500 feet from the MCC. The motor is expected to run constantly except during three weeks of maintenance downtime (approximately 8,200 hours per year). Using the tables in the NEC:

Full load current:	156 amps
Base wire size:	3/0 XHHW
Base conduit:	2"
Estimated loss @ 100% load:	2.8 kW, 23,000 kWh
Estimated loss @ 75% load:	1.6 kW, 13,000 kWh
Base cost, wire:	$2,500
Base cost, conduit:	$2,525
Base cost, Total:	$5,025

It is proposed that the wire size be increased to 4/0 XHHW, which can still use the 2" conduit:

Estimated loss @ 100% load:	2.3 kW, 19,000 kWh
Estimated loss @ 75% load:	1.3 kW, 11,000 kWh
Proposed cost, wire:	$2,800
Proposed cost, conduit:	$2,525
Proposed cost, total:	$5,325
Incremental cost:	$ 300
Simple payback @$.07/kWh, 100% load:	1.1 years
Simple payback @$.07/kWh, 75% load:	2.1 years

This second example has a much better payback than the first example because of the combination of longer operating hours (which produce larger savings) and the lower incremental cost, since the conduit size does not change when the wire size is increased. In addition to receiving a fast payback, the motor user would save $140 per year for the life of the installation if the motor ran at 75% load.

Drivetrains

As discussed in chapter 3, synchronous belts can be considerably more efficient than V-belts. Synchronous belts can be used effectively in both new and retrofit applications. They cost more than V-belts and require more costly cogged pulleys, but they last longer. V-belts are sometimes used for safety reasons, because they will slip if the equipment jams. Because synchronous belts do not slip, some applications might require safety equipment such as clutches or shear pins, which will add to the cost of the system.

There is no simple correlation between motor size and the cost of synchronous belts. The cost of the belts and pulley depend on the gear ratio between the motor and the equipment, the amount of torque that the belt will see, and the distance between the centers of the pulleys. In new construction, a conventional V-belt system will cost about 65% to 75% less than a synchronous belt system.

Although belt costs do increase with motor size, the increase is nonlinear, so retrofitting a belt on a larger motor is more cost-effective than doing the same retrofit on a smaller motor. Two examples follow.

Synchronous Belt
Example #1

A synchronous belt drive is being considered for a fan in an air handler that is driven by a 5-hp motor via a conventional V-belt. The fan operates ten hours per day, five days per week (2,600 hours per year). The system parameters are as follows:

Motor efficiency:	88.7% (efficient motor)
V-belt efficiency:	92%
Synchronous belt efficiency:	97%
Energy cost:	$.07/kWh
Motor load:	75%

Energy use for the V-belt system is:

$$(5 \text{ hp}) \times (.746 \text{ kW/hp}) \times (.75 \text{ load}) \times \frac{(2{,}600 \text{ hrs/yr})}{(.92 \times .887)} = 8{,}900 \text{ kWh/yr}$$

Operating cost is:

$$8{,}900 \text{ kWh/year} \times \$0.07/\text{kWh} = \$623/\text{yr}$$

The energy use for the synchronous belt system is:

$$(5 \text{ hp}) \times (.746 \text{ kW/hp}) \times (.75 \text{ load}) \times \frac{(2{,}600 \text{ hr/yr})}{(.97 \times .887)} = 8{,}450 \text{ kWh/yr}$$

Operating cost is:

$$8{,}450 \text{ kWh/yr} \times \$.07/\text{kWh} = \$592/\text{yr}$$

The dollars savings due to the use of the synchronous belt is

$31 per year. The cost as a retrofit is $275, and the incremental cost in new construction is $155. The simple paybacks are 8.9 years for retrofits and 5.0 years for new construction.

Example #2

A synchronous belt is being considered for a fan in an air-handler driven by a 75-hp motor. The fan operates 24 hours per day, five days per week (6,240 hours per year). The system parameters are as follows:

Motor efficiency:	95.1% (efficient motor)
V-belt efficiency:	92%
Synchronous belt efficiency:	97%
Energy cost:	$.07/kWh
Motor load:	75%

The energy use for the V-belt system is:

$$\text{(75 hp)} \times \text{(.746 kW/hp)} \times \text{(.75 load)} \times \frac{(6240 \text{ hr/yr})}{(.92 \times .951)} = 299,000 \text{ kWh/yr}$$

Operating cost is:
$$299,000 \text{ kWh/yr} \times \$0.07/\text{kwh} = \$20,900/\text{yr}$$

The energy use for the synchronous belt system is:

$$\text{(75 hp)} \times \text{(.746 kW/hp)} \times \text{(.75 load)} \times \frac{(6,240 \text{ hrs/yr})}{(.97 \times .951)} = 284,000 \text{ kwh/yr}$$

Operating cost is:
$$284,000 \text{ kwh/yr} \times \$0.07/\text{kwh} = \$19,900/\text{yr}$$

The dollars savings due to the use of the synchronous belt is $1,000 per year. The cost as a retrofit is $1,250 and the incremental cost in new construction is $710. The simple paybacks are 1.3 years for retrofits and 0.7 years for new construction.

The paybacks in the second example are far more attractive because of the larger equipment size and longer operating hours. In cases where synchronous belts are not economic, cogged V-belts should be considered. They fall between conventional V-belts and synchronous belts in cost and efficiency.

Equipment Manufacturers and Associations

This appendix is divided into seven sections:

1. Motor manufacturers
2. ASD manufacturers
3. Power-factor controller manufacturers
4. Vendors of power-factor correction systems
5. Test and repair equipment manufacturers
6. Software suppliers
7. Trade and professional associations related to motors and drives

1. Motor Manufacturers

Product characteristics are listed in parentheses as follows for those manufacturers who responded to our survey:

> (F) fractional horsepower
> (I) integral horsepower
> (L) large motors (over 350 hp)
> (1) single-phase
> (3) three-phase
> (p) permanent-magnet (brushless)
> (s) synchronous
> (i) induction

A.O. Smith Electric Motor Co.
(FI13i)
531 N. Fourth St.
Box 688
Tipp City, OH 45371
(513) 667-6800
(513) 667-5030 (fax)

ASEA–Brown Boveri
(FIL13psi; to European specifications)
Industrial Systems
P.O. Box 372
Milwaukee, WI 53201
(414) 785-3200

Baldor Electric Co. (FI13pi)
5711 S. 7th St.
Fort Smith, AR 72902
(501) 646-4711
(501) 648-5792 (fax)

Canadian General Electric Co.,
  Ltd. (L3si)
107 Park St., North
Peterborough, Ontario K9J7B5
(705) 748-8486

Delco Products Division
  (FIL3i)
General Motors Corporation
P.O. Box 1042, Mail Stop 4-09
Dayton, OH 45401-1042
(513) 455-9097

Emerson Electric Co.
  Emerson Motor Co. (F13i)
  P.O. Box 3946
  8000 W. Florissant Ave.
  St. Louis, MO 63136
  (314) 553-2000
  U.S. Electrical Motors
    (FIL13i)
  8100 W. Florissant Ave.
  St. Louis, MO 63136
  (314) 553-2000

Fasco Industries, Inc. (F1pi)
Motor Division
500 Chesterfield Center, #200
St. Louis, MO 63017
(314) 532-3505
(314) 532-9306 (fax)

Franklin Electric (FI13i)
400 E. Spring
Bluffton, IN 46714
(800) 437-6897

General Electric Co.
  (FIL13psi)
Commercial & Industrial
  Sales Dept.
1635 Broadway
P.O. Box 2222
Fort Wayne, IN 46801
(219) 428-2000

Grainger Engineering (Dayton)
  (FI13si)
5959 W. Howard
Chicago, IL 60648
(708) 647-8900

Leeson (FI13pi)
21000 Washington Ave.
Grafton, WI 53024
(414) 377-8810

Leroy-Somer (FIL13pi)
Division of Emerson Electric
  Co.
560 S. Hicks Rd.
Palatine, IL 60067-6996
(708) 359-2440
(708) 359-2156 (fax)

Lincoln Electric Co. (I3i)
22801 St. Clair Ave.
Cleveland, OH 44117-1199
(216) 481-8100

MagneTek Century Electric
  (FIL13i)
1881 Pine St.
St. Louis, MO 63103
(314) 436-7800

MagneTek Indiana General
  (FI; DC only)
1168 Barranca Dr.
El Paso, TX 79935
(915) 593-1621

✓ MagneTek Louis Allis (IL3si)
P.O. Box 2020
Milwaukee, WI 53201-2020
(414) 481-6000
(414) 481-8895 (fax)

MagneTek Universal Electric
  (F13i)
300 E. Main Street
Owosso, MI 48867
(517) 723-7866

Marathon Electric Mfg. Corp.
  (FIL13i)
P.O. Box 8003
Wausau, WI 54402-8003
(715) 675-3311

Morrill Motors (F1i)
3685 Northrup
Ft. Wayne, IN 46805
(219) 484-1519

Powertec Corp. (FIp)
13504 S. Point Blvd.
P.O. Box 410129
Charlotte, NC 28241-0129
(704) 588-1956
(704) 588-7315 (fax)

Reliance Electric Co.
  (FIL13psi)
24701 Euclid Ave.
Cleveland, OH 44117
(800) 245-4501
(216) 266-7536 (fax)

Siemens Energy & Automation
  (IL3psi)
Industrial Motor Div.
4260 Forest Ave.
Norwood, OH 45212-3396
(513) 841-3100

Sterling Electric (FIL13i)
16752 Armstrong Ave.
Irvine, CA 92714
(800) 654-6220
(714) 474-0543 (fax)

Teco American (FIL13si)
6877 Wynnwood
Houston, TX 77008
(713) 864-5980

Toshiba/Houston (IL3psi)
13131 W. Little York Rd.
Houston, TX 77041
(800) 231-1412

U.S. Motors
see *Emerson Electric*

Westinghouse Motor Co. (L3si)
P.O. Box 277
Round Rock, TX 78680-0277
(800) 451-8798

2. ASD Manufacturers

AC Technology Corp.
Douglas St.
Uxbridge, MA 01569
(508) 278-6781
(508) 278-7873 (fax)

Allen Bradley
Drives Division
1201 S. Second St.
Milwaukee, WI 53204
(414) 382-2000
(414) 382-4444 (fax)

ASEA–Brown Boveri
Industry Div.
1460 Livingston Ave.
North Brunswick, NJ 08902
(201) 932-6000

Autocon Industries, Inc.
995 University Ave.
St. Paul, MN 55104
(800) 328-3351

Baldor Electric Co.
P.O. Box 2400
Fort Smith, AR 72902
(501) 646-4711
(501) 648-5792 (fax)

BBC Brown Boveri
Industry Div.
1460 Livingston Ave.
North Brunswick, NJ 08902
(201) 932-6000

Eaton Corp.
Electric Sales Div.
3122 14th Ave.
Kenosha, WI 53141
(800) 322-4986

Emerson Electric Co.
Industrial Controls
3036 Alt Blvd.
Grand Island, NY 14072
(716) 773-2321

GEC Automation
Projects Incorporated
2871 Avondale Mill Rd.
Macon, GA 31206
(912) 784-5200

General Electric Co.
Drive Systems Dept.
1501 Roanoke Blvd.
Salem, VA 24153
(703) 387-8735

Graham Co.
8800 W. Bradley Rd.
Milwaukee, WI 53223
(414) 355-8800
(414) 355-6117 (fax)

Hampton Products Co.
Division of Danfoss
2995 Eastrock Dr.
Rockford, IL 61109
(815) 398-2770

Hitachi America Ltd.
50 Prospect Ave.
Tarrytown, NY 10591
(914) 332-5800

Lovejoy Electronics
2655 Wisconsin Ave.
Downers Grove, IL 60515
(800) 323-3534

MagneTek
Drives & Systems Div.
16555 W. Ryerson Rd.
New Berlin, WI 53151
(800) 541-0939
(414) 782-1283 (fax)

Mitsubishi Electric
Sales America
799 N. Bierman Circle
Mount Prospect, IL 60056
(800) 323-4216

Parametrics
Unit of ASEA
88 Marsh Hill Rd.
Orange, CT 06477
(203) 795-0811

Polyspede Electronics Co.
6770 Twin Hills Ave.
Dallas, TX 75231
(214) 363-7245

Relcon
80 Walker Dr.
Brampton, Ontario LGT 4H6
(416) 458-1100

Reliance Electric Co.
Electrical Drives Group
24703 Euclid Ave.
Cleveland, OH 44117
(216) 266-7000

Robicon Corp.
100 Sagamore Hill Rd.
Pittsburgh, PA 15239
(412) 327-7000

Rondo Motor Control
Motor Div. of ASEA–Brown
   Boveri
2150 W. 6th Ave.
Broomfield, CO 80020
(303) 469-1742

Ross Hill Controls Corp.
1530 West Belt Drive North
Houston, TX 77043
(713) 467-9888

Saftronics
5580 Enterprise Parkway
Ft. Myers, FL 33905
(800) 533-0031
(813) 693-2431 (fax)

Siemens Energy & Automation
150 Hembree Park Dr.
Roswell, GA 30077
(404) 442-2500

Southcon
3608 Rozzells Ferry Rd.,
Charlotte, NC 28216
(704) 393-1636

Square D Co.
P.O. Box 9247
Columbia, SC 29290-0247
(803) 776-7500

T.B. Wood's Sons Co.
440 N. 5th Ave.
Chambersburg, PA 17201
(717) 264-7161

Toshiba/Houston
13131 W. Little York Rd.
Houston, TX 77041
(800) 231-1412

Vee-Arc Corp.
50 Milk St.
Westborough, MA 01581
(617) 366-7451

Voith Transmissions
7 Pearl Ct.
Allendale, NJ 07401
(201) 825-8855

Westinghouse Electric Corp.
Control Div.
P.O. Box 819
Oldsmar, FL 33557
(813) 855-4621

Westinghouse Electric Corp.
I/C Projects Div.
2040 Ardmore Blvd.
Pittsburgh, PA 15221
(412) 636-3010

York Internat'l Corp.
P.O. Box 1592
York, PA 17405
(717) 771-7890

Zycron Systems
72 Acton St.
West Haven, CT 06516
(203) 932-8471
(203) 932-1459 (fax)

3. Power-Factor Controller Manufacturers

Commander Control
301 Chardonnay Circle
Clayton, CA 94517
(415) 672-3837
(415) 672-8272 (fax)

Electrical South
P.O. Drawer 21888
Greenboro, NC 27420
(800) 222-6662
(919) 375-7444

Motortronics
4241 114th Terrace N.
Clearwater, FL 34622
(813) 573-1819

Nordic Controls Co.
155 N. Van Nortwick Ave.
Batavia, IL 60510
(800) 323-5450
(312) 879-7500

Planum Technology Corp.
1413 Chestnut Ave.
Hillside, NJ 07205
(201) 923-3444
(800) 631-7854

Square D Co.
P.O. Box 9247
Columbia, SC 29290-0247
(803) 776-7500

4. Vendors of Power-Factor Correction Systems

ABB Control
Div. of ASEA–Brown Boveri
1206 Hatton Rd.
Wichita Falls, TX 76302
(800) 877-3232

A.C.M.
P.O. Box 607
Frankfort, IN 46041
(317) 659-4646

AEMC Corp.
99 Chauncy St.
Boston, MA 02111
(617) 266-8506
(800) 343-1391

American Switchgear Corp.
8967 Pleasantwood Ave. NW
North Canton, OH 44720
(216) 499-1210

Control Systems Engineering
1420 Providence Hwy.
Norwood, MA 02062-4691
(617) 762-2900

Cornell-Dubiller
1605 E. Rodney French Blvd.
New Bedford, MA 02741-9990
(201) 256-2000

Electrical Equipment Sales Co.
1338 Reliez Valley Rd.
Lafayette, CA 95459
(415) 939-2521

Fargo Mfg. Co.
130 Salt Point Rd.
Box 2900
Poughkeepsie, NY 12603
(914) 471-0600

General Electric Co.
One Winners Circle
Albany, NY 12205
(518) 438-6500

Horizon Capacitor Corp.
1378 N. Wolcott Ave.
Chicago, IL 60622
(312) 252-2211

Industrial Drives
201 Rock Rd.
Radford, VA 24141
(703) 639-2495

Joliet Equipment Corp.
Box 114
Joliet, IL 60433
(815) 727-6606

McGraw-Edison Power
    Systems
Cooper Industries
P.O. Box 2850
Pittsburgh, PA 15230
(412) 777-3274

M-C Products
Div. of Material Control
7720 E. Redfield Rd., #2
Scottsdale, AZ 85260
(602) 993-9577

Myron Zucker
708 W. Long Lake Rd.
Bloomfield Hills, MI 48013
(313) 643-2277

Planum Technology Corp.
Div. of Cynex Products
28 Sager Pl.
Hillside, NJ 07205
(201) 399-3334

Porter Co.
H.K., Electrical Div.
Porter Bldg.
Pittsburgh, PA 15219
(412) 391-1800

Queensboro Transformer Tech.
   Corp.
115-25 15th Ave.
College Point, NY 11356
(718) 461-5552

RTE Aerovox
Subsid. of RTE Corp.
740 Belleville Ave.
New Bedford, MA 02745
(617) 994-9661

Sangamo Energy Mgmt.
P.O. Box 48400
Atlanta, GA 30362
(404) 447-7300

Sprague Electric Co.
Power Capacitor Dept.
North Adams, MA 01247
(413) 664-4461

Square D Co.
1601 Mercer Rd.
Lexington, KY 40505
(606) 254-6412

Time Mark Corp.
11440 E. Pine St.
Tulsa, OK 74116
(918) 438-1220

Versatex Industries
P.O. Box 354
Brighton, MI 48116
(313) 229-5751
(313) 229-7863 (fax)

Yokogawa Corp. of America
2 Dart Rd.
Newnan, GA 30265
(404) 253-7000

5. Manufacturers of Motor System Test and Repair Equipment

Abbreviations for product categories are as follows:

(c) core loss testers
(d) dynamometers
(k) kW meters
(m) megohmmeters
(p) power quality analyzers
(r) rewind equipment
(t) tachometers

Note that while digital multimeters and surface thermometers are also commonly used for motor testing, they are widely available and thus not listed here.

AEMC Corp. (m)
99 Chauncy St.
Boston, MA 02111
(617) 451-0227

Ametek (t)
8600 Somerset Dr.
Largo, FL 33543
(813) 536-7831
(813) 539-6882 (fax)

Amprobe (m)
630 Merrick Rd.
Lynnbrook, NY 11563
(516) 593-5600
(516) 593-5682 (fax)

AW Dynamometer (d)
P.O. Box 428
Colfax, IL 61728
(800) 447-2511
(309) 723-4951 (fax)

Basic Measurement
  Instruments (BMI) (k,p)
335 Lakeside Dr.
Foster City, CA 94404
(800) 876-5355
(415) 574-2176 (fax)

Biddle Instruments (m)
510 Township Line Rd.
Blue Bell, PA 19422
(800) 424-3353
(215) 643-2670 (fax)

Dranetz Technologies (k,p)
1000 New Durham Rd.
P.O. Box 4019
Edison, NJ 08818-4019
(800) 372-6832
(201) 287-8627 (fax)

Dreisilker Electric Motors (r)
352 Roosevelt Rd.
Glen Ellyn, IL 60137
(312) 469-7510
(312) 469-3474 (fax)

Eaton Corp. (d)
Electric Drives Div.
3122 14th Ave.
Kenosha, WI 53141
(800) 322-4986

Esterline Angus Instrument
  Corp. (k)
P.O. Box 24000
Indianapolis, IN 46224
(317) 244-7611
(800) 543-0829
(317) 247-4749 (fax)

Inductor, Inc. (d)
5821 5th Ave.
Kenosha, WI 53141
(414) 657-0984
(414) 657-1200 (fax)

Lexington Sales and
  Engineering (Lexseco) (c)
304 Blankenbaker Ln.
Louisville, KY 40207
(502) 897-7005

Monarch Instruments (t)
Columbia Drive
Amherst, NH 03031
(603) 883-3390

Multi-Amp Corp. (m,c)
4271 Bronze Way
Dallas, TX 75237
(214) 333-3201
(214) 333-3038 (fax)

NLB (r)
29830 Berk Rd.
Wixom, MI 48096
(313) 624-5555

WOMA (r)
P.O. Box 6793
Edison, NJ 08818
(201) 417-0010

Yokogawa Corp. (t,m)
2 Dart Rd.
Newnan, GA 30265-1018
(404) 253-7000
(404) 251-2088 (fax)

6. Software Suppliers

The number of suppliers of software related to motors and their applications is rapidly growing. The following listings are intended to be a sample rather than a comprehensive compilation. Several periodicals (including ASHRAE Journal, Consulting-Specifying Engineer, and Energy Engineering) publish directories of software on a regular basis.

Fan, Pump, Mechanical Transmission, Duct, and Pipe Software

In addition to general-purpose analysis tools, an increasing number of manufacturers have incorporated computerized versions of their catalogs into software written to assist users in finding the proper fan, pump, or transmission equipment for a given application. Software designations in this subsection are as follows:

      (d) ducts
      (f) fans
      (mt) mechanical transmission components
      (pi) pipes
      (pu) pumps

APEC, Inc. (d,pi)
Miami Valley Tower, Suite 2100
40 W. 4th St.
Dayton, OH 45402
(513) 228-2602

Aurora Pump (pu)
Unit of General Signal
800 Airport Rd.
North Aurora, IL 60542
(708) 859-7074

Bayley Fan Group (f)
843 Indianapolis Ave.
P.O. Box 646
Lebanon, IN 46052-0646
(317) 482-3650

Browning Manufacturing (mt)
P.O. Box 687
Maysville, KY 41056
(606) 564-2011

Buffalo Forge Co. (f)
Box 985
Buffalo, NY 14240
(716) 847-5282

Carrier Corporation (d,pi)
Carrier Parkway/TR-1
Syracuse, NY 13221
(315) 433-4018

E. Jessup & Associates (d,pi)
4977 Canoga Ave.
Woodland Hills, CA 91364
(808) 884-3997

Elite Software (d,pi)
P.O. Drawer 1194
Bryan, TX 77806
(800) 648-9523
(409) 846-2340

Engineered Software, Inc.
(pi,pu)
1015 10th Ave. SE
Olympia, WA 98507
(206) 786-8545

Greenheck Fan Corporation (f)
Box 128
Schofield, WI 54476
(715) 359-6171

Hartzell Fan, Inc. (f)
P.O. Box 919
Piqua, OH 45356-0919
(800) 334-3267

ITT Bell & Gossett (pu)
ITT Fluid Technology Corp.
8200 N. Austin Ave.
Morton Grove, IL 60053
(708) 966-3700

MC-2 Engineering Software
(d,pi)
8107 SW 72nd Ave. #425
Miami, FL 33143
(305) 665-0100

MTS Software (pu)
5 Oak Forest Ct.
St. Charles, MO 63303
(314) 441-1022

National Planning and
Consulting Corp. (mt)
P.O. Box 140489
Coral Gables, FL 33114-0489
(305) 442-1133

O-I/Schott Process Systems,
Inc. (pu)
Box T
Vineland, NJ 08360
(609) 692-4700

Penn Ventilator Co., Inc. (f)
Red Lion & Gantry Roads
Philadelphia, PA 19115
(215) 464-8900

Taco, Inc. (pu)
1160 Cranston St.
Cranston, RI 02920
(401) 942-8000

The Trane Company (d,pi)
3600 Pammel Creek Rd.
La Crosse, WI 54601
(608) 787-3926

United McGill Corporation (d)
Sheet Metal Products
Engineers
200 E. Broadway
Westerville, OH 43801
(614) 882-7401

High-Efficiency Motor Savings Analysis

High-Efficiency Motor Financial Evaluation Program

Ontario Hydro
Technical Services and
   Development Dept.
Attn: Gord Latos
700 University Ave.
Toronto, Ontario M5G 1X6
(416) 592-5332

High-Efficiency Motor Rebate Software Package

BC Hydro
Industrial Energy Management
HEMA Software Package
1045 Howe St., 6th floor
Vancouver, BC V6Z 2B1
(604) 663-3761

NEWMOTOR.WK1

Walter Johnston
Industrial Extension Service
North Carolina State
   University
Box 7901, Page Hall
Raleigh, NC 27695
(919) 737-2356

Industrial Motor Database

Washington State Energy
   Office
Electric Ideas Clearinghouse
809 Legion Way SE, #FA-11
Olympia, WA 98504-1211
(206) 956-2148 or
(800) 872-3568 in OR, WA,
   MT and ID

Adustable-Speed Drives Savings Analysis

Adjustable-Speed Drives Economic Analysis

BC Hydro
Industrial Energy Management
Drives Software Package
1045 Howe St., 6th floor
Vancouver, BC V6Z 2B1
(604) 663-3761

ASCON I and ASCON II

Power Electronics Applications
   Center
10521 Research Dr., #400
Knoxville, TN 37932
(615) 675-9505

Building Energy Simulation

ASEAM:

Dale Stanton-Hoyle
ACEC Research Management
  Foundation
1015 15th Ave. NW
Washington, DC 20005
(202) 347-7474

DOE-2 (mainframe version):

Simulation Research Group
Bldg. 90-3147
Lawrence Berkeley Lab
Berkeley, CA 94720
(415) 486-5711

DOE-2 (PC version):

Acrosoft International
9745 E. Hampden Ave.
Denver, CO 80231

ADM Associates
3299 Ramos Cir.
Sacramento, CA 95827
(916) 363-8383

Trak-Load:

Morgan Systems
2560 9th St., #211
Berkeley, CA 94710
(415) 548-9616
(415) 548-7885 (fax)

7. Trade and Professional Associations Related to Motors and Drives

Air Conditioning
and Refrigeration Institute
1501 Wilson Blvd. Suite 600
Arlington, VA 22209
(703) 524-8800

Air Movement and Control
  Assoc.
30 W. University Dr.
Arlington Heights, IL 60004
(708) 394-0150

American Chain Assoc.
932 Hungerford Dr., #36
Rockville, MD 20850
(301) 738-2448

American Gear Manufacturers
  Assoc.
1500 King St., #201
Alexandria, VA 22314
(703) 684-0211

American Society of
  Lubrication Engineers
838 Busse Hwy.
Park Ridge, IL 60068
(312) 825-5536

American Society of
  Mechanical Engineers
345 E. 47th St.
New York, NY 10017
(212) 705-7800

Anti-Friction Bearing
  Manufacturers Assoc.
1101 Connecticut Ave. NW,
  #700
Washington, DC 20036
(202) 429-5155

Association of Energy
  Engineers
4025 Pleasantdale Rd., #420
Atlanta, GA 30340
(404) 447-5083

Bearing Specialists Association
800 Roosevelt Rd., Bldg. C,
  #20
Glen Ellyn, IL 60137
(312) 858-3838

Electrical Apparatus Service
  Assoc. (EASA)
1331 Bauer Blvd.
St. Louis, MO 63132
(314) 993-2220

Electric Power Research
  Institute
P.O. Box 10412
Palo Alto, CA 94303
(415) 855-2000

Electronic Motion Control
  Assoc.
230 N. Michigan Ave., #1200
Chicago, IL 60601
(312) 372-9800

Fluid Power Distributors
  Assoc.
1900 Arch St.
Philadelphia, PA 19103
(215) 564-3484

Institute of Electrical and
  Electronics Engineers
345 E. 47th St.
New York, NY 10017-2394
(212) 705-7557

Mechanical Power
   Transmission Assoc.
800 Custer Ave.
Elkhart, IN 46515
(219) 264-9421

National Electrical
   Manufacturers Assoc.
   (NEMA)
2101 L St. NW
Washington, DC 20037
(202) 457-8400

National Fluid Power Assoc.
3333 N. Mayfair Rd., #311
Milwaukee, WI 53222
(414) 778-3344

National Industrial Belting
   Assoc.
1900 Arch St.
Philadelphia, PA 19103
(215) 564-3484

National Industrial Distributors
   Assoc.
1900 Arch St.
Philadelphia, PA 19103
(215) 564-3484

National Lubricating Grease
   Inst.
4635 Wyandotte St.
Kansas City, MO 64112
(816) 931-9480

Power Electronics Applications
   Center
10521 Research Dr., #400
Knoxville, TN 37932

Power Transmission
   Representatives Assoc.
5845 Horton, #201
Shawnee Mission, KS 66202
(913) 262-4512

Power Transmissions
   Distributors Assoc.
100 Higgins Rd.
Park Ridge, IL 60068
(312) 825-2000

Rubber Manufacturers Assoc.
1400 K St. NW
Washington, DC 20005
(202) 682-4800

Society of Tribologists
and Lubrication Engineers
838 Busse Hwy.
Park Ridge, IL 60068

Small Motor Manufacturers
   Assoc.
P.O. Box 637
Libertyville, IL 60048
(312) 362-3201

Southern Industrial
   Distributors Assoc.
11 Corporate Sq., #200
Atlanta, GA 30329
(404) 325-2776

Glossary

Material for this glossary was taken in part from the following sources:

- *Cooling and Heating Load Calculation Manual,* Atlanta: American Society of Heating, Refrigerating and Air-Conditioning Engineers (ASHRAE), 1979.
- *1981 Fundamentals,* Atlanta: American Society of Heating, Refrigerating and Air-Conditioning Engineers (ASHRAE), 1981.
- *Terminology of Heating, Ventilation, Air-Conditioning, and Refrigeration,* Atlanta: American Society of Heating, Refrigerating and Air-Conditioning Engineers (ASHRAE), 1986.
- "Guide to HVAC Equipment," Sacramento: California Energy Commission, September 1980.
- "Energy Savings Potential in California's Existing Office and Retail Buildings," Staff Report, Sacramento: California Energy Commission, June 1984.
- "Glossary of Frequently Occurring Motor Terms," Wallingford, Conn.: EMS, Inc., 1983.
- "Guidelines for Saving Energy in Existing Buildings: Building Owners and Operators Manual, ECM 1," Washington, D.C.: Federal Energy Administration, 16 June 1975.
- *IEEE Standard Dictionary of Electrical and Electronics Terms,* New York: Institute of Electrical and Electronics Engineers, 1988.
- Nayler, J.L. and G.H.F. Nayler, *Dictionary of Mechanical Engineering,* New York: Hart Publishing Company, 1967.

Actuator: A device, either electrically, pneumatically, or hydraulically operated, which changes the position of a valve or damper.

Adjustable-speed drive (ASD): A motor accessory that enables the driven equipment (e.g., fan or pump) to be operated over a range of speeds. The two general categories of ASDs are mechanical units (installed between the motor and the driven load) and electronic units (installed in the electrical wiring to the motor).

Air transport system: System that distributes air to the various spaces in a building. Generally comprised of fans, ducts, dampers, registers, etc. Sometimes referred to as a "ventilation system," but air transport of warm or cool air for space conditioning may be separate from the mechanical ventilation system in some buildings.

Alternating current (AC): Electric current that is characterized by the electrons flowing back and forth along the conductors that make up the circuit. Normal building wiring in the United States is alternating current with a frequency of back-and-forth flow of 60 cycles per second. See also *Direct Current*.

Ambient: Surrounding (e.g., ambient temperature is the temperature in the surrounding space).

Amperes (amps):

   Full load amps (FLA): The amount of current the motor can be expected to draw under full load (torque) conditions when operating at the rated voltage. Also known as "nameplate amps."

   Locked-rotor amps (LRA): Also known as "starting inrush," this is the amount of current the motor can be expected to draw under starting conditions when full voltage is applied.

   Service-factor amps: The amount of current the motor will draw when it is subjected to a percentage of overload equal to the service factor on the nameplate of the motor. For example, many motors have a service factor of 1.15, meaning that the motor can handle a 15% overload. See also *Current*.

Apparent efficiency (motor): The product of a motor's efficiency and its power factor.

ASHRAE: American Society of Heating, Refrigerating and Air-Conditioning Engineers.

ASHRAE 90: Voluntary building standards for new buildings, developed by ASHRAE. These standards include minimum equipment efficiencies, building envelope characteristics, and required control strategies for nonresidential buildings.

Average efficiency (motor): See *Nominal efficiency*.

Avoided cost: Cost to the utility of the marginal kWh produced. When conservation or alternative supply allows a utility to reduce its own power production, the savings to the utility are its avoided cost. This quantity (which includes avoided operations and maintenance [O&M], transmission and distribution [T&D], and capacity costs) varies depending on a wide range of factors, including fuel cost, generation type (which may vary over the course of the day and the year), etc.

Bearings: The supports for holding a revolving shaft in its correct position. In the context of motors, the two rotor shaft bearings (mounted in the motor frame) allow rotary motion of the shaft relative to the enclosure, while preventing axial or radial motion. Bearings come in a wide variety of types. Most integral-hp motors use ball bearings, which use rolling steel balls contacting the two main parts ("races") of the bearing to allow the relative motion. Many fractional-hp motors (especially in the smaller sizes) use sleeve or journal bearings, which depend on the bearing lubricant to keep the spinning shaft from contacting the stationary bearing.

Belt: A band of flexible material (usually rubber or plastic reinforced with fabric or steel) for transmitting power from one shaft to another by running over flat, grooved, or toothed pulleys. See Figure 3-15 for illustrations. The common belt types include

  Flat belts: Smooth belts with a flat cross section, riding on corresponding smooth pulleys. Flat belts are thinner and wider than V-belts for the same application.

  Synchronous belts: Belts with a flat cross section and teeth formed in the inner belt surface. The belt teeth engage teeth in the pulleys, preventing any slippage (hence the name).

  V-belts: Belts with a V-shaped cross section; inside surfaces may be smooth or cogged (i.e. having teeth formed in the inner surface of the belt for greater flexibility). V-belts ride in pulleys (sheaves) with corresponding smooth, V-shaped grooves.

Bipolar transistor: Three-terminal electronic switch in which the current between two terminals (the collector and the emitter) is controlled by the third terminal (the base). The base current is typically 50–100 times smaller than the output current.

Brushes: Conductors, usually composed in part of carbon, serving to maintain an electrical connection between the stationary and rotating parts of a motor. Brushes contact either slip rings (in AC wound-rotor motors) or the contacts of the commutator (in DC motors).

Capacitor: A component containing a dielectric (non-conducting) material sandwiched between two metallic layers. Capacitors are widely used for power-factor compensation and filters. See also *Power factor.*

Centrifugal chiller: A machine that produces cold water by using centrifugal action in its compressor to raise the pressure level of the refrigerant gas. Centrifugal chillers are commonly used in large commercial buildings to supply chilled water to cooling coils in the buildings' HVAC systems. Chiller unloading (operating at cooling loads below maximum) is generally regulated by varying the flow of refrigerant gas with variable-inlet vanes on the input side of the compressor.

Centrifugal fan: Device for propelling air by centrifugal action. Forward-curved fans have blades that are sloped forward relative to direction of rotation. Backward-curved fans have blades that are sloped backward relative to direction of rotation. Backward-curved fans are generally more efficient at high pressures than forward-curved fans.

Chiller: A refrigeration machine that produces cooled water, generally at a temperature of 40–55°F. Types include reciprocating, screw, centrifugal (named for the type of compressor used in the motor-driven compression-expansion cycle), and absorption (for the heat-driven absorption cycle).

Chopper: A device that converts DC power into a square wave. When used with an output filter, a chopper can be used with a constant-voltage input to create a variable-voltage output by varying the ratio of on-time to off-time in the square wave.

Code letter: An indication of the amount of locked rotor (inrush) current required by the motor when it is started. See also *Amperes, Locked-rotor.*

Coefficient of performance (COP): A measure of the efficiency of cooling or refrigeration equipment, COP is defined as the ratio of cooling output to energy input, with both quantities in the same units of measure (kW or BTU/h). Electric cooling equipment has COPs ranging between about 2 and 6. See also *Energy efficient ratio.*

Compressor: A mechanical device that increases the pressure, and thereby the temperature, of a gas. Refrigerant compressors are the most common in building applications, followed by air compressors.

Condenser: A heat exchanger in which a refrigerant is condensed from a vapor to a liquid. Common types are air-cooled (either by natural air flow, as in the coil on the back of many residential refrigerators, or fan-forced, as on air conditioners); water-cooled (as in most large chillers for commercial buildings); and evaporative (in which water is sprayed on the outside of the refrigerant tubes and a fan forces air to evaporate some of the water, providing a cooling effect).

Cooling load: The heat and moisture that accumulate in a building and that must be removed in order to maintain comfortable temperature and humidity conditions.

Cooling tower: Device that cools water directly by evaporation. Typically used to reject heat from one or more condensers.

Current: The flow of electrons in an electrical circuit. Current is measured in amperes (or just "amps"); an ampere is equal to a flow of 6.25×10^{18} electrons per second, or one coulomb per second. See also *Amperes.*

Current-source inverter (CSI): A type of electronic ASD that works by converting the AC input to controlled-current DC, then synthesizes the variable-frequency AC output by using a DC-to-AC inverter. See also *Adjustable-speed drive, Variable-frequency drive,* and *Voltage-source inverter.*

Cycloconverter: An AC converter in which the AC supply from the grid is converted directly into another AC voltage waveform with a lower frequency, without an intermediate DC stage. The output frequency ranges between 0% and 50% of the input frequency.

Damper: A restrictive device used to vary the volume of air passing through an air outlet, inlet, or duct.

Demand charge: The amount charged by the utility per kW of peak power used (demanded) by the customer. Demand charges are usually billed per month; the peak demand is measured by a special demand meter that records the highest average demand (typically over a 15- or 30-minute interval) during the month. The charge may be fixed or variable according to the time of day, season, and level of demand.

Design: The design letter on a motor nameplate is an indication of the shape of the torque-speed curve. Figure 2-8 (page 20) shows the typical shape of the most commonly used NEMA design letters (A, B, C, and D). Design B is the standard industrial-duty motor which has reasonable starting torque with moderate starting current and good overall performance for most industrial applications. Design C is used for hard-to-start loads and is specifically designed to have high starting torque. Design D is the so-called high-slip motor, which tends to have very high starting torque but has high-slip rpm at full load torque. Design D motors are particularly suited for low-speed punch press, hoist, and elevator applications.

Generally, the efficiency of Design D motors at full load is rather poor, and thus they are normally used on those applications where the torque characteristics are of primary importance. Design A motors are not commonly specified, but specialized motors used on injection molding applications have characteristics similar to Design A. The most important characteristic of Design A is that the pull-out torque is somewhat higher than Design B's; otherwise A and B are quite similar.

Direct current (DC): Electrical current characterized by electrons flowing in one direction only. See also *Alternating Current.*

Discharge dampers: Dampers that regulate the flow of air on the outlet side of a fan on variable-air-volume (VAV) systems. Dampers are the least efficient method of regulating air flow.

Eddy (or eddy-current) losses: See *Magnetic losses.*

Efficiency (motor): In general, this is the ratio of the mechanical power output to the electrical power input. See other efficiencies: *Apparent, Minimum,* and *Nominal.*

Electromagnetic interference (EMI): Impairment of a transmitted electromagnetic signal by an electromagnetic disturbance. Particularly relevant to data processing and communications applications.

Energy charge: The amount charged by the utility for each kWh of energy used by the customer. The energy charge may be fixed or variable, depending on the time of day, season, and level of usage.

Energy efficiency ratio (EER): A measure of cooling equipment efficiency, defined as (cooling output in Btu/h)/(electric input in watts). EER = COP $\times$ 3.412. SEER is seasonal energy efficiency ratio, measured in a standard test that averages over different part-load ratios of equipment throughout a simulated cooling season.

Explosion-proof (EXP): A type of motor package ("enclosure") designed to withstand the explosion of a specified gas or vapor within it and to prevent ignition of a specified external gas or vapor by sparks, flashes, or explosions that may occur within the motor casing.

Forced commutation inverter: Inverter in which a special commutation circuit is required to turn off the thyristor, making the inverter design more complex.

Fractional horsepower (hp): Motors with a rated output power of less than 1 hp. See also *Horsepower* and *Integral horsepower.*

Frame size: Motors come in various physical sizes to match the requirements of the application. In general, the frame size gets larger with increasing hp or with decreasing speed. In order to promote standardization, NEMA prescribes standard frame sizes for certain hp, speed, and enclosure combinations. Frame size specifies the mounting and shaft dimensions of standard motors. For example, a motor with a frame size of 56 will always have a shaft height above the base of 3½ inches. Frame sizes are usually listed as a combination of a number and a letter, with the number indicating the relative size, and the letter the general frame type (such as T, U, etc.). See also *Frame type.*

Frame type: The general characteristics of a motor's size and mounting configuration; generally given by a letter. For example, NEMA T-frame motors (base-mount, single-ended shaft) are the most commonly made three-phase frame type; the similar but larger U-frame motors were most common until the 1960s. U-frame and T-frame motors have the same shaft size for the same power and speed. Another early design of the same type, A-frame motors, differ from T-frames in both motor size and shaft size. C- and J-frame motors are end-mounted, designed to be bolted directly to the driven equipment. L-frame motors are similar to C-frame, except they are designed to mount vertically above the load (usually a pump). Fractional-hp motors generally do not have a letter designation. See also *Frame size.*

Free rider: A participant in a promotional conservation program who would have performed the conservation action even without the program.

Frequency: Rate of oscillation of alternating current, expressed in cycles per second (or Hz). In North America, the predominant frequency of AC power is 60 Hz.

Full load speed: The approximate speed that the motor will run when it is operating at full rated output torque or hp.

Gate turn-off thyristor (GTO): Electronic switch with the same properties as the thyristor, but in which it is possible to turn off the device by applying a small control signal in the gate. This is in contrast to standard thyristors, which must have the voltage across the main terminals brought close to zero in order to turn off (requiring the use of such techniques as forced commutation).

Gears: A mechanical system for transmitting rotation through the use of toothed wheels in direct engagement. Gears are used to change the speed, direction, or orientation of rotation from one shaft to another. There are a great many types and combinations of gears; four of the most common types of gears are bevel gears, helical gears, worm gears, and spur gears. Helical and worm gears are shown and described in Figure 3-12; spur gears are cylindrical gear wheels in which the teeth are parallel to the shaft. They are used for transmitting power between parallel shafts. Bevel gears are beveled in order to transmit rotation between nonparallel shafts; they are commonly used to transmit shaft power at 90 degrees, that is, between shafts with intersecting axes at right angles.

Harmonics: Electrical signals with frequencies that are integral multiples of the fundamental frequency. For example, in a 60-Hz application, a 180-Hz component is called the third harmonic.

Hertz (Hz): Frequency of AC power in cycles per second. The predominant frequency of power in North America is 60 Hz and in most other countries, 50 Hz. See also *Frequency*.

High-inertia load: A load that has a relatively high flywheel effect (or moment of inertia). Large fans, blowers, punch presses, centrifuges, industrial washing machines, and similar loads can be classified as high-inertia loads. See also *Inertia*.

Horsepower: A unit of power equal to 746 watts or 33,000 ft-lb/minute. In the United States, horsepower is used to indicate the rated output (shaft) power of a motor. One horsepower = torque (ft-lb) × speed (rpm)/5,252. In compressor sizing, it is the full load output rating of the electric motor driving the compressor.

HVAC system: A system that provides one or more of the functions of heating, ventilation, and air conditioning (cooling) for a building.

Hysteresis losses: See *Magnetic losses.*

Inductance: The property of an electrical device or circuit by virtue of which a varying current induces an electromotive force in that circuit, thereby resisting the change in current.

Induction motor: The most common type of AC motor, in which a primary winding on one member (usually the stator) is connected to the power source, and a secondary winding (in the case of wound-rotor induction motors) or a squirrel cage of metal bars (in the case of squirrel cage induction motors) on the other member (usually the rotor) carries induced current. The changing magnetic field created by the stator induces a current in the rotor conductors, which in turn creates the rotor magnetic field. The interaction of the stator and rotor magnetic fields causes the motor to rotate.

Inductors: Generally devices with a magnetic core around which windings of wire are wrapped, a construction which results in high *inductance* relative to the size of the device. An electromagnet is a type of inductor.

Inertia: That property of a body by which it tends to resist a change in its state of rest or uniform motion. Inertia is measured by mass (equivalent to weight) when linear accelerations are considered. In the context of motor systems, where rotational acceleration is the primary concern, inertia is measured by the moment of inertia, about the axis of rotation. The moment of inertia is the sum Σmr^2, where m is the mass of a part of the rotating equipment and r its perpendicular distance from the axis of rotation. That is, the moment of inertia depends on the weight of the rotating system, and on how far the weight is from the axis of rotation (the farther it is, the more effect the same weight will have).

Inlet vanes: Variable vanes on the inlet side of a fan that regulate air flow in a variable-air-volume system. Inlet vanes are also used in *centrifugal chillers.*

Insulated gate transistor (IGT): Three-terminal electronic switch whose input stage is an *MOS transistor* and output stage is a *bipolar transistor.* In this way the IGT combines the best properties of both transistors (negligible input power to control the transistor and low losses in the conduction state when the IGT is fully on).

Insulation class: A measure of the resistance of the insulation components of a motor to degradation from heat. The four major classifications of insulation used in motors are, in order of increasing thermal capabilities, class A, B, F, and H. Class A is no longer used in integral hp motors; the designations C through E and G were never used.

Integral horsepower (hp): Motors with output power ratings of 1 hp or above. See also *Fractional horsepower* and *Horsepower.*

Inverter: Device or system that changes DC power to AC power.

Isolation transformer: A transformer with primary and secondary windings physically separated, thus preventing primary circuit voltage from being forced on the secondary circuits. Isolation transformers are often used with large ASDs to reduce the power quality degradation caused by the ASD.

kVA (kilovolt-amperes): The product of the voltage (in volts) and current (in amperes) in an electrical circuit, divided by one thousand. In DC circuits, kVA equals kW flowing. In AC circuits, the kVA equals the kW if the power factor equals one; otherwise the kVA is higher than the kW. See also *kW* and *Power factor.*

kW (kilowatt): A unit of (usually) electrical power equal to one thousand watts, or the flow of one thousand joules of energy per second. Equivalent to 3,412 British thermal units (Btu) per hour of thermal power or 1.34 hp. Other than in the United States, commonly used to indicate motor output (shaft) power. See also *Horsepower* and *kVA*.

kWh (kilowatt-hour): A unit of electrical energy equal to one kilowatt of power flowing for one hour, i.e. 3,600,000 joules of energy. Equivalent to 3,412 British thermal units (Btu) of thermal energy or 1.34 hp-hr. kWh is the most common unit used for metering electricity. See also *kW*.

Laminations: Thin steel sheets stacked together and used in electromagnetic devices. In motors, they form the core of the stator and rotor magnets. In inductors and transformers, laminations provide the magnetic core around which the windings of wire are placed.

Leakage reactance: The motor reactance associated with that fraction of the magnetic flux generated by the stator winding that does not cross the air gap and so does not reach the rotor (and, vice versa, from the rotor to the stator). The leakage reactance is a trade-off value, as, for example, a high degree of leakage reactance results in lower starting current (a desirable result), but with undesirable reductions in steady-state motor performance. The leakage reactance increases with the air gap size and is also a function of other motor design parameters such as slot design, saturation of the magnetic circuit, and winding configuration.

Load profile: Time distribution of the heating, cooling, ventilation, electrical, or any other loads of a building or process. Load profile is usually expressed on an hourly basis over a day but may also be expressed on a seasonal basis over a year.

Load types:

   Constant horsepower: Loads where the torque requirement decreases as the speed increases, and vice versa. Constant-hp loads are usually associated with applications such as traction (in electric vehicles, for example) and metal removal (drill presses, lathes, milling machines).

   Constant torque: Loads where the amount of torque required to drive the machine is constant regardless of the speed at which it is driven. For example, most conveyors and many reciprocating compressors are constant-torque loads.

   Variable torque: Loads that require low torque at low speeds and increasing torque as the speed is increased. Centrifugal fans and pumps are typical examples of variable-torque loads.

Magnetic losses: When the iron core in the motor is subjected to a changing magnetic field, as it is during normal operation, there are two types of losses: eddy current and hysteresis. Eddy-current (or simply "eddy") losses are due to the currents induced in the iron by the change in the magnetic flux, with losses growing with the square of the flux density and the square of the frequency. Eddy losses can be minimized by using thinner laminations and silicon steel with a higher electric resistivity. Hysteresis losses are due to the rotation of groups of iron atoms as they are excited by the changing magnetic field. Hysteresis losses are proportional to the square of the flux density and to the frequency. Hysteresis losses can be decreased by using high-performance silicon steel with high permeability and a narrow hysteresis cycle. Both types of magnetic losses can be decreased by using a lower magnetic flux density, which means using larger cross sections in the magnetic circuit—that is, more iron in the motor.

Mechanical cooling: Cooling by energy-using cooling equipment such as chillers and air conditioners. Cooling accomplished through use of outside air or by evaporative coolers is generally *not* considered mechanical cooling.

Microelectronic: Electronic devices characterized by highly integrated circuits (many semiconductor devices on one chip of silicon), usually used for computation and control, and generally operating at currents well below one ampere and voltages below ten volts. See also *Power electronic.*

Minimum efficiency (motor): The minimum level of efficiency for a group of motors of the same specification. Up to 5% of motors can have an efficiency lower than the minimum efficiency. Minimum efficiency is sometimes guaranteed by the motor manufacturer. The NEMA minimum efficiency levels are set at two standard increments of efficiency below the NEMA nominal efficiency. See also *Nominal efficiency.*

Minimum efficiency standard: A standard requiring a particular type of equipment to meet a minimum level of operating efficiency. In the case of motors, such standards generally set different minimum levels of nominal motor efficiency according to the motor size (in hp output rating). See also *Efficiency* and *Nominal efficiency.*

MOS transistor: Three-terminal electronic switch in which the conduction between the two main terminals (the drain and the source) is controlled by the voltage applied between the third terminal (the gate) and the source. The input current in the gate is almost zero, and the input power required to control the transistor is negligible. This leads to simple control circuits and improved efficiency.

Motor (electric): Machine that converts electrical power into mechanical power in the form of a rotating shaft. See also *Induction motor* and *Synchronous motor.*

Natural commutation: Circuit in which the voltage applied to the thyristors reverses in polarity leading to the turnoff of the device when the voltage crosses zero.

NEMA: National Electrical Manufacturers Association.

Nominal efficiency (motor): The average expected efficiency for a group of motors of the same specification. Half of the motors are expected to fall below the nominal value, half above. NEMA nominal efficiency (a rating indicating that the motor's nominal efficiency falls within a certain range) is now being stamped on the nameplate of most domestically produced integral-hp electric motors. See also *Minimum efficiency.*

Open drip-proof (ODP): A type of motor package ("enclosure") in which cooling is provided by internal fan(s) forcing air through the motor. The ventilation openings are positioned to keep out liquid or solid particles falling at any angle from 0° to 15° from the vertical.

Participation rate: Fraction (or percentage) of the eligible customers taking part in a program.

Part-load ratio: The ratio of instantaneous output from a piece of equipment to the equipment's rated output. For example, if a piece of cooling equipment is putting out 60% of its full cooling capacity, the part-load ratio is 0.6.

Peak cooling load: The maximum rate of cooling occurring in a building during the year.

Penetration rate: Degree to which a technology has become the standard in a marketplace. For example, if energy-efficient motors are sold for 10% of the general-purpose motor applications, then they have achieved a 10% penetration rate in that market. The market context must be clarified for penetration rate to be meaningful. For example, one needs to know if the market is for new applications or the existing stock.

Phase: The indication of the type of power supply for which the motor is designed. The two main categories are single phase and three phase (sometimes referred to as polyphase).

Poles: The ends of a magnet. Poles are always present in a pair consisting of a north and a south pole. Thus, the number of poles is always even. Poles may be located on permanent magnets or electromagnets. In AC motors, the synchronous speed is determined by the frequency of the power supply and the number of poles: four different motors operating at 60 Hz with two, four, six, and eight poles will have synchronous speeds of 3,600, 1,800, 1,200, and 900 rpm, respectively.

Positive displacement: Term used to describe mechanical equipment (such as compressors, pumps, and blowers) characterized by a reduction of the internal volume of a chamber, usually by a piston.

Power electronic: Electronic devices used for the direct control of electrical power to various types of equipment, including motors. Power electronic devices are available with ratings up to about 5,000 volts and 5,000 amperes, with a trend toward ever higher ratings.

Power factor: The ratio between the real power (measured in watts or kW) and apparent power (the product of the voltage times the current measured in volt-amperes or kVA). Power factor is expressed either as a decimal fraction (from zero to one) or a percentage (0% to 100%). In the case of pure sinusoidal waveforms (those not distorted by harmonics), the power factor is equal to the cosine of the phase angle between the voltage and current waves in an AC circuit. This value is known as the displacement power factor because it deals with the time displacement between the voltage and current. Since cosine values range from 0 to 1, the apparent power is always greater than or equal to the real power. If the power factor is less than 1, more current is required to deliver a unit of real power at a certain voltage than if the power factor were 1. In the case of waveforms that include harmonics, the harmonic current adds to the total current without contributing to the real power, so the power factor is reduced. Many power electronic devices (such as ASDs) have high displacement power factors (over 90%) but overall power factors that are significantly lower depending on design and operating conditions (see *Current* and *Voltage*). This higher current is undesirable because the energy lost to heat in the wires supplying power is proportional to the square of the current. In motors and other inductive loads operating in AC circuits, the current wave lags behind the voltage wave. When a capacitive load is applied to an AC circuit, the voltage wave lags behind the current wave. Since these are opposite effects, they can be used to cancel each other. Thus capacitors can be (and very commonly are) used to correct low power factor. In DC circuits, the power factor is always 1. See also *kVA* and *kW*.

Pulley: see *Sheaves*.

Reciprocating compressor: A machine that uses positive displacement pistons for compression. The pistons move back and forth within their cylinders, much as in a standard automobile engine. Common applications of reciprocating compressors are refrigeration, air conditioning (including reciprocating chillers), and compressed-air systems.

Rectifier: Two-terminal (the positive = anode and the negative = cathode) electronic device that conducts current in one direction with low resistance and blocks the current flow in the opposite direction. The main application of rectifiers is to convert AC power into DC power. The most common rectifiers made are solid-state silicon devices. In the past, mercury rectifiers (using liquid mercury in a vacuum tube) were commonly used in high-current applications.

Regeneration capability (also called regenerative braking): The return of energy to the supply system when a motor is braking, in which case the motor is working as a generator. The input stage of the ASD must have the capability to work as an inverter to pump back the energy to the AC supply.

Resistance: A property of electrical conductors that, depending on their dimensions, material, and temperature, determines the current produced by a given voltage difference across the resistance. Resistance is the property of a material that impedes current and results in the dissipation of power in the form of heat. Resistance is measured in ohms; one ohm is the resistance through which a voltage difference of one volt will produce a current of one ampere.

Resistor: A device connected into an electrical circuit to introduce a specified *resistance*.

RMS (root mean square): The constant value of a periodic time-changing current or voltage that when applied to a resistance would produce the same amount of power. RMS is also known as equivalent DC. The RMS value of a periodic quantity is equal to the square root of the average of the squares of the instantaneous values of the quantity taken throughout the period. For example, the mathematical expression of the RMS value of a current is

$$I_{RMS} = \sqrt{\frac{1}{\tau} \int_{t_1}^{t_2} I^2(t)dt}$$

where τ = the period of time for one cycle
  t_1 = the time measurement starts
  t_2 = the time measurement ends
  $I^2(t)$ = the square of instantaneous value of the current at a time t between t_1 and t_2.

A similar expression applies to the RMS voltage and power values. If the quantity is a sine wave (the nominal form for voltage and current in AC circuits), the RMS value is 0.707 times the peak value of the wave.

Rotary compressor: A positive-displacement compressor that changes internal volume of its compression chamber(s) by the rotary motion of a positive-displacement member(s). Two common types of rotary compressor are

    Rolling-piston compressor: A small rotary compressor having a rotor aligned eccentrically within the stator and used in domestic refrigerators and some room air conditioners.

    Screw compressor: A rotary compressor that produces compression with two intermeshing helical rotors. Applications include medium-to-large refrigeration, and HVAC (including screw chillers) and compressed-air systems.

Rotor: The part of the motor that rotates.

SCR: See *Thyristor.*

Screw compressor, screw chiller: See *Rotary compressor.*

Self-commutation: Circuits that use electronic devices such as transistors and gate-turnoff thyristors, in which those devices turn off by applying a small control signal at their input.

Service factor: The service factor is a multiplier that indicates the amount of overload a motor can be expected to handle. For example, a motor with a 1.0 service factor cannot be expected to handle more than its nameplate hp on a continuous basis. Similarly, a motor with a 1.15 service factor can be expected to safely handle continuous loads amounting to 15% beyond its nameplate hp.

Servomotor: A low-power electric motor that performs a positioning function. Examples include actuators for dampers, valves, and adjustable pulleys.

Sheaves: Grooved wheels attached to the motor shaft and to the shaft of the driven equipment, such as a fan. Sheaves transmit mechanical power by means of one or more belts that ride in the grooves of the pair of sheaves. Another name for sheave is pulley.

Silicon-controlled rectifier (SCR): See *Thyristor.*

Slip: The difference between motor operating speed and the synchronous motor speed, expressed either directly in rpm or as a percentage of synchronous speed (see *Synchronous speed*). For example, an 1,800-rpm motor operating at a full load speed of 1,725 rpm is running at a slip of 75 rpm or 4.2%. Most standard induction motors run at a full load slip of 2% to 5%.

Slip rings: In an AC motor, a set of metal rings that are mounted on the rotor shaft and that conduct current into or out of the rotor through stationary brushes.

Space conditioning loads: A building's heat losses and gains that need to be counteracted by heating or cooling in order to maintain comfortable temperature and humidity.

Squirrel cage induction motor: A type of induction motor with a squirrel cage winding consisting of a number of conducting bars connected at each end by metal rings that are located in slots in the rotor core. The bars are parallel to the motor shaft; the rings are concentric with the axis of the shaft. This motor is the most common type in use. In order to deliver torque to a load, its shaft must run with slip, or below synchronous speed. See also *Induction motor, Slip,* and *Synchronous speed.*

Stator: The nonrotating magnetic section of a motor. In most induction motors, the stator contains the windings.

Synchronous motor: An AC motor in which the speed of operation is exactly proportional to the frequency of power to which it is connected (the motor operates with no slip). Synchronous motors generally have the rotor electromagnets supplied with DC power through slip rings. Since these motors produce little torque except at speeds near to the synchronous speed, they need special methods for starting.

Synchronous speed: The speed at which the motor's magnetic field rotates. It approximates the speed of no-load operation. A four-pole motor running on 60-cycle-per-second power will have a synchronous speed of 1,800 rpm; a two-pole motor at the same frequency will have a synchronous speed of 3,600 rpm. See also *Slip.*

TEFC (Totally enclosed fan-cooled): A type of motor package ("enclosure") in which there is no air exchange between the inside and outside of the motor. The fan is located in a cover opposite the driving (power output) shaft and is driven by an extension of the motor shaft through the housing.

Temperature, ambient: The maximum safe room temperature surrounding the motor if the motor is going to be operated continuously at full load. In most cases, the standardized ambient temperature rating is 40°C (104°F). Certain types of applications, such as on ships and in boiler rooms, may require motors with a higher ambient temperature capability such as 50°C or 60°C. Note that this definition is specific to motors, in contrast to the general definition of *ambient.*

Temperature rise: The amount of temperature increase that can be expected within the winding of the motor from nonoperating (cool condition) to its temperature at full load and continuous operation. Temperature rise is normally expressed in degrees Celsius.

Throttle: A device that regulates the flow of a gas or liquid by directly restricting the flow. Discharge dampers, inlet vanes, and valves can all be throttles.

Thyristor (also called silicon-controlled rectifier [SCR] or phase-controlled rectifier): Electronic devices that have both the same capabilities of rectifiers and a third terminal (the gate). The gate allows control of conduction from 0% to 100% when the polarity applied to the main terminals is positive. If the polarity is negative, the thyristor blocks the current like a rectifier.

Time rating: Most motors are rated for continuous duty, meaning that they can operate at full load torque continuously without overheating. Motors used in certain applications such as waste disposers, valve actuators, hoists, and other intermittent loads will frequently be rated for short-term duty such as 5 minutes, 15 minutes, 30 minutes, or 1 hour.

Torque: The twisting force exerted by the motor shaft on the load. Torque is measured in units of length times force in foot-pounds or inch-pounds, or, on small motors, inch-ounces. For an illustration of the following types of torque, see Figure C-1.

  Breakdown torque: See *Pull-out torque.*

  Full load torque: The rated continuous torque that the motor can support without overheating within its time rating.

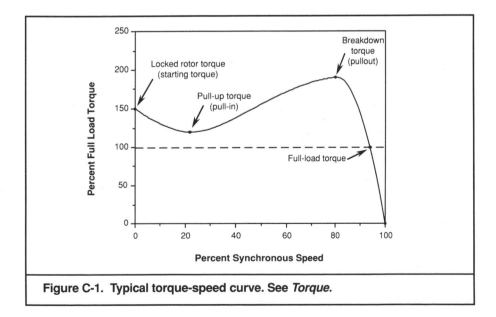

Figure C-1. Typical torque-speed curve. See *Torque.*

Peak torque: Many types of loads, such as reciprocating compressors, have cycling torques where the amount of torque required varies depending on the position of the machine. The actual maximum torque requirement at any point is called the peak torque requirement. Peak torques are involved in types of loads (such as punch presses) with an oscillating torque requirement. A motor's pull-up torque must be greater than the load's peak torque requirement to prevent stalling the motor.

Pull-out torque: Also known as "breakdown torque," the maximum amount of torque that is available from the motor shaft when the motor is operating at rated voltage and running at full speed.

Pull-up torque: The lowest point on the torque-speed curve for a motor accelerating a load up to speed. Pull-up torque limits a motor's ability to accelerate its load and to meet a load's peak torque requirement. Some motor designs (typically NEMA Designs A and B) do not have a separate value for pull-up torque because the lowest point may occur at the locked rotor (starting) point. In this case, pull-up torque is the same as starting torque.

Starting torque: Also known as locked-rotor torque, the amount of torque the motor produces when energized at full rated voltage with the shaft locked in place. It is the amount of torque available when the motor is energized to break the load away (start it moving) and begin accelerating it up to speed.

Transformer: In its most common form, a device to increase or decrease the voltage in an AC system. The primary side of the transformer is connected to the source of power; the secondary side, to the load. A step-down transformer (the most common type in transmission and distribution systems) reduces the primary voltage to the secondary voltage. A step-up transformer (used, for example, at power plants to increase the generation voltage to the transmission voltage) increases the primary voltage to the secondary voltage. Transformers work by using the current in the primary winding to create a changing magnetic field, which is used to induce a voltage (and thus current, when connected to a load) in the secondary winding. Another common transformer type is the *Isolation transformer.*

Transistor: See *Bipolar transistor, Insulated gate transistor,* and *MOS transistor.*

Variable-air-volume (VAV): An HVAC system in which the amount of cooling is controlled by changing the air flow rate; VAV heating systems are also used, as well as VAV control of room pressurization.

Variable-frequency drive (VFD): Another name for the most common type of electronic adjustable-speed drive. This type of drive uses an electronic package between the fixed-frequency AC input and the motor. The speed is varied by supplying the motor with synthesized AC power of changing frequency. See *Adjustable-speed drive*.

Variable-speed drive: See *Adjustable-speed drive*.

Ventilation: The introduction of fresh air into a building specifically for the purpose of maintaining good air quality. Air is usually drawn from outdoors but can also be purified, recirculated air. Often, the term "ventilation" is used loosely to include transport of any air, not just of fresh air. See *Air transport system*.

Voltage: The rated voltage for which a motor or related electrical equipment is designed to operate. In general, voltage is the electrical potential at any point relative to some reference point in a circuit. The voltage represents the energy level of a quantity of electrical charge (electrons) at that point in the circuit. In the Systeme Internationale system of measurement, the unit of voltage is the volt, which equals one joule of energy per one coulomb of charge (see *Current*). When there is a flow of charge at a given voltage, this flow of energy is electrical power. This power is measured in watts (joules per second), which at any instant is equal to the product of the voltage and the current in the circuit.

Voltage-source inverter (VSI): A type of electronic ASD that converts the AC input to controlled-voltage DC, then synthesizes the variable-frequency AC output by using a DC-to-AC inverter. See *Adjustable-speed drive*, *Current-source inverter*, and *Variable-frequency drive*.

Watt: Unit of (usually) electrical power equal to one joule of energy flowing per second. See also *kW*.

Winding: In motor stators, transformers, inductors, and electromagnets, a number of turns of insulated wire (usually copper) wrapped around the core of steel laminations. The stator windings are generally connected to the power supply. In squirrel cage motor rotors, the "winding" is composed of several bars of uninsulated aluminum or copper, arranged in a cylinder and connected together at both ends by rings of the same material. The windings of wound-rotor motors are similar to those of the motor stator. When a motor is rewound, the insulated wire is removed and replaced with new wire.

Wk^2: Symbol used for moment of inertia and measured in lb-ft^2. See also *Inertia*.

Annotated Bibliography

Books and Reports

This Annotated Bibliography describes some of the books, reports, journals, and other periodicals that are most useful for obtaining additional information on motor systems.

Andreas, John. *Energy-Efficient Electric Motors.* New York: Marcel Dekker, 1982. This reference, written in simple language, provides guidelines for selecting and applying electric motors on the basis of life-cycle costs. Particular emphasis is given to single- and three-phase motors from 1 to 125 hp. The book covers the economics of energy-efficient motors in detail, and it discusses some of the interactions between the power supply and the motor. There is a brief section on adjustable-speed drives.

Argonne National Laboratory. *Classification and Evaluation of Electric Motors and Pumps.* DOE/TIC-11339. Springfield, VA: National Technical Information Service, 1980. This study describes the motor and pump markets of the late 1970s and analyzes whether efficiency standards and labeling requirements for motors and pumps are desirable. The report contains many detailed breakdowns on the motor and pump populations, but the accuracy of some of the numbers is questionable (due to limitations in the underlying data). Unfortunately, more accurate or recent data are often not available, so for lack of competition, Argonne's numbers are sometimes the best available. The report contains a politically biased conclusion that neither efficiency standards nor labeling requirements are desirable (an issue discussed in chapter 9 of this book). An earlier version of the report (DOE/CS-0147, same title and publisher) concluded that efficiency standards and labeling may be advantageous. This earlier version also contains some data that did not make it into the final report.

Baldwin, Sam. *Energy-Efficient Electric Motor Drive Systems,* Working Papers 91-94. Princeton, NJ: Princeton University Center for Energy and Environmental Studies, 1988. Working Paper 91, *High-Performance Materials, Adjustable-Speed Drives, and System Design,* is a highly readable

75-page summary, for non-engineers, of how motors work, how motor systems are designed, the types of adjustable-speed drives, and the role of materials advances in the evolution of motors. (Later versions of this paper appeared in *Annual Review of Energy,* Volume 13 [1988], and in *Electricity: Efficient End-Use and New Generation Technologies and Their Planning Implications,* Lund University Press, 1989.) Working Paper 92, *Electric Motor Fundamentals—A Tutorial,* summarizes electromagnetic theory as it pertains to motors. Packed with equations, this 90-page paper is aimed at a technical audience. Working Paper 93, *Motor and System Diagnostics,* is a 45-page treatment of diagnostic theory and methods for an engineering audience. Working Paper 94, *A Field Study of the Jamaican Sugar Industry,* examines the savings potential from high-efficiency motors, ASDs, and optimized sizing and design practice. Although the specifics pertain to the sugar industry, the study's methodology is widely applicable. Each of these papers contains extensive references.

Bose, Bimal. *Power Electronics and AC Drives.* Englewood Cliffs, NJ: Prentice Hall, 1986. This textbook covers the basics of the main semiconductor devices for drivepower systems, explains the fundamentals of induction and synchronous machines, and features a comprehensive description of AC electronic drives, including the use of microcomputers to control AC drives. Conventional and vector control of AC machines are discussed in detail. Readers should have some background in electronics and machines.

British Columbia Hydro. *High-Efficiency Motors.* Vancouver, British Columbia: British Columbia Hydro, 1988. This well-written 26-page booklet is designed to educate end-users about the advantages of high-efficiency motors. It includes discussion of high-efficiency motor economics, of noneconomic criteria to consider when buying high-efficiency motors, and of simple steps to increase motor life. The booklet concludes with answers to common questions on high-efficiency motors.

Comnes, G.A., and R. Barnes. *Efficient Alternatives for Electric Drives.* ORNL/TM-10415. Oak Ridge, TN: Oak Ridge National Laboratory, 1987. This 60-page report summarizes information on motor and ASD sales for 1982–83 and describes opportunities for energy savings with energy-efficient motors, new motor designs, and adjustable-speed drives. Economics and behavioral factors that affect decision-making about industrial drives are also discussed. The report contains some useful information but is becoming a bit dated.

De Almeida, Anibal, Steven Greenberg, and Carl Blumstein. "Demand-Side Management Opportunities through the Use of Energy-Efficient Motor Systems." *IEEE Transactions on Power Systems* 5 (August). Parsippany, NJ: IEEE, 1990. This paper summarizes the key issues related to efficient motor system operation from a demand-side management perspective. Particular emphasis is given to ASDs, including potential savings in the residential, commercial, and industrial sectors. Research, development, demonstration, and training options are also described.

De Almeida, Anibal, Steve Greenberg, Carl Bauer, and Carl Blumstein. *Applications of Adjustable-Speed Drives for Electric Motors: Technology Assessment.* RP-1966-24. Palo Alto, CA: Electric Power Research Institute, 1991. This report describes conventional speed controls as well as the various types of electronic ASDs and their characteristics and fields of application. The impact of ASDs on power and communications networks is covered, along with economic analyses and options for further research.

Dreisilker, Henry. "Safe Stator and Rotor Stripping Method." Undated 10-page typescript. Glen Ellyn, IL: Dreisilker Electric Motors. Henry Dreisilker, president of a large motor distribution and repair business, has waged a one-man campaign for thirty years against the use of burnout-oven stripping. He maintains that conventional motor repair practice damages motors and that the low-temperature, mechanical technique he uses and markets does a better job without damaging the motors. Dreisilker has an extensive collection of testimonials in support of his method and of case studies of the damage caused by conventional practice.

Electrical Apparatus Service Association. *Core Iron Study.* St. Louis: Electrical Apparatus Service Association, 1985. This widely cited study by the trade association of motor repair shops sought to resolve the question of whether conventional burnout-oven stripping degrades motor cores. Although the study concludes that no damage should occur when burnout ovens are set no higher than 650°F, the data from EASA's tests do show some damage and suggest that lower temperature limits may be warranted. This important issue thus remains unresolved.

Greenberg, Steve, Jeffrey Harris, Hashem Akbari, and Anibal de Almeida. *Technology Assessment: Adjustable-Speed Motors and Motor Drives (Residential and Commercial Sectors).* #LBL-25080. Berkeley, CA: Lawrence Berkeley Laboratory, 1988. This report describes the technologies applicable for speed control in residential and commercial buildings. Particular emphasis is placed on ASDs, including cost-effectiveness, savings potential, and possible adverse effects. Additional implementation options, major uncertainties, and recommendations for future work are covered.

Lawrie, Robert J., ed. *Electric Motor Manual: Application, Installation, Maintenance, Troubleshooting.* New York: McGraw-Hill, 1987. This book is a collection of useful articles from the "Motor Facts" department of *Electrical Construction and Maintenance* Magazine. As the title implies, there are articles covering motor selection, installation, and care. The articles are written for the technician doing the work and thus cover practical details frequently lacking in more theoretical references.

Lobodovsky, Konstantin, Ramesh Ganeriwal, and Anil Gupta. "Field Measurements and Determination of Electric Motor Efficiency." *Energy Engineering 86(3).* Atlanta: Association of Energy Engineers, 1987. This paper explains how to conduct field measurements of motor efficiency. The authors use IEEE Standard Method E/F to achieve results similar to laboratory testing. The results of a field testing survey showing widespread oversizing are presented. Limitations of the method are also explained.

Lovins, Amory, Joel Neymark, Ted Flanigan, Patrick Kiernan, Brady Bancroft, and Michael Shepard. *The State of the Art: Drivepower.* Snowmass, CO: Rocky Mountain Institute (RMI), 1989. This 419-page encyclopedia on the technical and economic aspects of motors, controls, mechanical equipment, electrical tune-ups, and other motor maintenance issues includes a long discussion on motor rewind issues and an estimate of the total drivepower savings potential in the United States (the authors' estimate being somewhat higher than the estimate made in chapter 7 of this book). This report, which is updated at least annually, contains a wealth of information, making it a useful reference volume. It is available primarily to subscribers to RMI's *Competitek* series—which includes *State of the Art* volumes on lighting, appliances, and other topics—although individual copies may be purchased. Contact RMI for ordering information. Individual copies are expensive.

Marbek Resource Consultants, Ltd. *Energy-Efficient Motors in Canada: Technologies, Market Factors and Penetration Rates.* Ottawa: Marbek Resource Consultants, 1987. This excellent report examines the market and opportunities for energy-efficient motors in Canada. It contains much useful information and data on how the motor market is structured and on market constraints that need to be overcome. This information is generally applicable to the United States as well as Canada. The report concludes with specific policy recommendations for Canadian utilities and governments.

Montgomery, David. "Testing Rewinds to Avoid Motor Efficiency Degradation." *Energy Engineering* 86(3). Atlanta: Association of Energy Engineers, 1989. Drawing on extensive field data, this reference analyzes the impact of rewinding on motor efficiency. The breakdown of losses is explained. Methods of economic analysis of rewinds, quality control measures, and core loss testing are presented.

Nailen, Robert. *Motors.* Vol. 6, *Power Plant Reference* series. Palo Alto: Electric Power Research Institute, 1987. This motor manual is directed mainly at power plant engineers, although most of the information is useful in other fields. Techniques for matching a motor to an application are described in relation to the load characteristics, environment, and power systems. Motor industry standards and maintenance practice are also covered.

Power Electronics Application Center. *ASD Directory.* PEAC.00.5.87. Knoxville: Power Electronics Application Center, 1987. The *ASD Directory* contains a listing of manufacturers and their products (by type of electronic ASD technology) in size categories from 7.5 to 20,000 hp. Case studies, average equipment list prices, and average installation costs for the range of sizes are also included. There are brief sections on how ASDs work and how to determine savings. Glossary and references are included. The directory is updated about every three years (a new edition is not available at press time).

Ryan, Maura. *Electric Variable-Speed Drive Reference Guide.* Toronto: Ontario Hydro, 1988. This small-format handbook is the fourth in Ontario Hydro's "product knowledge" series, written for the utility's customer service representatives. It offers brief, but comprehensive, coverage of the topic: basic definitions, selection and economics, and power quality issues and performance standards.

Seton, Johnson & Odell, Inc. *Energy Efficiency and Motor Repair Practices in the Pacific Northwest.* Portland, OR: Bonneville Power Administration, 1987. This detailed report on motor rewind practices includes results of end-user and rewind shop surveys. It discusses the perspective of end-users facing rewind-versus-replacement decisions. Rewind-practices and their effect on motor efficiency are covered. Opportunities for improving motor efficiencies are discussed, including improvements in core stripping techniques and ways to identify and repair or replace motors with damaged cores.

Seton, Johnson & Odell, Inc. *Report on Lost Conservation Opportunities in the Industrial Sector.* Portland, OR: Bonneville Power Administration, 1987. This report examines opportunities for obtaining efficiency improvements at low cost when new equipment is purchased or existing equipment is being replaced. It discusses several motor-related industrial energy efficiency opportunities, including motors, pumps, and piping. Extensive data on motor sales, costs, and efficiencies are included.

Stout, Timothy, and William Gilmore. "Motor Incentive Programs: Promoting Premium Efficiency Motors." Paper presented at Electric Council of New England National Conference on Demand-Side Management, November 16 and 17, 1989. Westborough, MA: New England Power Service Company. This paper discusses the lessons one utility has learned from offering several types of motor rebate programs. Program designs and results are discussed, including information on marketing and on setting eligibility and rebate levels.

Wilke, Kenneth, and T. Ikuenobe. "Guidelines for Implementing an Energy-Efficient Motor Retrofit Program." *Proceedings of the 10th World Energy Engineering Conference.* Atlanta: Association of Energy Engineers, 1987. This paper summarizes a successful program at Stanford University to replace 73 in-service HVAC motors with energy-efficient models. Many cases had paybacks of less than two years; the payback of the whole project averaged three years with a utility rebate, five years without the rebate. Nearly half the motors were downsized upon replacement. This little known project deserves greater scrutiny; it suggests that targeted group motor replacements may be a cost-effective option in many settings.

Journals and Periodicals

ASHRAE Journal, ASHRAE Transactions. American Society of Heating, Refrigerating and Air-Conditioning Engineers, 1791 Tullie Circle NE, Atlanta, GA 30329. The *Journal* is the monthly magazine of ASHRAE, covering the topics the name implies in articles, advertisements, and product listings; it is of primary interest to mechanical engineers designing or retrofitting HVAC and refrigeration systems. *Transactions* is published twice each year and contains the research papers presented at each of two annual ASHRAE meetings.

Consulting-Specifying Engineer. The Cahners Publishing Co., 275 Washington Street, Newton, MA 02518-1630. This magazine is published monthly and is addressed to mechanicial and electrical engineers working in the building construction industry. The articles, advertising, and product listings cover a wide range of technologies, including those related to motors.

Design News. Cahners Publishing Co., 275 Washington Street, Newton, MA 02158. This magazine is published twice monthly and is written for electrical and mechanical engineers designing components and systems for buildings, industry, and transportation. It includes articles, advertising, and product listings.

Electrical Construction and Maintenance. Intertec Publishing, 888 7th Avenue, 38th Floor, New York, NY 10106. This monthly magazine covers the installation, maintenance, and repair of a range of electrical technologies. Each issue includes "Motor Facts," covering a variety of issues in the selection, installation, and care of motors.

Energy Engineering. Association of Energy Engineers, 700 Indian Trail, Lilburn, GA 30247. This bimonthly publication is the journal of the Association of Energy Engineers. Each issue concentrates on a single topic, such as motor systems, energy management control systems software, and so forth, and includes several articles plus a product directory.

Energy User News. The Chilton Co., 7 E. 12th Street, New York, NY 10003. This magazine, published monthly, is targeted at facility managers in commercial and institutional buildings. Articles include case studies, interviews, and surveys and report trends in energy costs. The magazine includes advertising and product directories on a variety of energy-efficiency technologies, including motors and drives.

Engineered Systems. Business News Publishing Co., P.O. Box 7016, Troy, MI 48007. This magazine, published bimonthly, "provides information to assist people who specify, install, buy, and maintain commercial-industrial-institutional HVAC/R systems." Articles, advertisements, and product directories cover a wide range of topics in the areas of both mechanical and electrical technologies, including motors and motor systems.

Heating, Piping, and Air Conditioning. Penton Publishing, Inc., 1100 Superior Avenue, Cleveland, OH 44114. This monthly magazine is addressed to mechanical engineers working in the building trade. Articles, advertising, and product listings cover a variety of topics, including pumps, fans, piping, and ductwork.

IEEE Transactions on Industry Applications, IEEE Transactions on Power Systems. Institute of Electrical and Electronics Engineers, 345 E. 47th Street, New York, NY 10017-2394. *Transactions on Industry Applications,* published six times a year, covers a variety of motor-related technologies of interest to industry—including recent developments in adjustable-speed drives and their applications—and publishes papers presented at conferences of the IEEE Industry Applications Society. *Transactions on Power Systems* focuses on topics of interest to electric utilities, including new types of motors and the interaction of motor systems with the utility. This quarterly publication contains papers presented at conferences of the IEEE Power Engineering Society.

Plant Engineering. Cahners Publishing Co., 275 Washington Street, Newton, MA 02158. This magazine is published twice monthly and is written for engineers working in industry. It includes articles, advertising, and product listings.

References

Abbate, G. 1988. "Technology Developments in Home Appliances." In *Demand-Side Management and Electricity End-Use Efficiency,* ed. A. de Anibal and A. Rosenfeld, 435–448. NATO Advanced Science Institutes Series E, vol. 149, Applied Sciences. Norwell, MA: Kluwer Academic Publishers.

A.D. Little, Inc. 1976. *Energy Efficiency and Electric Motors.* Report PB-259 129 (August), U.S. Federal Energy Administration, Office of Industrial Programs. Springfield, VA: National Technical Information Service.

———. 1980. *Classification and Evaluation of Electric Motors and Pumps.* Report DOE/CS-1047 (February), U.S. Department of Energy, Office of Industrial Programs. Springfield, VA: National Technical Information Service.

Alliance to Save Energy. 1983. *Industrial Investment in Energy Efficiency: Opportunities, Management Practices, and Tax Incentives.* Washington, D.C.

American Society of Heating, Refrigerating and Air-Conditioning Engineers, Inc. 1990. *ASHRAE/IES Standard 90.1-1989: Energy Efficient Design of New Buildings Except Low-Rise Residential Buildings.* Atlanta: American Society of Heating, Refrigerating and Air-Conditioning Engineers.

———. 1991. "ASHRAE/IES 90.1c — Addendum to ASHRAE/IES 90.1-1989." Atlanta, GA: American Society of Heating, Refrigerating and Air-Conditioning Engineers.

Anderson, K., and N. Benner. 1988. "The Energy Edge Project: Energy Efficiency in New Commercial Buildings." Paper presented to American Society of Mechanical Engineers. Portland, OR: Pacific Power & Light.

Andreas, J. 1982. *Energy-Efficient Electric Motors: Selection and Application.* New York: Marcel Dekker.

Argonne National Laboratory. 1980. *Classification and Evaluation of Electric Motors and Pumps.* Report DOE/TIC-11339 (September), U.S.

Department of Energy. Springfield, VA: National Technical Information Service. (This is largely a rewrite of A.D. Little, Inc. 1980.)

Baldwin, A., and N. Planer. 1982. *Evaluation of Electrical Interference to the Induction Watthour Meter.* Report EL-2315. Palo Alto: Electric Power Research Institute.

Baldwin, S. 1986. "New Opportunities in Electric Motor Technology at U.C. Berkeley." *IEEE Technology and Society Magazine,* March, 11–18.

———. 1988a. *Energy-Efficient Electric Motor Drive Systems.* Working Papers 91-94, Princeton University Center for Energy and Environmental Studies. These reports have extensive bibliographies.

———. 1988b. "The Materials Revolution and Energy-Efficient Electric Motor Drive Systems." *Annual Review of Energy* 13: 67–94.

———. 1989. "Energy-Efficient Electric Motor Drive Systems." In *Electricity: Efficient End-Use and New Generation Technologies, and Their Planning Implications,* edited by T.B. Johansson, B. Bodlund, and R.H. Williams. Lund, Sweden: Lund University Press.

BC Hydro. See British Columbia Hydro.

Benner, N. 1988. Portland Energy Conservation, Inc., 2950 SE Stark, Portland, OR 97214, (503) 248-4636. Meeting with Steven Nadel, April.

Benner, N., R. Christle, J. McFerran, and K. Miller. 1989. "Lessons Learned in Demand-Side Planning for Connecticut Light and Power's New Building Program: Commercial Sector." In *Demand-Side Management Strategies for the 90s, Proceedings: Fourth National Conference on Utility DSM Programs.* EPRI CU-6367. Palo Alto: Electric Power Research Institute.

Bodine, C., ed. 1978. *Small Motor, Gearmotor, and Control Handbook,* 4th edition. Chicago: Bodine Electric Company.

Bonneville Power Administration. 1989. "The Energy $avings Plan." Portland, OR. Photocopy.

Bose, B. 1986. *Power Electronics and AC Drives.* Englewood Cliffs, NJ: Prentice Hall.

Boston Edison Co. et al. 1990. *The Power of Service Excellence: Energy Conservation for the 90's.* Boston: Boston Edison.

Bowles, H.E. 1989. The Electrification Council, 1111 19th Street, Washington, DC 20036, (202) 778-6900. Phone conversation with Steven Nadel, July.

BPA. See Bonneville Power Administration.

British Columbia Hydro. 1988. *High-Efficiency Motors.* Vancouver, B.C.: British Columbia Hydro.

———. 1989a. *Adjustable-Speed Drives Economic Analysis.* Vancouver, B.C.: British Columbia Hydro.

———. 1989b. "Electric Motor Database System." Vancouver, B.C.: British Columbia Hydro.

British Columbia Hydro, Ontario Hydro, and Hydro Quebec. 1990. "Motor Manufacturers Meeting, October 25, 1990: High-Efficiency Motors, a Coordinated Utility Approach." Toronto: Ontario Hydro.

Buckley, T. 1990. Burlington Electric Department, 585 Pine Street, Burlington, VT 05401, (802) 658-0300. Meeting with Steven Nadel, September.

Bureau of the Census. 1988. *1986 Annual Survey of Manufacturers*. Report M86(AS)-1. Washington, D.C.: U.S. Government Printing Office.

———. 1990. *Current Industrial Reports: Motors and Generators, 1989*. Report MA36H(89)-1. Washington, D.C.: U.S. Government Printing Office.

Burley, L., and J. Leber. 1981. "Analysis of the September 1980 Department of Energy Motors Report." Sacramento: California Energy Commission.

Burrell, C. 1990. Ontario Hydro, 700 University Avenue, Toronto, Ontario M5G 1X6, (416) 231-4111. Phone conversation with Steven Nadel, May.

Calhoun, R. 1984. "The Great PG&E Energy Rebate," In *Doing Better: Setting an Agenda for the Second Decade*. Vol. 1. Washington, D.C.: American Council for an Energy-Efficient Economy.

Calhoun, R. 1987. Morgan Systems Corp., 2560 9th Street, Suite 211, Berkeley, CA 94710, (415) 548-9616. Phone conversation with Steven Nadel, October.

California Energy Commission. 1984. "Energy-Savings Potential in California's Existing Office and Retail Buildings." Staff analysis, Technology Assessments Project Office (P. Gertner and T. Tanton). Sacramento: California Energy Commission.

Canadian Standards Association. 1991. *Energy Efficiency Test Methods for Three-Phase Induction Motors (Efficiency Quoting Method and Permissible Efficiency Tolerance)*. Rexdale, Ontario: Canadian Standards Association. July.

Carolina Power and Light. 1986. Printout of motor data compiled by Carolina Power & Light, Raleigh, NC; Contact: Carl Castellow.

Castelow, C. 1989. Carolina Power and Light, P.O. Box 1551, OHS-9B5, Raleigh, NC 27602, (919) 546-7078. Phone conversation with Steven Nadel, October.

Central Maine Power. 1988. "Efficiency Buy Back Program." Augusta, ME: Central Maine Power. Photocopy.

———. *1989 Energy Management Report*. Augusta, ME: Central Maine Power.

Clarkson, J. 1990. Southwire Company, Carrollton, GA, (404) 832-4608. Phone conversations and meetings with Steven Nadel and Michael Shepard, September.

Clarkson, J., and N. Deese. 1987. "Reducing Electrical Losses." In *Proceedings of the1987 World Energy Engineering Congress*. Atlanta: Association of Energy Engineers.

Clippert, P. 1989. "Commercial/Industrial/Farm—Smart Money Energy Program, Total Completed Status—17 March 1989." Wisconsin Electric Power Company. Printout.

CMP. See Central Maine Power.

Cohen, R. 1992. Canadian Standards Association, 178 Rexdale Boulevard, Rexdale, Ontario M9W 1B3, (416) 747-4000. Phone conversation with Steven Nadel, May.

Colby, R., and D. Flora. 1990. "Measured Efficiency of High Efficiency and Standard Induction Motors." Paper presented at IEEE Industry Applications Society Annual Meeting, October, in Seattle.

Commonwealth of Massachusetts. 1990. House 6077, *An Act Requiring Minimum Efficiency Standards for Lighting Fixtures, Lightbulbs, Floor Lamps, Table Lamps and Electric Motors.* Boston.

Comnes, G.A., and R. Barnes. 1987. *Efficient Alternatives for Electric Drives.* Report ORNL/TM-10415. Oak Ridge, TN: Oak Ridge National Laboratory.

Connors, D., and D. Jarc. 1983. "Application Considerations for AC Drives." *IEEE Transactions on Industry Applications* IA-19 (May/June): 455–460.

Consolidated Edison Company. 1989. *Demand-Side Management Program Plan.* New York: Consolidated Edison.

Cornell Pump Company. 1987. "Model 6NHP-Various Speed." Pump curve number 770-130. Portland, OR: Cornell Pump Company.

Cowern, E. 1989. Baldor Motors, 5711 S. 7th Street, Fort Smith, AR 72902, (501) 646-4711. Phone conversation with Michael Shepard, August.

CP&L. See Carolina Power and Light.

Davies, J. 1989. City of Palo Alto Utilities Department, P.O Box 10250, Palo Alto, CA 94303, (415) 329-2695. Phone conversation with Steven Nadel, February.

De Almeida, A., S. Greenberg, C. Bauer, and C. Blumstein. 1991. "Applications of Adjustable-Speed Drives for Electric Motors: Technology Assessment." Final report for EPRI Research Project (RP) 1966-24. To be published in 1991.

De Almeida, A., S. Greenberg, and C. Blumstein. 1990. "Demand-Side Management Opportunities Through the Use of Energy-Efficient Motor Systems." *IEEE Transactions on Power Systems.* PWRS-5 (August): 852–861.

De la Moriniere, O. 1989. "Energy Service Companies: The French Experience." In *Electricity: Efficient End-Use and New Generation Technologies, and Their Planning Implications,* edited by T.B. Johansson, B. Bodlund, and R.H. Williams. Lund, Sweden: Lund University Press.

Desmond, J. 1989. Taunton Municipal Lighting, 55 Weir Street, Taunton, MA 02780, (508) 824-5844. Phone conversation with Steven Nadel, September.

Dewan, S., G. Slemon, and A. Straughen. 1984. *Power Semiconductor Drives.* New York: John Wiley.

DeWitt, L., and D. Wolcott. 1986. "Innovative Approaches to Facilitate Energy Efficiency Improvements in New York State's Public Sector." In *Proceedings of the 1986 Summer Study on Energy Efficiency in Buildings.* Vol. 4, *Incentives.* Washington, D.C.: American Council for an Energy-Efficient Economy.

DOE. See U.S. Department of Energy.

Donaldson, J. 1989. Gainesville Regional Utilities, P.O. Box 490, Station 52, Gainesville, FL 32602, (904) 374-2834. Phone conversation with Steven Nadel, June.

Donnelly, B, and L. Gudbjargsson. 1990. British Columbia Hydro, 1045 Howe Street, 6th Floor, Vancouver, B.C. V6Z 2B1. (604) 663-3969. Phone conversation with Steven Nadel, April.

Dreisilker, H. 1985. "Effect of Voltage Unbalance on Motors." *Electrical Construction and Maintenance,* 84 (August).

———. 1987. "Modern Rewind Methods Assure Better Rebuilt Motors." *Electrical Construction and Maintenance* 86 (August): 30–36.

———. "Safe Stator and Rotor Stripping Method." Undated 10-page typescript. Dreisilker Electric Motors, Inc., 352 Roosevelt Road, Glen Ellyn, IL 60137, (312) 469-7510.

EASA. See Electrical Apparatus Service Association.

Eaton Corporation. 1988. *Dynamatic Industrial Drives Catalog.* Kenosha, WI: Eaton Corporation.

Edison Electric Institute. 1987. *Statistical Yearbook 1987.* Washington, D.C.: Edison Electric Institute.

Eggars, M. 1989. New York State Energy Office, 2 Rockefeller Plaza, Albany, NY 12223, (518) 473-2007. Phone conversation with Steven Nadel, August.

Electric Power Research Institute. 1982. *Evaluation of Electrical Interference to the Induction Watthour Meter.* Report EL-2315. Palo Alto, CA: Electric Power Research Institute.

———. 1985. "Electronic Adjustable-Speed Drives for Boiler Feedpumps." *First Use* document FS5414B/D/E, *Results* series. Palo Alto, CA: Electric Power Research Institute.

———. 1988a. *Electrotechnology Reference Guide.* Report EM-4527. Palo Alto, CA: Electric Power Research Institute.

———. 1988b. *Energy Utilization Catalog.* Palo Alto, CA: Electric Power Research Institute.

———. 1989. *Proceedings: Advanced Adjustable Speed Drive R&D Planning Forum.* Report CU-6279. Palo Alto, CA: Electric Power Research Institute.

Electrical Apparatus Service Association. 1985. "Core Iron Study." St. Louis: Electrical Apparatus Service Association.

Emmett, E., and P. Gee. 1986. "Achieving Energy Efficiency in Government Operations: The Local Energy Officer Project." In *Proceedings of the ACEEE 1986 Summer Study on Energy Efficiency in Buildings.* Vol. 4, *Incentives.* Washington, D.C.: American Council for an Energy-Efficient Economy.

Englander, S., and L. Norford. 1988. "Fan Energy Savings: Analysis of a Variable-Speed Drive Retrofit." In *Proceedings of the 1988 ACEEE Summer Study on Energy Efficiency in Buildings.* Vol. 3, *Commercial and Industrial Building Technologies.* Washington, D.C.: American Council for an Energy-Efficient Economy.

EPRI. See Electric Power Research Institute.

Estey, D. 1989. "Bidding Conservation Against Cogeneration: The Level Playing Field." In *Demand-Side Management Strategies for the 90s, Proceedings of the Fourth National Conference on Utility DSM*

Programs. Report CU-6367. Palo Alto: Electric Power Research Institute.

Eto, J., and A. de Almeida. 1987. "Saving Electricity in Commercial Buildings with Adjustable-Speed Drives." *IEEE Transactions on Industry Applications.* 24 (3): 439–43.

Fenno, S. 1989. New York State Energy Office, 2 Rockefeller Plaza, Albany, NY 12223, (518) 473-2007. Phone conversation with Steven Nadel, August.

Ferraro, R. 1989. Power Electronics Application Center, 10521 Research Drive, Suite 400, Knoxville, TN 37932, (615) 675-9505. Phone conversation with Michael Shepard, November.

Fitzgerald, A., C. Kingsley, and S. Umans. 1983. *Electric Machinery.* New York: McGraw-Hill.

Fitzpatrick, D. 1992. Pacific Gas & Electric, 444 Market Street, San Francisco, CA 94177, (415) 972-5404. Phone conversation with Steven Nadel, May.

Flora, D. 1990. North Carolina Alternative Energy Corp., P.O. Box 12699, Research Triangle Park, NC 27709, (919) 361-8000. Phone conversation with Steven Nadel, May.

France, S. 1989. Puget Power, P.O. Box 97034, Bellevue, WA 97034, (206) 462-3742. Phone conversation with Steven Nadel, August.

Furugaki, I. 1988. "The Energy Conservation Policy System in Japan." In *Energy Efficiency Strategies for Thailand.* Ed. D. Bleviss and V. Lide. Lantham, Md.: University Press of America.

Futryk, R., and J. Kaman. 1987. "Variable-Speed Control of Lorain Assembly Plant Boiler Fans." Paper presented at the 1987 Ford International Energy Conference, October 5–7, in Cologne, Germany.

Garay, P. 1990. *Pump Application Desk Book.* Lilburn, GA: Fairmont Press.

Geller, H. 1987. "Energy and Economic Savings from National Appliance Efficiency Standards." Washington, D.C.: American Council for an Energy-Efficient Economy. Photocopy.

——. 1990. American Council for an Energy-Efficient Economy, 1001 Connecticut NW, Suite 535, Washington, D.C. 20036, (202) 429-8873. Letter to Steven Nadel, June. Data from PROCEL, the Brazilian national electricity conservation program.

General Electric. 1989. *G.E. Motors Energy Saver Line.* Publication GEA-11818. Fort Wayne, IN: General Electric.

——. 1991. *G.E. Stock Motors.* Publication GEP-500G. Fort Wayne, IN: General Electric.

Gilmore, W. 1989. Walco Electric Co., 303 Allens Avenue, Providence, RI 02901, (401) 467-6500. Phone conversations with Steven Nadel and Michael Shepard, September.

——. 1990. Bryant College, 1150 Douglas Pike, Smithfield, RI, (401) 232-6425. Letter to Steven Nadel, October.

Goldman, C., and E. Hirst. 1989. *Key Issues in Developing Demand-Side Bidding Programs.* Report LBL-27748. Berkeley: Lawrence Berkeley Laboratory.

Gordon, F., M. McRae, M. Rufo, and D. Baylon. 1988. "Use of Commercial Energy Efficiency Measure Service Life Estimates in Program and Resource Planning." In *Proceedings of the 1988 ACEEE Summer Study on Energy Efficiency in Buildings*. Vol. 3, *Commercial and Industrial Building Technologies*. Washington, D.C.: American Council for an Energy-Efficient Economy.

Gordon, G. 1990. Walco Electric Co., 303 Allens Avenue, Providence, RI 02901, (401) 467-6500. Phone conversation with Steven Nadel, August.

Greenberg, S., J. Harris, H. Akbari, and A. de Almeida. 1988. *Technology Assessment: Adjustable-Speed Motors and Motor Drives (Residential and Commercial Sectors)*. Report LBL-25080. Berkeley: Lawrence Berkeley Laboratory.

Greenheck Fan Corporation. 1986. Curve and table for model 30BISW. *Backward-Inclined Centrifugal Fans*. Publication No. BISW/BIDW-3-86 R. Schofield, WI: Greenheck Fan Corporation.

Gudbjargsson, L. 1990. British Columbia Hydro, 1045 Howe Street, 6th Floor, Vancouver, B.C. V6Z 2B1, (604) 663-1898. Phone conversation with Steven Nadel, April.

Gunn, R. 1989. Northern States Power, 414 Nicollet Mall, Minneapolis, MN 55401, (612) 330-7821. Phone conversation with Steven Nadel, July.

Gustafson, G., and J. Peters. 1987. *Process Evaluation of the Industrial Test Program*. Portland, Ore.: Bonneville Power Administration.

Guttman, M., and A. Stotter. 1984. "The Influence of Oil Additives on Engine Friction and Fuel Consumption." In *Proceedings of the 39th Annual Meeting of the American Society of Lubrication Engineers*. Chicago: American Society of Lubrication Engineers.

Habart, J. 1990. British Columbia Hydro, 1045 Howe Street, 6th Floor, Vancouver, B.C. V6Z 2B1, (604) 663-2219. Phone conversation with Steven Nadel, August.

Harris, J., and R. Diamond. 1989. *Energy Edge Impact Evaluation, Findings and Recommendations from the Phase One Evaluation*. Bonneville Power Administration.

Hawley, T. 1990. Wisconsin Electric, 333 West Everett, Milwaukee, WI 53201, (414) 221-3887. Phone conversation with Steven Nadel, May.

Henriques, D. 1989. British Columbia Hydro, 1045 Howe Street, 6th Floor, Vancouver, B.C. V6Z 2B1, (604) 663-3286. Phone conversation with Steven Nadel, June.

Hicks, E. 1989. "Third-Party Contracting Versus Customer Programs for Commercial/Industrial Customers." In *Energy Conservation Program Evaluation: Conservation and Resource Management, Proceedings of the August 23-25, 1989 Conference*. Argonne, Ill.: Argonne National Laboratory.

Hotrum, E. 1990. Alberta Power, 10035 105th Street, Edmonton, Alberta T5J 2V6. (403) 420-7091. Phone conversation with Steven Nadel, August.

Hudson, W. 1989. Penn Dower Petroleum Co., 6412 Dower House Road, Upper Marlboro, MD 20772, (301) 599-6500. Phone conversation with Steven Nadel, October.

Ibáñez, P. 1978. "Electromechanical Energy." In *Efficient Electricity Use*, ed. C. Smith et al. Elmsford, NY: Pergamon Press. 369–409.

IEEE. See Institute of Electrical and Electronics Engineers.

Institute of Electrical and Electronics Engineers. 1981. *IEEE Guide for Harmonic Control and Reactive Compensation of Static Power Converters*. IEEE Standard 519. New York: Institute of Electrical and Electronics Engineers.

Jackson, J. 1987. "New England Power Pool Commercial Energy Demand Model Systems." Jerry Jackson and Associates, Sandwich, Mass. Photocopy.

Johnston, W. 1990. North Carolina Industrial Extension Service, North Carolina State University, Box 7901, Page Hall, Raleigh, NC 27695, (919) 737-2356. Phone conversations and meetings with Steven Nadel, September.

Jordan, H. 1983. *Energy Efficient Electric Motors and Their Application*. New York: Van Nostrand Reinhold.

Katz, G. 1990. Momentum Engineering, Portland, OR. Phone conversation with Steven Greenberg, June.

Kellum, V. 1989. North Carolina Alternative Energy Corp., P.O. Box 12699, Research Triangle Park, NC 27709, (919) 361-8000. Phone conversation with Steven Nadel, July.

Keneipp, M., R. Ciliano, F. Stern, and K. Osvatic. 1990. "Developing Demand-Side Programs for Commercial/Industrial New Construction Projects." In *Proceedings of the 1990 ACEEE Summer Study on Energy-Efficiency in Buildings*. Vol. 8, *Utility Programs*. Washington, D.C.: American Society for an Energy-Efficient Economy.

Kent, J. 1989. Kent Oil Co., Lakeland, FL, (813) 665-0070. Phone conversation with Steven Nadel and Michael Shepard, October.

Kirsch, F. 1989. *Energy Management Assistance for Small and Medium-Size Manufacturers: Manufacturers' Evaluations of EADCs' Services 1987–88*. Philadelphia: University City Science Center.

Kochensparger, J. 1987. "Applying Predictive Maintenance Testing to Minimize Motor Failure Downtime." *Plant Engineering*, March 12.

Krupp-Widia. 1987. Krupp-Widia Magnet Engineering, Essen, Germany. Information extracted from an advertisement.

Lann, R., B. Riall, J. McMenamin, N. Khosla, and Q. Looney. 1986. *The COMMEND Planning System: National and Regional Data and Analysis*. EM-4486. Palo Alto: Electric Power Research Institute.

Lannoye, M. 1988. Letter to A. Lovins of Rocky Mountain Institute, 22 September 1988, regarding Teamwork Incentive Program, Washington State Energy Office, 809 Legion Way SE, #FA-11, Olympia, WA 98504, (206) 586-5000.

Lawrie, R.J., ed. 1987. *Electric Motor Manual*. New York: McGraw-Hill.

Leonard, W. 1984. *Control of Electrical Drives*. New York: Springer-Verlag.

——. 1986. "Microcomputer Control of High Dynamic Performance AC Drives—A Survey." *Automatica,* 22(1):1–19.

——. 1988. "Electro-Mechanical Energy Conversion by Controlled Electrical Drives." In *Demand-Side Management and Electricity End-Use Efficiency,* ed. A. T. de Anibal and A. H. Rosenfeld, 269–297. NATO Advanced Science Institutes Series E, vol. 149, Applied Sciences. Norwell, MA: Kluwer Academic Publishers.

Lihach, N. 1984. "Pacing Plant Motors for Energy Savings." *EPRI Journal,* March, 22–38.

Linn, C. 1987. "Calculating Daylight for Successful Retail Design." *Architectural Lighting,* January, 29–34.

Linn, J. 1990. Central Maine Power, Edison Drive, Augusta, ME 04336, (207) 623-3521. Phone conversation with Steven Nadel, February.

Litman, T. 1990. Washington State Energy Office, 809 Legion Way SE, #FA-11, Olympia, WA 98504, (206) 586-5000. Phone conversation with Steven Nadel and Michael Shepard, May.

Little, S. 1989. Bonneville Power Administration, P.O. Box 3621, Portland, OR 97208, (503) 230-3973. Phone conversation with Steven Nadel, April.

Lloyd, T.C. 1969. *Electric Motors and Their Applications.* New York: Wiley Interscience.

Lobodovsky, K. 1989. Pacific Gas and Electric, 77 Beale Street, San Francisco, CA 94106, (415) 972-7000. Phone conversation with Amory Lovins and related to Michael Shepard, March.

Lobodovsky, K., R. Ganeriwal, and A. Gupta, 1983. "Field Measurements and Determination of Electric Motor Efficiency." In *Procedings of the Sixth World Energy Engineering Congress.* Atlanta: Association of Energy Engineers.

——. 1989. "Field Measurements and Determination of Electric Motor Efficiency." *Energy Engineering* 86(3): 41-53. Atlanta: Association of Energy Engineers.

Lovins, A., J. Neymark, T. Flanigan, P. Kiernan, B. Bancroft, and M. Shepard. 1989. *The State of the Art: Drivepower.* April. Snowmass, CO: Rocky Mountain Institute (Competitek).

Lovins, A., R. Sardinsky, P. Kiernan, T. Flanigan, B. Bancroft, and J. Neymark. 1988. *The State of the Art: Lighting.* March. Snowmass, CO: Rocky Mountain Institute (Competitek).

Magnusson, D. 1984. "Energy Economics for Equipment Replacement." *IEEE Transactions on Industry Applications.* IA-20 (March/April): 402–6.

Mahany-Braithwait, C., and M. Mauldin. 1988. "The Ontario Hydro Motor Rebate Pilot: An Evaluation." Toronto: Ontario Hydro. Photocopy.

Marbek Resource Consultants, Ltd. 1987. *Energy Efficient Motors in Canada: Technologies, Market Factors, and Penetration Rates.* Ottawa: Marbek Resource Consultants.

Mayo, B. 1989. Southern California Edison, 2244 Walnut Grove Avenue, Rosemead, CA 91770, (213) 491-2263. Phone conversation with Steven Nadel, July.

McAteer, M. 1990. "Design 2000, New England Electric's New Construction Program." Paper presented at the New England Environmental Expo, April, in Medford, MA.

McDonald, W.J., and H.N. Hickok. 1985. "Energy Losses in Electrical Power Systems." *IEEE Transactions on Industry Applications.* IA-21 (May/June): 803–19.

Messenger, M. 1989. California Energy Commission, 1516 9th Street, Sacramento, CA 95814, (916) 324-3259. Phone conversation with Steven Nadel, September.

Miller, P., J. Eto, and H. Geller. 1989. *The Potential for Electricity Conservation in New York State.* Albany: N.Y. State Energy Research and Development Authority.

Milton, B., and E. Carter. 1982. "Fuel Consumption and Emission Testing of an Engine Oil Additive Containing PTFE Colloids." In *American Society of Lubrication Engineers Transactions.* 39(2):105–10.

Mohan, N. 1981. *Techniques for Energy Conservation in AC Motor-Driven Systems.* EPRI Report EM-2037. Palo Alto: Electric Power Research Institute.

Mohan, N., K. Brooks, and W. Montgomery. 1988. *Energy Conservation in AC Motor Driven Systems by Means of Solid-State Adjustable-Frequency Controllers.* Minneapolis: University of Minnesota.

Monroe, E., and C. Moscarrillo. 1989. Edison Electric Institute Statistical Department, 1111 19th Street NW, Washington, D.C. 20036. Phone conversation with Michael Shepard, January.

Montgomery, D. 1984. "The Motor Rewind Issue: A New Look." *IEEE Transactions on Industry Applications.* IA-20(5):1330–36.

———. 1989. "Testing Rewinds to Avoid Motor Efficiency Degradation." *Energy Engineering* 86(3): 24–40.

Moore, T. 1988. "The Advanced Heat Pump: All the Comforts of Home . . . and Then Some." *EPRI Journal,* March, 4–13.

Morante, P. 1990. Northeast Utilities, 100 Corporate Place, Rocky Hill, CT 06067, (203) 721-2707. Phone conversation with Steven Nadel, June.

Nadel, S. 1990a. *Lessons Learned: A Review of Utility Experience with Conservation and Load Management Programs for Commercial and Industrial Customers.* Albany: New York State Energy Research and Development Authority.

———. 1990b. *Testimony in Support of an Act Requiring Minimum Efficiency Standards for Lighting Fixtures, Lightbulbs, Floor Lamps, Table Lamps, and Electric Motors* (revised). Washington, D.C.: American Council for an Energy-Efficient Economy.

Nadel, S., and H. Tress. 1990. *The Achievable Conservation Potential in New York State from Utility Demand-Side Management Programs.* Washington, D.C.: American Council for an Energy-Efficient Economy.

Nailen, R. 1987. *Motors.* Vol. 6, *Power Plant Electrical Reference Series.* Palo Alto: Electric Power Research Institute.

National Electrical Code. 1990. Document ANSI/NFPA 70. Quincy, MA: National Fire Protection Association.

National Electrical Manufacturers Association. 1982. *Tests and Performance of AC Fractional and Integral-Horsepower Motors.* Standards Publication MG 1-1 (Standards 12.30 to 14.44). Washington, D.C.: National Electrical Manufacturers Association.

———. 1988. *Energy Management Guide for Selection and Use of Single-Phase Motors.* Standards Publication MG 11-1977, rev. 1982, 1987. Washington, D.C.: National Electrical Manufacturers Association.

———. 1989a. *Energy Management Guide for Selection and Use of Polyphase Motors.* Standards Publication MG 10-1983, rev. 1988. Washington, D.C.: National Electrical Manufacturers Association.

———. 1991a. *Motors and Generators.* Standards Publication MG 1-1987 with Revisions No. 1 and 2. Washington, D.C.: National Electrical Manufacturers Association.

———. 1991b. "Proposal for NEMA Design E Motor Based on IEC Locked Rotor Current and Starting Torque and Higher Efficiency Limits." Washington, D.C.: National Electrical Manufacturers Association.

NCAEC. See North Carolina Alternative Energy Corp.

NEC. See National Electrical Code.

NEES/CLF. See New England Electric System and Conservation Law Foundation of New England.

NEMA. See National Electrical Manufacturers Association.

NEPSCO. See New England Power Service Co.

New England Electric System. 1990. "Motor Price Analysis, July 1990." Westborough, MA: New England Electric System.

New England Electric System and Conservation Law Foundation of New England. 1989. *Power by Design: A New Approach to Investing in Energy Efficiency.* Westborough, MA: New England Electric System.

New England Power Service Co. 1989. "Rhode Island Motor Survey." Westborough, MA: New England Power Service Co.

New York State Energy Office. 1989. *New York State Energy Plan.* Vol. 5, *Electricity Supply Assessment.* Albany: New York State Energy Office.

———. 1991. *New York State Energy Construction Code.* Albany: New York State Energy Office, Bureau of Codes and Standards.

Niagara Mohawk. 1987. "Niagara Mohawk Power Corporation Motor Retrofit Program, an Industrial Customer Rebate Demonstration Program, Project No. C1P-17." Syracuse: Niagara Mohawk. Photocopy.

Nisson, N. 1988. "A Quantum Leap in Gas Furnace Technology." *Energy Design Update,* April, 12–14.

North Carolina Alternative Energy Corp. 1989. "Energy-Efficient Electric Motor Systems." Project proposal. Research Triangle Park, NC: North Carolina Alternative Energy Corp. Photocopy.

Northern States Power. 1988. *Conservation Improvement Program, Annual Report and Evaluation.* Minneapolis: Northern States Power.

NSP. See Northern States Power.

Oliver, J., and M. Samotyj. 1989. *Lessons Learned from Field Tests of Large Induction Motor Adjustable Speed Drives, 1984–1989.* Palo Alto: Electric Power Research Institute.

Ontario Hydro. 1988. *Marketing High-Efficiency Motors.* Toronto: Ontario Hydro.

———. 1989. "High-Efficiency Motors Evaluation Software Program." Ontario Hydro, Toronto. Photocopy.

———. 1992. "Hydro Motor Efficiency Levels." Toronto, Ontario: Ontario Hydro.

Oviatt, V. 1989. Association of Energy Engineers, 4025 Pleasantdale Road, Suite 420, Atlanta, GA 30340, (404) 447-5083. Phone conversation with Steven Nadel, July.

Paco Pumps. 1983. "Pump Curve Number RC-2010." Oakland, Calif.: Paco Pumps Company.

———. 1985. "A1b End Suction Centrifugal Pumps Type L; Pump Selection Charts." Oakland, CA: Paco Pumps Company.

Payton, R. 1988. Reliance Electric Company, 411 Borel Avenue, Suite 405, San Mateo, CA 94402, (415) 574-8860. Phone conversation with Steven Greenberg, February.

PEAC. See Power Electronics Applications Center.

Peat, Marwick, Main and Company. 1987. *Program Evaluation of the New York State Energy Advisory Service to Industry Program.* Albany: New York State Energy Office.

Peddie, R. 1988. "Smart Meters." In *Demand-Side Management and Electricity End-Use Efficiency,* ed. A. de Anibal and A. Rosenfeld, 171–80. NATO Advanced Science Institutes Series E, vol. 149, Applied Sciences. Norwell, MA: Kluwer Academic Publishers.

Peele, T., and R. Chapman. 1986. "Automating Maintenance Management." *Plant Engineering,* October 23, 76–78.

Pendleton, D. 1989. "Commercial Conservation Report: 1988 Summary." Everett, WA: Snohomish Public Utility District.

Perkins, R. 1989. Compaq Computer, P.O. Box 692000, Houston, TX (713) 370-0670. Phone conversation with Steven Nadel, October.

Peters, J. 1989. *Interim Process Evaluation of the Bonneville Power Administration's Energy Savings Plan (E$P) Program.* Portland, OR: Bonneville Power Administration.

Poole, J. 1989. CRS Sirrine, Inc., 5511 Capital Center Drive, Raleigh, NC 27695, (919) 549-6205. Phone conversation with Michael Shepard, August.

Poole, M., J Moran, D. Seitzinger, T. Johnson, G. Stengl, S. Salib, and D. Wangerin. 1990. *Commercial and Industrial Applications of Adjustable-Speed Drives.* CU-6883. Palo Alto: Electric Power Research Institute.

Power Electronics Applications Center. 1987. *ASD Directory.* 2nd ed. Knoxville: Power Electronics Applications Center.

Raba, J. 1990. National Electrical Manufacturers Association, 2101 L Street NW, Washington, DC 20037, (202) 457-8423. Letter to and phone conversation with Steven Nadel, September.

Resource Dynamics Corporation. 1985. *Adjustable Speed Drives: Directory of Manufacturers and Applications.* McLean, VA.

Ryan, M. 1988. *Electric Variable-Speed Drive Reference Guide.* Toronto: Ontario Hydro.

Schwartz, K. 1990. British Columbia Hydro, 1045 Howe Street, Vancouver, B.C. V6Z 2B1, (604) 663-3761. Phone conversation with Steven Nadel, April.

Seattle Dept of Construction & Land Use. 1991. *1991 Seattle Energy Code.* Seattle, WA: Seattle Dept of Construction & Land Use.

Seton, Johnson & Odell, Inc. 1983. Summary data from proprietary industrial motor drive study prepared for Seattle City Light, June 1983.

———. 1987a. *Lost Conservation Opportunities in the Industrial Sector.* Portland, OR: Bonneville Power Administration.

———. 1987b. *Energy Efficiency and Motor Repair Practices in the Pacific Northwest.* Portland, OR: Bonneville Power Administration.

Shepard, M., A. Lovins, J. Neymark, D. Houghton, and R. Heede. 1990. *The State of the Art: Appliances.* August. Snowmass, CO: Rocky Mountain Institute (Competitek).

Smeaton, R. 1987. *Motor Application and Maintenance Handbook.* New York: McGraw-Hill.

Smith, W. 1990. Electric Power Research Institute, 3412 Hillview Avenue, Palo Alto, CA 94304, (415) 855-2000. Phone conversation with Steven Nadel, April.

Sperber, R. 1989. "Maintenance Software System Yields Two-Year Payback." *Food Processing,* October, 72-78.

Stout, T. 1990. New England Electric Service, 25 Research Drive, Westborough, MA 01582, (508) 366-9011. Phone conversation with Steven Nadel, June.

Stout, T., and W. Gilmore. 1989. "Motor Incentive Programs: Promoting Premium Efficiency Motors." Paper presented at the *Electric Council of New England National Conference on Demand-Side Management,* November 16-17. Westborough, MA: New England Power Service Company.

Strohs, R. 1987. "Application of Variable Speed Drive Pumping Systems for Energy Savings." Paper presented at the 1987 Ford International Energy Conference, October, in Cologne, Germany.

Thompson, E. 1990. Inland Rome Lumber Co., Rome, GA, (404) 232-0851. Phone conversation with Gail Katz, January.

Toshiba. 1987. *Squirrel Cage Induction Motors.* Publication No. 9116. Houston: Toshiba/Houston.

Treadle, S. 1987. "The Interaction of Lighting and HVAC Systems." *Lighting Design and Application,* May.

Tyre, R. 1989. Nevada Power, P.O. Box 230, Las Vegas, NV 98151, (702) 367-5113. Phone conversation with Steven Nadel, July.

Umans, S. 1988. "Unity Plus Motor Tests." Unpublished report available from author, 5 Regent Road, Belmont, MA 02178, (617) 253-7351.

Umans, S., and H. Hess. 1983. "Modelling an Analysis of the Wanlass Three-Phase Induction Motor Configuration." *IEEE Transactions on Power Apparatus and Systems.* PAS-102 (September): 2912–26.

U.S. Department of Energy. 1989a. *Annual Energy Review 1988.* DOE/EIA-0384(88). Washington, D.C.: Energy Information Administration.

——. 1989b. "Energy Conservation Voluntary Performance Standards for Commercial and Multi-Family High-Rise Residential Buildings; Mandatory for New Federal Buildings; Interim Rule." *Federal Register* 54 (18) (January 30): 4538–4720.

U.S. Senate. 1992. *S.2166. National Energy Security Act of 1992.* Washington, D.C.: U.S. Government Printing Office.

Van Son, D. 1989. Baldor Motors, 5711 S. 7th Street, Fort Smith, AR 72902, (501) 646-4711. Phone conversation with and letter to Michael Shepard, August.

Wanlass, C. 1978. U.S. Patent 4063135, December 13.

——. 1980. U.S. Patent 4187457, February 5.

Weedall, M., and F. Gordon. 1990. "Utility Demand-Side Management Incentive Programs: What's Been Tried and What Works to Reach the Commercial Sector." In *Proceedings of the 1990 ACEEE Summer Study on Energy-Efficiency in Buildings.* Vol. 8, *Utility Programs.* Washington, D.C.: American Society for an Energy-Efficient Economy.

Wilke, K., and T. Ikuenobe. 1987. "Guidelines for Implementing an Energy-Efficient Motor Retrofit Program." In *Proceedings 10th World Energy Engineering Congress,* 399–406. Atlanta: Association of Energy Engineers.

Willis, P. 1990. British Columbia Hydro, 1045 Howe Street, 6th Floor, Vancouver, B.C. V6Z 2B1, (604) 663-2303. Phone conversation with Steven Nadel, May.

Wisconsin Electric. 1989. *1988 Smart Money Energy Program Evaluation, Final Report.* Milwaukee: Wisconsin Electric.

Xenergy. 1989. Motor inventory for Wisconsin Electric. Burlington, MA: Xenergy.

Zacheral, L. 1988. Brown-Boveri Corporation, 1460 Livingston Avenue, North Brunswick, NJ 08902, (201) 932-6000. Phone conversation with Brady Bancroft and related to Michael Shepard, Rocky Mountain Institute, June.

About the Authors

Anibal T. de Almeida, Professor of Electrical Engineering at the University of Coimbra, Portugal, holds a Ph.D. in Power Systems from the University of London. He has conducted research on drive-power systems in Europe and the U.S. and was Chairman of the 1987 NATO Advanced Study Institute on Demand-Side Management and Electricity End-Use Efficiency. He is involved in several international projects on efficient electricity use and industrial automation.

Steve Greenberg, Research Associate in Lawrence Berkeley Laboratory's Energy Analysis Program, holds a bachelor's degree in Mechanical Engineering and a master's in Energy and Resources, both from the University of California at Berkeley. A licensed professional engineer, Greenberg has conducted extensive research and engineering analysis on drivepower technologies in commercial, industrial, and institutional facilities since 1980. He has participated in several studies on the application of adjustable-speed drives and is the technical editor of *Home Energy* magazine.

Gail Katz held degrees in electrical and mechanical engineering from Portland State University. She spent seven years at the engineering firm of Seton, Johnson & Odell, developing energy conservation plans for Pacific Northwest industries, before founding her own firm, Momentum Engineering, in 1988. Her consulting practice grew to include extensive work with industrial facilities and utilities throughout the country, national laboratories, and other groups. In 1990 she was invited to join a National Academy of Sciences delegation to help Poland address its environmental problems. She died at her home of heart and kidney failure on August 14, 1990.

Steven Nadel is Senior Associate with the American Council for an Energy-Efficient Economy (ACEEE). He holds an M.A. in Environmental Studies from Wesleyan University and an M.S. in Energy Management from New York Institute of Technology, and has over ten years of experience in the field of energy conservation. At ACEEE he directs projects dealing with utilities, lighting, and motor systems, and is the author of a recent study examining lessons learned from utility C&LM programs over the past decade. Before joining ACEEE, he planned and evaluated conservation and load management programs for the New England Electric System, including programs to improve the efficiency of motor systems.

Michael Shepard is Director of the Energy Program at Rocky Mountain Institute and Manager of RMI's Competitek information service. He holds a master's degree from the Energy and Resources Group at U.C. Berkeley, and a bachelor's in Natural Resource Conservation from Cornell University. He has written extensively on energy issues for over a decade and was a co-author of the Competitek report, *The State of the Art: Drivepower.* Shepard formerly was Senior Feature Writer for the *EPRI Journal* and Publications Director of the New Mexico Solar Energy Association.

Index

Magnetic losses. *See* Core losses
Maintenance, 94-97
 cleaning, 96-97
 end-user practices, 187-188
 importance of, 5
 lubrication, 94-95
 national savings potential from
  improvements, 178
 painting, 97
 periodic checks, 96
 policy and program
  recommendations, 248
 seminars for maintenance staff,
  202
 software for tracking, 208
Managers. *See* Energy managers
Manufacturers
 control equipment, 196-197
 motor, 190-191
 original equipment, 193-195
Market. *See* Motor market
Massachusetts, efficiency
 standards proposed by, 221,
 224
Mechanical equipment
 representatives and distributors,
 197
Mechanical losses, 39
Megohmmeter manufacturers,
 307-308
Microprocessors, ASD technology
 and, 119-120
Minimum efficiency standards
 described, 59-60
 need for, 9
 recommendations, 249
 See also Standards
Monitoring. *See* Testing
Motor applications, 131-151
 for adjustable-speed drives,
  113-120, 149-150
 air compressors, 148-149
 centrifugal compressors, 149
 chillers, 149
 conveyors, 150-151
 fans and pumps, 131-148
  characteristics of, 135-140

fluid flow fundamentals,
  131-135
 system control and
  optimization techniques,
  140-146
 system optimization case
  studies, 146-148
 replacing DC drives with AC
  motors and ASDs, 149-150
 summary table for motor types,
  37
 synchronous motor
  applicability, 31
 See also specific applications
Motor cycling, 83-86
 allowable cycling, 84-85, 86
 candidates for, 83-84
 starting controls for, 85
 thermal and mechanical stresses
  of, 84
Motor distributors, 191-193
Motor efficiency, 36-43
 cube law and, 138
 defined, 56
 load and, 72-76
 loss-minimizing features, 39,
  40
 losses, 38-39
 oversizing and, 72-76
 testing standards, 56-63
  comparison of, 59
  field measurements, 62-63
  nameplate efficiency index,
   58-62
 See also Drivepower efficiency;
  Energy-efficient motors
Motor efficiency statistics
 induction motors vs. brushless
  DC motors, 33
 NEMA Design B motors, 21, 41
 ODP motors, 41, 215
 shaded-pole motors, 19
 single-phase squirrel cage
  induction motors, 18
 TEFC motors, 41
 See also Energy use
Motor-generator sets, 104, 176-177